MANAGING HOUSEKEEPING AND CUSTODIAL OPERATIONS

Edwin B. Feldman, P.E.

PRENTICE HALL
Englewood Cliffs, New Jersey 07632

Prentice-Hall International (UK) Limited, *London*
Prentice-Hall of Australia Pty. Limited, *Sydney*
Prentice-Hall Canada, Inc., *Toronto*
Prentice-Hall Hispanoamericana, S.A., *Mexico*
Prentice-Hall of India Private Limited, *New Delhi*
Prentice-Hall of Japan, Inc., *Tokyo*
Simon & Schuster Asia Pte. Ltd., *Singapore*
Editora Prentice-Hall do Brasil, Ltda., *Rio de Janeiro*

10 9 8 7 6 5 4 3 2 1

Library of Congress Cataloging-in-Publication Data

Feldman, Edwin B.
Managing housekeeping and custodial operations / Edwin B. Feldman.
p. cm.
Includes index.
ISBN 0-13-378159-3 : $69.95
1. Buildings—Cleaning—Handbooks, manuals, etc. 2. Industrial housekeeping—Handbooks, manuals, etc. 3. Plant maintenance—Handbooks, manuals, etc. I. Title.
TH3361.F45 1992
648′.068—dc20 91-33677
CIP

ISBN 0-13-378159-3

PRENTICE HALL
Business Information & Publishing Division
Englewood Cliffs, NJ 07632
Simon & Schuster, A Paramount Communications Company

Printed in the United States of America

To one of the great joys of my life, Marissa, with love.

About the Author

EDWIN B. FELDMAN has more than forty years' experience in custodial maintenance and plant engineering. This includes ten years as plant engineer, plant manager, and director of engineering for a chemical specialties manufacturing firm and thirty years operating a firm specializing in housekeeping and maintenance consulting and training. He is currently an independent consultant, based in Atlanta, Georgia.

Feldman has written ten books and more than 250 articles in technical and trade magazines, all on various aspects of building maintenance. He is a frequent speaker for the American Institute of Plant Engineers, the Association of Physical Plant Administrators, the Cleaning Management Institute, the International Facilities Management Association, and other organizations. He has a bachelor's degree in industrial engineering from Georgia Institute of Technology. He is a registered Professional Engineer and is certified by the National Council of Engineering Examiners. He has received the Outstanding Service Award of the National Society of Professional Engineers.

Foreword

I first understood that there was more to housekeeping than met the eye more than forty-five years ago, when I obtained a part-time job in a factory that manufactured sanitation supplies.

When I graduated from Georgia Institute of Technology as an Industrial Engineer, I became a Plant Engineer of that company. Over a period of ten years, and through various job positions, I also spent a good deal of time with the company's sales staff—they asked me questions about time requirements for cleaning, the proper equipment or chemicals to use, procedures, and the like. It was apparent that this information was not available to them or to their customers on a commercial basis without relationship to the sale of products.

In order to satisfy this need, I founded an independent custodial consulting firm in 1960. I now consult as an independent, with no ties to contract cleaning or supply firms of any kind.

This eventually led to the writing of this book, which is based on consulting experiences for clients all over the United States and Canada, as well as exposure to thousands of housekeeping managers and supervisors at seminars that I regularly conduct in various cities.

The book has had a great deal of input from various sources, including material from seminar participants. I am grateful to them.

Edwin B. Feldman, P.E.

What This Book Will Do for You

Managing Housekeeping and Custodial Operations brings the varied aspects of cleaning buildings of all types into one complete volume.

This book is the culmination of the author's thirty-year career as a housekeeping consultant, seminar leader, and writer. The book distills the experience of consulting in all types of facilities; researching and writing 250 magazine articles and ten books; and conducting seminars for thousands of cleaning department managers.

Useful not only as a daily operational aid and a planning resource, the book also is an invaluable training resource and reference work.

It will directly appeal to the cleaning department head (whether it be Custodial Manager, Executive Housekeeper, Director of Sanitation, Environmental Services Manager, or one of a dozen other titles); to involved administration (such as Physical Plant Director, Facilities Manager, Building Manager, Assistant Administrator, Operations and Maintenance Manager, Plant Engineer); and to the first line supervisor as well.

This book is equally valuable for use in schools, hospitals, office buildings, factories, airports, nursing homes, colleges and universities, hotels, retail establishments and shopping malls, as well as in government and contract-cleaned properties; in fact, in any structure that is to be cleaned.

The book is written in the author's best-selling, no-nonsense, easily readable style. The numerous charts, illustrations, forms and graphs make the book even more useful.

The arrangement of the material makes for an easy flow of information, as well as ready research. The use of the book is further simplified by its division into three sections: Management (ten chapters), Supervision (seven chapters), and Operations (thirteen chapters).

Throughout, the book provides numerous opportunities for cost saving and quality improvement, in such categories as staffing, organization, equipment, chemicals, cleaning methods, motivation, supervisory development, worker training, and contracted services.

This is a book of problem-solving chapters . . . cost-savings and quality improvement chapters . . . chapters of opportunity for you.

You will easily find—

In *Chapter 1,* that improvement begins with pinpointing your strengths and weaknesses and identifying your costs and problems. You are given 13 benefits with which to sell your concept to management, and 11 steps you can take to optimize your cleaning efficiency.

In *Chapter 2,* you will learn the easiest way to measure your needs and to set your objectives. You are given guidelines for making a no-nonsense custodial survey.

In *Chapter 3,* you are given the basis for establishing the seven key types of standards: personnel, frequency, methods, materials, equipment, time, and quality.

Chapter 4 tells you how to organize your department, including program format, job descriptions, personnel requirements, projects and relief, policing, and job assignments.

You get the four key steps to successfully present your program to management and staff and see how to coordinate implementation in *Chapter 5.*

Chapter 6 shows how to get buildings designed for effective and economic cleaning, in four critical areas, and gives you a design checklist as well.

The controversial question of contract cleaning is simplified in *Chapter 7,* so you can make valid comparisons with in-house services, and get your money's worth if you use contract cleaning.

Chapter 8 shows how to save money through night cleaning, and it explains its benefits where applicable.

In *Chapter 9,* you will learn how to select and use consultants, and how to decide when and if to use them.

A housekeeping self-audit that targets dozens of opportunities for cost and quality improvement is provided in *Chapter 10.*

Chapter 11 begins the supervision section of the book, with a recipe for workable, day-by-day leadership. It tells how to determine how many supervisors are needed and how to select a supervisor.

You are told in *Chapter 12* how to select and evaluate housekeeping workers, and you are given typical rules to follow.

In Chapter 13, you will find proven techniques for developing and operating an effective training program, including orientation, on-the-job training, and classroom training.

Chapter 14 takes the theory out of motivation and gives you a practical system that works. Public relations is shown as an aspect of this.

You are faced with the universal problem of absenteeism, and *Chapter 15* tells you how to handle it.

Safety, fire prevention, and body mechanics are covered for you in *Chapter 16,* as well as waste disposals and handling chemicals.

Chapter 17 gives you money-saving tips for purchasing cleaning supplies and equipment.

Part Three of the book begins with the important subject of soil prevention—keeping dirt out of your building, rather than cleaning it up inside, in *Chapter 18.*

Chapter 19 tells you fourteen ways to control the graffiti problem, and how to remove the markings.

Chapter 20 explains the care of rest rooms, including choosing supplies and tools, controlling odors, and gives you thirty-two tips for efficient cleaning.

You will see in *Chapter 21* how to organize your storage and closet areas for maximum productivity.

The selection and maintenance of powered equipment is covered in *Chapter 22.* You are given the four benefits of mechanization, and the eleven steps used to assess your needs.

Chapter 23 gives you the latest information on manual equipment, ranging from mops to soap dispensers, from waste receptacles to custodial carts.

You are given in *Chapter 24* the means of choosing the best chemicals for your needs, both frequent and specialized.

Floor care guidelines are presented to you in *Chapter 25,* including selection, cleaning, and maintenance.

In *Chapter 26,* you are given the solutions to floor-care problems for the six basic types of floors, as well as tips on removing nine stubborn stains.

Chapter 27 clarifies for you the controversy of carpets versus resilient tile, tells you how to choose the right carpet, how to clean it and destain it, and gives you a summary care checklist.

How to clean during construction, including how to isolate the area, is given to you in *Chapter 28.*

In *Chapter 29,* you will find information on controlling germs and preventing cross-infection, how to choose the proper chemical, and how to handle isolation and critical areas.

The book concludes, in *Chapter 30,* with a discussion of how you can avoid the costly common mistakes that are so frequently made in cleaning.

Contents

Part ONE

MANAGEMENT

1

How to "Clean Up" Your Cleaning Department: Self-Help Tips for Increased Cost Efficiency

Billions of dollars are spent each year on housekeeping and sanitation. The efforts of millions of custodial workers are involved. Unfortunately, a good part of this effort is wasted. This chapter will help you target specific areas for reducing cost and improving overall efficiency.

HOW TO IDENTIFY YOUR TOTAL HOUSEKEEPING COSTS AND MAJOR PROBLEM AREAS

For too long, housekeeping has been overlooked as a potential source of immediate and considerable cost improvement. Focus has been growing on indirect costs associated with custodial cleaning, but management often is not aware of the total costs involved or aware that the costs being spent may be out of line. Housekeeping and associated costs are not minor items! To determine where your housekeeping costs are being expended, you should analyze your total housekeeping program. Include in your cost analysis the following:

- Direct wages of workers
- Salaries of supervisors
- Overtime costs
- Cost of cleaning performed on contract
- Cost of housekeeping materials
- Amortized cost of housekeeping equipment
- Costs borne by personnel outside the housekeeping department in cleaning operations.

A more detailed picture can be obtained by considering such items as:

- Space for housekeeping storage and facilities
- Overhead costs for housekeeping (utilities, services, etc.)
- Purchasing and stock control costs
- Value of management time consumed in housekeeping
- Fringe benefits.

Unfortunately, many managers believe their company has a more effective or economical housekeeping operation than the evidence warrants. Most housekeeping operations can be 10 to 15% more effective and economical than they currently are. For every improvement suggested by your own internal housekeeping staff, numerous other improvements go unchecked. Housekeeping is an annually recurring expense with constantly rising costs; improvements tend to be recurrent.

The housekeeping operations of hundreds of organizations have been analyzed carefully over the years by experts. Two points are clear: (1) Conditions, for the most part, remain the same. (2) The problems, however, are more serious because of the increased costs involved due to higher wages—wages account for all but 7 to 8% of direct housekeeping costs. In today's highly competitive business world, *every* road to cost improvement must be explored.

Keep in mind that problems are opportunities in disguise. The following list of common housekeeping problems can help you reduce costs and improve the quality of your housekeeping departments. The solution to each of these 22 problems is the same: Positive communication with management. You must educate management concerning the opportunities that are available for you to optimize the department. For example:

- Make a monthly written report to management, always including your recommendations.
- Follow up this written report (such as three days later) with a meeting. This should be a standing appointment, where it is not necessary for you to request it each time.
- Provide illustrations and charts to support your recommendations.
- Improve your ability to "sell" your ideas to management; you might want to read a book or attend a seminar on this subject.
- Be persistent—remember that most ideas are sold after many discussions.
- Remember that silence indicates your approval of existing conditions—do not be silent if you want conditions to change.

Here are your 22 opportunities for improvement:

1. *Top Management Indifference.* Top management often views housekeeping as a "necessary evil." Management is not conversant with the hazards and costs of poor housekeeping or the advantages of good housekeeping. Demands are made to "get the place cleaned up" without any conception of how such general results are to be obtained, or their costs.

2. *Insufficient Budget.* Often, the size of the housekeeping budget is directly proportioned to such factors as square footage, current production levels (in a manufacturing plant), enrollment (in a school), or bed capacity (in a hospital), whereas the requirements of housekeeping do not relate directly at all to these factors.

3. *Lack of Management Direction.* Management has failed to detail the objectives and responsibilities of the housekeeping department and of other personnel relating to housekeeping.

4. *Low Status of the Housekeeping Department.* The housekeeping department is not given the same organizational integrity and status as other departments. It may suffer arbitrary budget cuts and yet be given no opportunity or procedure to outline the consequences or alternative possibilities to management. Promising personnel are pulled out of the housekeeping department to be placed in "important" work.

5. *Lack of Recognition.* Often management does not recognize the housekeeping function as requiring a separate identity, although related functions such as safety and transportation have been so identified and given special attention. These very services cannot be completely effective without the support of the housekeeping operation. Contributing to this situation is the fact that housekeeping can be dispersed throughout the various departments, in part if not fully. Thus is lost the opportunity for standardization and effective supervision.

6. *Negative Objectives.* The housekeeping department is often given a negative objective, with descriptions and understandings of its function expressed as bad conditions which it should prevent or avoid. Custodians often perceive their jobs as being devoted to keeping things from happening. Rather, housekeeping should be a positive concept, with the housekeeping department considered as providing conditions of benefit.

7. *Lack of Career Path.* Cleaning workers often consider themselves "at the bottom of the ladder" with nowhere to go. This attitude is amply reinforced—even justified at times—by existing conditions which require correction.

8. *Insufficient Supervision.* Supervision is spread too thinly. Many times a single housekeeping foreman is responsible for the direct supervision of dozens of personnel working on three different shifts.

9. *Responsibility for Miscellaneous Work.* The housekeeping department has to perform nonhousekeeping duties which other departments do not want, or which just do not seem to fit anywhere else. This can include such tasks as mail delivery, running errands, and chauffeuring. (Because of the indivisibility of labor, this situation may sometimes be justified.)

10. *Lack of Training.* The housekeeping workers are untrained in their jobs despite the fact that they are custodians of valuable property. In addition, they are dealing with ever more complicated materials and equipment, and the effect of their work on productivity and quality control is of considerable importance.

11. *Poor Service from Suppliers.* Suppliers may not give suitable services. The average sanitary-supply firm is not organized or financially able to give comprehensive field service.

12. *Poor Image.* Little has been done to correct the image of the "janitor" of 20 years ago, who is depicted as shuffling along with a soiled mop and bucket of dirty water. This contributes markedly to the problems of morale and employee and management attitudes.

13. *Worker Turnover.* Labor contracts sometimes create difficult situations for housekeeping, the most prevalent example of which is company-wide seniority. Since it is mistakenly believed that anyone can perform a housekeeping task, the turnover in the housekeeping department becomes significant during periods of changing employment levels. In addition, the practice of using group leaders (which can offer many advantages) cannot be followed because the best qualified worker may not get the job.

14. *Decentralization of Housekeeping.* Housekeeping operations are decentralized, which further compounds the problems of training, supervision, status and morale, standardization, etc.

15. *Lack of Work Standardization.* There may be little standardization in housekeeping levels, materials, methods, and equipment. This is often true even in different parts of the same facility.

16. *Irreputable Contractors.* Some organizations perform their basic cleaning operations on a contract basis. A comparison of the objective of the organization (obtaining the most cleaning for the housekeeping dollar) with that of the contractor (retaining as much of the contract fee as possible for a profit) indicates the problems inherent in such an arrangement if the contractor is not reputable.

17. *Lack of Preventive Maintenance.* Housekeeping is performed on a corrective rather than a preventive basis—a matter of "putting out fires." As a result, projects that should be performed on a periodic basis are not adequately programmed and carried out, and some important tasks are *never* performed.

18. *Poor Work Conditions.* Poor appearance, hazardous conditions, and bad odors often indicate housekeeping situations that should not be permitted to exist.

19. *Lack of Mechanization.* Although there are a number of timesaving housekeeping equipment items now on the market, and others being developed, mechanization in the housekeeping department has not kept pace with mechanization in other fields.

20. *Poor Worker Attitudes.* One of the fundamental problems, particularly in larger facilities, is the attitude of the employees toward their housekeeping responsibilities. Very often, no responsibility is felt at all—and since all housekeeping work is performed because of human activities, such attitudes seriously affect the work load with which the custodian is faced each day. The situation becomes more serious because the fruits of the custodian's effort deteriorate almost overnight. In a given facility, the proportion of custodians to other personnel may be one to a hundred. Consider the difference between the custodian working *with* the other workers toward better housekeeping conditions and the custodian struggling with the conditions created by a hundred careless workers.

21. *Poor Building Design for Maintenance.* Architects may build structures without regard to the maintenance which must follow. (This subject is pursued further in Chapter 6.) Design engineers and decorators may fail to obtain the viewpoint of the housekeeping department and housekeeping consultants on the method and cost of maintaining new or renovated structures.

22. *Poor Attitude of Labor Organizations.* Labor organizations may fail to convey to their members a proprietary attitude about sanitation and housekeeping where they work.

13 MONEY-SAVING BENEFITS OF GOOD HOUSEKEEPING

A sound housekeeping program will provide you with these benefits, all of which ultimately reduce costs:

1. Better hygienic conditions, leading to improved health and decreased absenteeism.
2. Improved safety.
3. Less fire hazards.
4. Material protected from damage from soiling or contamination.
5. Improved morale, because of better overall appearance of the facility and general conditions.

6. Less wear and resulting premature repair and replacement of surfaces.
7. Promotional benefits to the organization by public acceptance and aesthetic appreciation.
8. Enhanced quality control and reduced inspection time.
9. Reduced training and orientation time for new or transferred employees.
10. Greater flexibility to handle emergencies, special assignments, changing weather conditions, etc.
11. Less turnover and absenteeism in the housekeeping department, because of improved status and morale.
12. Better internal communications between the housekeeping department and other departments.
13. More accurate budgets.

The future undoubtedly will bring surprising developments in the technology of housekeeping equipment. You may see devices that use ultrasonic, electrostatic and spray cleaning, so that a device placed in the center of a room would loosen all soil from every surface in that area and collect it on or within the machine for disposal.

Housekeeping must not fall behind in this rapidly changing world. You must be prepared to use new materials and equipment and new ideas and methods, as they are developed. In the meantime, there is a more urgent problem: to take advantage of the tools already at hand and to apply existing knowledge and experience to obtain the yet unrealized returns from your housekeeping operations.

The size of your housekeeping staff should be at that point where the additional cost of another custodian would not provide the organization with a suitable return on the investment. In other words, where the decrease of one custodian will cause a dollar loss considerably beyond the total cost of maintaining that custodian's job.

HOW TO PINPOINT THE STRENGTHS AND WEAKNESSES OF YOUR HOUSEKEEPING OPERATIONS IN SEVEN MAIN AREAS

Where does your facility stand with respect to housekeeping? Check your answers to these simple but revealing questions. The results should not only prove interesting but may also help you to pinpoint the strengths and weaknesses of your present operation.

Your ability to answer all of these questions positively relates directly to your communications with management—not once on each subject, but continually on many subjects. Do not let your silence indicate your disinterest.

1. General Management

- Is management satisfied with present levels of housekeeping?
- Does top management show an active interest in housekeeping?
- Has management voiced any objectives to housekeeping work?
- Are you fully aware of, and adhering to, all regulatory statutes affecting housekeeping?
- Have you ever undertaken an employee awareness program to build up interest in, and cooperation with, better housekeeping?
- Have reasonable standards been established as goals for the housekeeping program?
- Have you ever evaluated the general morale of the housekeeping staff?
- Have you taken any steps to improve morale, or maintain it at existing levels?
- Is the housekeeping department given the opportunity to select materials, etc., for new construction or renovation?

2. Cost Control

- Have you recently calculated your total annual housekeeping costs, including all costs relating to the cleaning function?
- Do you have some way to compare your housekeeping costs with those of similar facilities?
- Has housekeeping been costed out by department or by job?
- If yes, are these figures adequate to reveal unusual situations?
- Could some housekeeping jobs, currently performed by outside contractors, be handled at a lower cost by your own housekeeping department?
- Could certain specialized cleaning jobs, such as window washing, be performed more economically or more safely on contract?
- Have you studied current housekeeping work methods and procedures to see if savings could be effected?
- Has your facility taken full advantage of the economies of mechanizing housekeeping?

3. Work Standards

- Have you made a time study of the various housekeeping jobs?
- Do you have a basis for comparing your frequencies of housekeeping job performance with those found at other well-run establishments in the same type of facility for the same general types of areas?
- Do you know whether the labor-hours invested in housekeeping are consistent with the labor-hour requirements for the job to be performed?
- Are both daily and project cleaning performed on a scheduled basis?
- Have acceptable performance levels been established for housekeeping workers?
- Do cleaning schedules allow for seasonal variations?

4. Organization

- Is housekeeping supervised on a centralized basis wherever practical?
- Have you studied your housekeeping structure recently to determine whether a reorganization might bring about better results or savings?
- Have definite lines of authority and reporting functions been established for the housekeeping department?
- Are custodians performing tasks other than housekeeping (e.g., serving as messengers or chauffeurs, performing errands, etc.)?
- Have you ever considered applying the group-leader principle to reduce supervisory costs and improve supervisory efficiency?
- Do you use blueprints and floor plans to help assign housekeeping tasks?
- Does your housekeeping program permit variations in the size of the staff (due to changing production levels, seasonal variations, etc.) without disturbing the basic scheduling?
- Can relief be provided for absent housekeeping workers without disrupting cleaning schedules?
- If a new building or wing is put up, do you have an accurate means of predetermining housekeeping requirements and costs?

5. Custodial Efficiency

- Do you have a formal plan to improve workers' skills and efficiencies through a continual training program in housekeeping methods and procedures?

- Have you fully outlined the extent of each custodian's responsibility to him or her?
- Does each custodian fully understand these responsibilities and their importance?
- Have you ever tested custodians to determine their understanding of their jobs and their duties?
- Do you adhere to specific qualifications for the hiring of housekeeping personnel?
- Do your housekeeping workers consider themselves as part of a team of specialists, responsible to a large degree for the safety of coworkers and the care of facilities?
- Are your workers provided with the latest developments in housekeeping materials?
- When purchasing supplies, are you avoiding "the cheapest and the least"?
- Are physical examinations given to prospective housekeeping employees?

6. Record Keeping

- Are adequate records maintained to ensure proper protection of surfaces at proper intervals (for example, the resealing of floors)?
- Are records maintained of the suggestions and complaints about housekeeping, the cost or amount of time necessary to remedy the situation, and the source?
- If so, is the complaint record studied periodically to determine how the general situation can be improved?
- Are regular reports made to enable management to check on housekeeping progress?
- Are regular reports made to enable management to check on supervisory performance?
- Are regular reports made to enable supervisors to check on custodial performance?

7. Facility Operations

- Is your accident record affected to any appreciable extent by faulty housekeeping procedures?
- Do you have fire hazards which better housekeeping would remove?

- Do you conduct regularly scheduled inspections of housekeeping and sanitation to avoid safety, health, and fire hazards?
- Has the safety department been consulted concerning housekeeping practices?
- Have housekeeping schedules been arranged to permit maximum efficiency along with minimum disruption of other work?
- Do you have an efficient system for stocking, inventorying, and distributing housekeeping equipment and supplies?
- Are custodial storage facilities adequate from the standpoints of size, number, location, ventilation, lighting, furnishing, and plumbing?
- Do employees complain about hand cleaners, odors in rest rooms, and other personnel items?
- Do you operate your housekeeping system as *preventive* maintenance rather than *corrective* maintenance?

Your answers to these questions may indicate that your housekeeping operation is more fraught with potential improvements than you might at first have suspected.

Management is now fixing its eye on the target of improving indirect costs. Housekeeping is the bull's-eye in that target.

Good housekeeping is being accepted as good economy by serious businesspeople. Most successful facilities are clean facilities, not because the fruits of success have permitted good housekeeping to follow, but rather because the viewpoint of management that provided a clean facility in the first place led to success. This growing realization brings many worthwhile changes.

ELEVEN STEPS TO OPTIMIZE THE EFFICIENCY OF YOUR CLEANING DEPARTMENT

You must develop a comprehensive program encompassing all variables relating to sanitation. The following eleven steps will help you improve the efficiency of your cleaning department.

Step 1: Use Work Measurement Techniques to Determine Your Cleaning Needs

To determine how many people you need in your cleaning department for a given quality level, or to determine what is a reasonable day's work for an individual custodian, you have two choices: guesswork or measurement.

Guesswork relies on experience, national averages (you do *not* have an average department!), experience exchanges with other organizations, trial and error, etc. Many failures in custodial management can be attributed to guesswork.

Measurement is the only approach that management, supervisors, and workers will believe. Using measurements for housekeeping, as described in Chapter 2, is the only way of avoiding the custodian's biggest complaint: that he or she was given more work to do than someone else.

Measurement also provides the only possible defense against arbitrary staffing cuts, because you can demonstrate the frequency and therefore the quality changes that are required by such a cut.

Chapter 2 provides details on how to determine your organization's cleaning needs.

Step 2: Train Cleaning Workers

Custodial workers obtain their information on how to do their work in one or more of these ways, in the absence of a formal training program:

- Television commercials promoting cleaning products
- Observing other custodians at work on various jobs
- Trial and error

All of the above systems lead to numerous errors, and even to property damage and accidents. Most cleaning commercials and advertisements showing people at work depict the work being done in a difficult or incorrect way. For example, the cover of a recent magazine showed a custodian mopping a lobby, with the worker bent far over (straining the back) and using the push-pull mopping technique, which is very ineffective. To overcome these problems, you need to provide a positive program of training, involving these three steps:

1. Orienting the new worker, an extremely important activity because this is a first impression.
2. On-the-job training, possibly provided by a group leader.
3. Classroom training, which is often overlooked or mishandled.

Classroom training not only provides useful information through slides or videos, but also improves the workers' morale by making them feel important—especially if a certificate is given for complete attendance. Guest speakers can enhance such a program.

For a classroom training program to be effective, it needs to be held at fixed dates, not catch-as-catch-can.

One of the greatest benefits of a classroom program is the development of the speaker. The speaker for "how to do it" subjects should be the first-line supervisor for a given work group. Such people have often risen through the ranks, and the self-confidence and separation from the workers that training classes provide is especially helpful. Chapter 13 discusses training in detail.

Step 3: Stimulate Management Involvement

Management must be educated to the relationship that exists between *quality and cost* in housekeeping. Many executive housekeepers or supervisors have had the experience of having someone in management come to them and say "we must get this place cleaned up," with a resulting crash program of hiring additional personnel, letting supplementary cleaning contracts, performing overtime work, and the like, until things seem to look the way management wants them. Yet, some months later, that same manager may return to the housekeeping department with a demand that "you've got to save a good deal of money; we're spending too much for cleaning!" The fact is, in general, management has no method of determining just how much cleaning is obtainable for a dollar spent in this direction.

It is up to you to prepare the figures to demonstrate, for example, that the cost of cleaning is *not* directly proportional to the production level or number of employees in an industrial plant, nor to the number of patients in a hospital, nor to the number of students in a school, nor to the number of guests in a hotel. There are relationships, but not *direct* proportions.

Further, you must be able to demonstrate mathematically the quality effect of either an increase or decrease in dollars spent for cleaning, through the required changes in frequency resulting from fluctuations in labor-hour provisions. Such information must concern itself with both daily cleaning, as well as the less frequent periodic cleaning. Where a fluctuating staff is unavoidable, it becomes your responsibility to arrange the work so there is the least disturbance to the long-term quality result.

Management must define interdepartmental responsibilities. An example of this is re-lamping, where in a given organization the housekeeping department might be required to re-lamp a given fixture on one occasion, yet the re-lamping may be done by electricians on another occasion. In such a case, the responsibility is not fixed and frictions and complaints will occur. "Gray areas" of this type should be eliminated by written directives from management.

It is management's responsibility to serve itself as well as its housekeeping department by better planning for construction and renovation, so that new surfaces, fixtures, or equipment are not installed that are overly difficult to maintain. Unfortunately, architects cannot always be relied on to show an interest in the maintainability of the items they specify. Thus, management is required to provide the architects and interior designers with a checklist of preferred surfaces, colorations, specifications, and the like. The potential savings here are impressive.

The housekeeping department badly needs official management recognition and support for the service that it is providing. A very positive way in which management can become involved is by developing an interdepartmental cooperation and employee awareness program. Although such a program requires a little thought, it involves only nominal amounts of money; yet the results can be significant. The program can secure benefits to a housekeeping department equivalent to the addition of several personnel to that department—at almost no cost.

Only management can provide *organizational integrity.* Avoid "leaks" to the cleaning department; if not controlled they can lead to destruction. Don't pull people out of the cleaning department to perform every other conceivable job that no other department wants to do or has no time to do. For example, a cleaning operator running a power sweeper may be interrupted from his work during a given day to assist in unloading a truck; then to deliver a package to another plant; then to run an errand for the plumber. These interruptions tend to prove to the custodian that, if he can be interrupted to handle anything else that comes up, then his cleaning job must not be important! If utility jobs are to be done, then utility personnel should be utilized for doing them; where cleaning work is to be done, then custodians should be used.

Management should consider contract cleaning as a legitimate alternative consideration, but not as a panacea or an easy way out.

The cleaning department should have its activities coordinated with other departments, such as grounds and maintenance. It is difficult to motivate one custodian to perform a satisfactory policing job in an office area when he has but to look out the window to see grounds that are badly littered; another custodian cannot understand the need to remove spots from a wall that obviously requires plastering or painting.

Finally, management will never achieve its optimum cleaning objectives without a realization that housekeeping represents a legitimate *investment* toward the achieving of the organization's fundamental objectives. Sometimes this requires selling management by emphasizing the benefits of a sound housekeeping program:

- Protects health
- Promotes safety
- Eliminates fire hazards
- Improves morale
- Avoids damage to materials
- Extends surface life
- Protects equipment
- Improves productivity
- Improves quality control
- Provides good public relations
- Brings optimum cost

Step 4: Develop a Workable Organization

The basic requirement for any organization is fixed lines of authority and responsibility, wherein each worker knows to whom he or she reports, and each supervisor knows the limit of his or her authority. When more than one supervisor can give orders to an individual cleaner, you have a condition worse than where the cleaner would have no supervisor at all; the worker is demoralized through conflicting instructions, and is soon looking for another job. A good test of this is to attempt to draw an organization chart, not representing the *desired* organization, but the actual *existing* conditions. If this chart is confused and complex, this can be a good clue to the need for reorganization.

Figure 1-1 shows a sample organization chart for a university hospital.

You should develop the department structure by analyzing all variables, rather than by simply trying to improve a few facets of the operation. You will achieve the best results by using the synthetic approach, wherein you develop a hypothetical situation to solve observed problems and conditions, without reference as to how the work might have been done in the past. Then, try to install this ideal operation, making whatever changes are necessary because of personality problems and other conditions.

No organization can function properly without open lines of communication. Custodial workers must be able to approach their supervisor—and go on to successively higher levels of management when they are not able to obtain satisfaction—with problems, observations, and suggestions.

Step 5: Schedule for Preventive Cleaning

The department should perform its cleaning functions through a *preventive* maintenance program rather than a *corrective* one. If you wait until floors apparently need stripping before taking any action, you will have waited too long. Floors need stripping *before* they appear to need it, as judged by trained personnel. This approach requires a scheduled frequency of special projects, utilizing a calendar or data-handling system. Since special assignments and crash programs are often inefficient, the preventive program, while providing a considerably higher quality level, generally costs no more than a corrective system.

The ideal housekeeping department will have an optimum number of cleaning workers; this means that if one additional worker were added to the payroll, the cost would exceed the benefit—direct and indirect—to be gained from the additional services, while, conversely, the reduction of the staff by one person would cost the organization more than that worker's total wages, benefits, and support cost in terms of damage to surfaces, decrease of production, quality control, and the like. No serious consideration can be given to optimum number until all aspects of the productive work—

Figure 1–1. Sample Organization Chart for a University Hospital

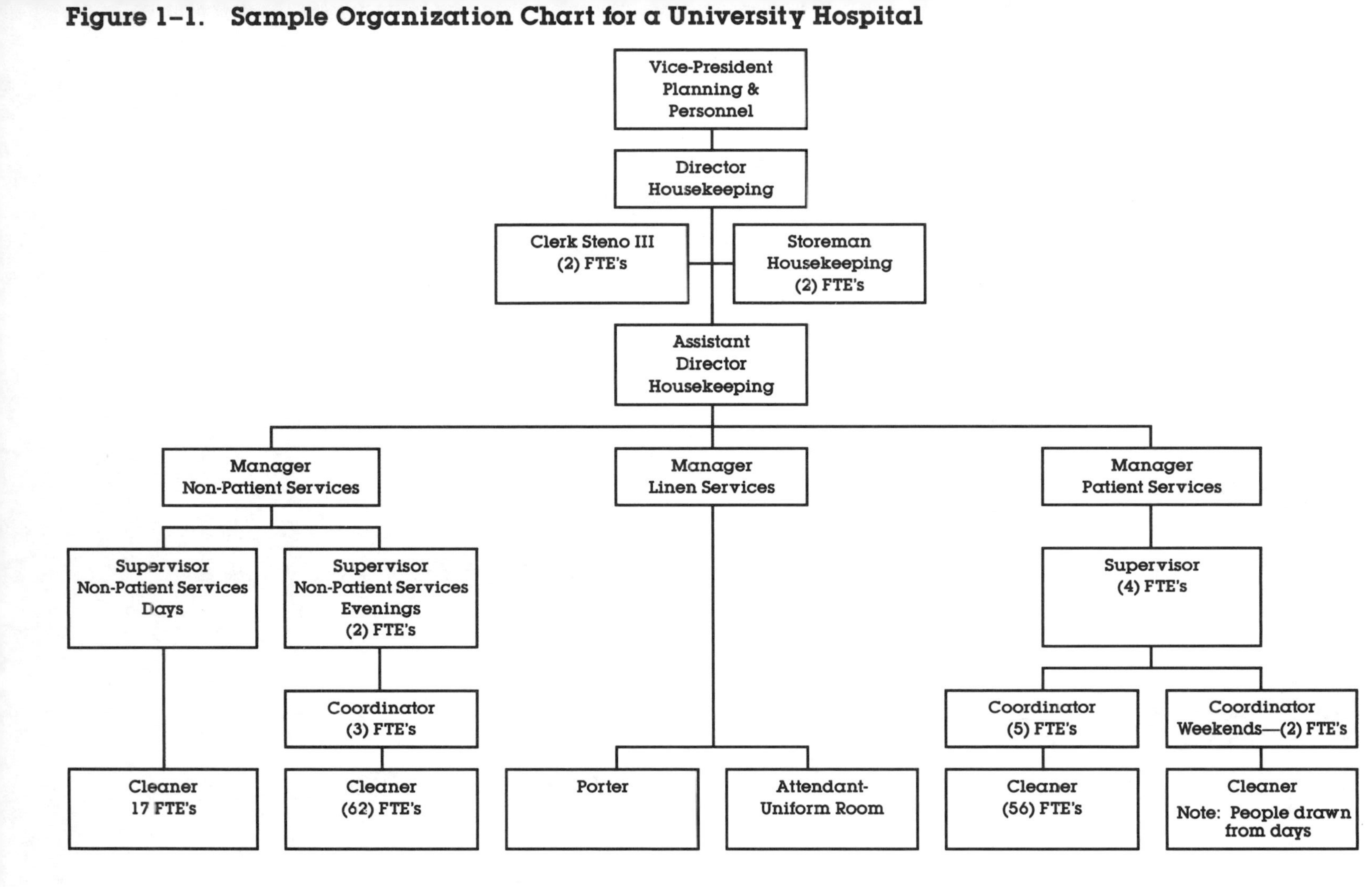

method, equipment, chemicals, frequency, custodial facilities, and so on—are considered. You want to limit the work done by each person to a reasonable day's work over a reasonable period of the day.

Complaints about excessive work load often indicate an *imbalanced* work load. When Henry, the custodian, complains "Boss, you've given me too much work to do!" what he probably means, possibly without quite realizing it himself, is "Boss, you've given me more work to do than you gave Jane!" Balanced work loads do not mean that each person should clean three rest room fixtures, twenty-three desks, and two and a half corridors; the effort required to perform each job assignment should be essentially equivalent, within the physical limitations of the worker. The department should be structured, and the cleaning program designed, to utilize the proper balance of men and women.

In staffing the department, pay attention to the recruiting and selection procedures. It is as bad to hire over-qualified workers as it is to hire those who are under-qualified; in the first case, the workers will not remain on the payroll very long, and a considerable turnover will be experienced, while in the latter case the workers will not be able to perform properly. You must place limits on the type of people who will be hired into the department, within organization rulings, of course. It is far better to concentrate on improving the morale and productivity of existing workers, when the department is understaffed, than to hire people where experience indicates that they will not stay around very long. In staffing, consider the use of part-time workers, special shifts for those who might work only four or five hours a day, handicapped personnel, and other relatively untapped sources of labor.

In scheduling work, it is first necessary to determine frequency, the basic determinant of cleaning quality. A carpet vacuumed daily will of course be at a higher quality level than one vacuumed weekly; although the time requirement for each performance of a cleaning job often increases as the frequency is lessened.

Finally, the work should be performed at the proper time. Typically, this is during those hours when the facility is at its lowest occupancy level. In an office building, this should be a late evening shift; in an industry, the third shift; in a school, evening shift if there are no evening classes, but the third shift if there are; in a hospital, an evening or night shift for administrative areas, but the day shift for patient areas. The times must also be selected with a consideration for transportation problems.

Step 6: Provide Effective Supervision

The housekeeping supervisor is the backbone of the cleaning operation. Although the supervisor is not expected to perform any of the cleaning work, it is his or her responsibility to motivate everyone else in the department to do the work in the way that management has developed it.

In larger organizations, a pyramid of supervision is required. With a cleaning force of one hundred or more, for example, you may have a department head to whom several supervisors report; to each of them might report two or three group leaders, as well as all the actual cleaners. In a still larger organization you might have assistant department heads on each shift, or assistant supervisors reporting to each area supervisor.

In smaller operations, a full-time department head, who is also the supervisor, is certainly justified for any department of ten or more personnel. For less than this number, the supervisor might also have additional responsibilities, such as security, grounds, maintenance, and the like. But even in a small operation, it is desirable to have some position of advancement for cleaners, such as to "group leader" or even to a position which might be called "senior custodian," which while not providing any authority, does give an additional compensation reward and title for exceptionally good work and service, and might also be used as a training aide for new workers.

The first requirement of the supervisor is a high motivational ability; this is more important than technical proficiency, experience, age, or any other such factor, since all of these things can be overcome in time, but without motivational ability the supervisor becomes a figurehead.

Don't overrate experience. Much previous experience in housekeeping involves methods that are now obsolete or will soon be obsolete, and a lengthy experience may merely indicate a repetitive history of use of these old techniques. On the other hand, a history of experience exchange is desirable, where the supervisor has taken the opportunity to visit other facilities, takes membership in such desirable organizations as the Environmental Management Association, the National Executive Housekeepers Association, or the Canadian Administrative Housekeeping Association, has read materials and attended seminars on cleaning, and the like. After such a person has been hired, this experience exchange should be continued and even intensified. So much is being changed in the cleaning field that it is impossible to keep up to date within the confines of a single physical facility.

In time, the supervisor must become quite knowledgeable about the techniques of cleaning, not only to be able to schedule the work and estimate the time requirement for it, but also to be able to command the fullest respect of custodial workers.

Part II discusses supervision in detail.

Step 7: Improve Cleaning Methods

Custodial workers do not instinctively find the best way to do a job. Therefore, you have a simple choice: either prescribe a method and see that it is utilized on a standardized basis, or abdicate this responsibility and let the workers determine for themselves what is the proper method. It is the very same principle that applies to time standards: either you determine the rate at which jobs will be performed, or, in the absence of this direction, the workers will determine it for themselves. This leads us directly to Parkinson's Law, where "work expands so as to fill the time allowed for its completion."

The utilization of effective methods can best be guaranteed when a theme of "let's work smarter, not harder" (assuming a reasonable effort) pervades all training and methods improvement. Workers are principally interested in the effect of change on them rather than how the company can make another few thousand dollars.

It is extremely important to provide a number of personnel who will not be assigned on a repetitive basis to fixed jobs, but who are available for relief of absent workers so that each job assignment will be performed daily; and when not required for this relief, these special personnel will be able to handle project and utility work.

Step 8: Provide Adequate Custodial Facilities

The first thing a construction contractor does when building any type of structure is to prepare a temporary housing for office, personnel, and storage requirements; in other words, the contractor provides work facilities. In a housekeeping department, however, you may give little thought to the provision of custodial facilities. Thus, you may find the central storage area in an inaccessible part of the basement, or the custodial closet under a back stairwell. Where adequate area is provided in original construction, it is often appropriated by other "more important" departments. With serious morale and motivation problems to begin with, it is doubly poor practice to require custodians to do their work without supplying the means to do it.

The first requirement is a central storage area, where the bulk chemicals, large equipment and spare parts can be stored, and where such duties as dust mop treatment; lubrication, repair, and cleaning of equipment; venetian blind cleaning; and other items can be performed. Having adequate space and storage racks for purchase of chemicals in quantity can save considerably in purchase price.

The basic work station is the custodial closet—or "janitor closet," shortened to "J.C.," as seen on blueprints—where the custodian's supplies are stored. Janitor closets are still being provided in new buildings with a typical floor area of ten square feet; this is supposed to take care of the utility sink, all chemicals, supplies, parts, mopping outfits, and motorized equipment! *No more than one percent of the architects in this country have any conception of how cleaning is being done in the buildings they design, and what provision for these supplies and equipment should be made in custodial closets, and where these closets should be located.*

Where inadequate provision for custodial closets has been made, you can purchase prefabricated cabinets, which are better than nothing. Some of these are well designed, and even have water and drain provisions.

A specialized aspect of these custodial facilities is the cleaning cart, which becomes a mobile work station to carry chemicals, manual equipment, a waste receptacle, and the like, thus avoiding numerous trips from the work area to the custodial closet. These carts are inexpensive and last a number of years, especially with repainting. The custodial closet should be designed to contain these carts when not in use.

Custodial facilities should include a distribution system for expendable materials and supplies. Avoid the need—as well as the excuse—for custodians to return to the housekeeping department office at various times during the day for additional supplies. These should be placed in their custodial closets or cabinets, perhaps on a weekly basis, to fulfill all their expected needs for a reasonable period of time.

Figure 1-2 provides a checklist for the condition and contents of a custodial closet.

For older facilities that have not been planned well for custodial operations, or where custodial facilities are inadequate, provide bibb faucets adjacent to fixtures in rest rooms. According to code, these bibb faucets must be fitted with a siphon breaker to avoid sucking dirty water into the clean system, and should also have a bracket so that a bucket may be held on the fixture without damage.

Chapter 21 discusses custodial facilities in detail.

Step 9: Supply Effective Cleaning Equipment

You wouldn't hire someone to paint your house and then provide a toothbrush to do the job; yet this is done regularly in housekeeping operations. A custodian is often required to clean a large floor area with a small bucket, an inadequate mop, and the wrong type of wringer with which to do this job; or mechanical equipment which is often the wrong size or type. Providing adequate custodial equipment can dramatically improve your cleaning operations in a relatively short period of time.

First look at the *number* of items of equipment. If you have scheduled assignments for each worker, this schedule will include certain equipage; if this is not provided, then the assignment should be changed, which in turn will change the method and frequency, and thus the quality. The companionship of equipment and assignment is inescapable.

Providing equipment of inadequate *size* is the mistake most often made in its selection. A 17″ single-disk floor machine is *not* essentially the same size equipment as a 19″ machine; calculations of the area of the brush in this case will indicate that the machine which is two inches larger in diameter will be about 25% more efficient. Purchasing agents, naturally, are anxious to save their organizations as much money as possible, but must be acquainted with what does and does not constitute an "or equal" situation. You must also bear in mind the advantages of standardization. If you have a dozen floor machines, and they come from six different manufacturers in seven different sizes, you have created a considerable problem in stocking various brush attachment plates, pad sizes, replacement switches, and the like. In pricing

Figure 1-2. Custodial Closet Checklist

Checklist for Assignment ________________ E-SA-SW ________________

CLOSET (location or room # ________________)

- ☐ Clean & Orderly
- ☐ No Unauthorized Items
- ☐ Securable
- ☐ Sink ☐ Clean ☐ Hose
- ☐ Shelves Adequate
- ☐ Tool Holder
- ☐ Light ☐ Bulb protected
- ☐ Odor Free
- ☐ Measuring Cup
- ☐ Funnel
- ☐ Pail ☐ Clean

- ☐ Gallon of Glass Cleaner
- ☐ Dust Mop Heads
- ☐ Gallon of Germicidal Detergent Concentrate
  - ☐ Labeled ☐ Jug pump
- ☐ Gallon of Germicidal Detergent Solution
  - ☐ Labeled ☐ Mixed correctly
- ☐ Container of Lotion Cleanser
- ☐ Container of Stainless Steel Cleaner and Polish
- ☐ Supply of Liners for Cart
- ☐ Supply of Liners for Trash Receptacles
- ☐ Supply of Disposable Cleaning Cloths
- ☐ Gallon of Neutral Detergent Concentrate
  - ☐ Labeled ☐ Jug pump
- ☐ Gallon of Neutral Detergent Solution
  - ☐ Labeled ☐ Mixed correctly
- ☐ Rest Room Supplies

- ☐ Folding Cart No. ________ ☐ Clean & Orderly ☐ Fireproof Ash Receptacle
- ☐ Tool Caddy ☐ Clean
- ☐ Spray Bottle Neutral Detergent ☐ Mixed correctly ☐ Labeled
- ☐ Spray Bottle Glass Cleaner ☐ Labeled
- ☐ Spray Bottle Germicidal Detergent ☐ Mixed correctly ☐ Labeled
- ☐ Lotion Cleanser
- ☐ Stainless Steel Cleaner and Polish
- ☐ Lambswool Duster ☐ Clean
- ☐ Putty Scraper
- ☐ Disposable Cleaning Cloths
- ☐ Dustpan ☐ Clean
- ☐ Rubber Gloves ☐ Safety Goggles
- ☐ Sponges ☐ Clean
- ☐ Scrub Pads
- ☐ Utility (Scrub) Brush ☐ Clean
- ☐ Toy Broom
- ☐ "Closed for Cleaning" Sign
- ☐ Large Foam Eraser ☐ Clean

WET DRY TANK VACUUM NO. ______

- ☐ Clean ☐ No Repair needed ☐ Hose & Wand
- ☐ Crevice Tool ☐ Brush Attachment
- ☐ Floor Tool ☐ Floor Squeegee

FOR REST ROOMS

- ☐ Acid-Type Bowl Cleaner
- ☐ Bowl Mop ☐ Clean
- ☐ Soap Film Remover
- ☐ Sponge ☐ Different color
- ☐ Clean ☐ Inspection Mirror

SPRAY-BUFFING

- ☐ Floor Machine No. ______
- ☐ Clean ☐ No repair needed
- ☐ Pads ☐ Clean
- ☐ Spray buff solution

Figure 1-2. (continued)

NON-CARPETED AREAS	CARPETED AREAS	
☐ Small Dust Mop ☐ Clean	☐ Upright Carpet Vacuum No._____	☐ Clean
☐ Large Dust Mop ☐ Clean		☐ No repair needed
☐ Extra Dust Mop Covers	☐ Hose & Wand ☐ Crevice tool	☐ Brush Attachment
☐ Push Broom ☐ Clean	☐ Stain Remover ☐ Gum Remover	
☐ Corner Scrub Brush ☐ Clean		
☐ "Doodle Bug" Brush ☐ Clean		
☐ 2 "Caution - Wet Floor" Signs ☐ Clean		
☐ Wet Mop ☐ Clean ☐ Deck Brush (Grouted tile only)		
☐ Mop Bucket & Wringer ☐ Clean ☐ Extra Mop Heads		

competitive bids for such equipment, give preference to a quotation that represents a move toward standardization.

There is also the question of *type* of equipment. An example of this is with carpet vacuums, where the type ranges from a light-duty surface vacuum, through a very aggressive pile-lifting vacuum, with various grades of surface and pile-lifting vacuums in between. In some cases, more than one type of vacuum will be needed, perhaps a surface vacuum for daily use and a pile-lifting vacuum for intermittent use, especially after the dry-cleaning process, for example.

Regarding preventive versus corrective maintenance, a specialized type of equipment is entrance matting. Providing suitable mats or runners at entrance areas can trap a good deal of soil and water, where it can be quickly and easily removed, rather than have this soil and water tracked throughout the building. Both rental type matting systems and those which are owned and cleaned internally can be useful. An extension of this program involves carpeting elevators and stair landings, to trap soil in these areas. Naturally, the policing program must include attention to these devices.

Chapters 22 and 23 discuss equipment in detail.

Step 10: Purchase Efficient Sanitation Chemicals

Cleaning chemicals are just another aspect of the "tools for the job." Returning to the analogy of the painter painting your house, you recognize that if a good deal of labor is going to be spent in this work, you will want a good quality of paint, so you don't have to do it over again very soon; this also applies to floor finish. It is demoralizing to cleaning workers to be required to put down a finish which they recognize is going to require a good deal of buffing, and have to be stripped much more frequently than otherwise would be required; that it might not be scuff resistant, and might be too slippery, thus presenting an endless source of complaints.

The chemical must be safe, both from the standpoint of avoiding hazards to the public (such as a slippery wax), as well as hazards to the user (such as a harsh cleaner that is damaging to the skin). It should be easy to use, and not require excessive stirring or working. Convenient packaging is helpful, so that excessive time is not used for decanting and measuring.

Although chemicals amount to just a percent of the total cleaning costs, their consumption should be controlled, not only to save money, but also to provide proper use dilutions. Over-use of a detergent, for example, can cause so much foaming that an extra rinsing operation is required, or a detergent film is left on the surface which looks dingy and attracts soil. Measuring cups and dispensing pumps, as well as portion packaging, are all useful consumption-control devices.

Chapter 24 discusses cleaning chemicals in detail.

Step 11: Develop Positive Morale

There are many approaches which you can take to alleviate a morale problem, many of which involve little or no expenditure. A number of these have already been discussed in terms of management's responsibilities: the development of inter-departmental cooperation; publicity programs; employee cooperation drives; recognition of the value of the work being done by the cleaning personnel. Morale is discussed in Chapter 14.

SUMMARY

Whether you are running the housekeeping department in an office building, industry, university, hospital, public school, or other facility, or whether you are a resident manager in any of these organizations for a contract cleaner, as the housekeeping manager you are, in a sense, in business for yourself. You have a given amount of money to spend and an objective to attain in terms of quality. Like a business, you are beset by competition on all sides: from individuals who want your job, both inside and outside your organization, to external organizations who want to disband your whole department. In any case, you can develop your own security and serve your organization best, by professionalizing your work, avoiding the building of empires (which are so vulnerable) and by developing an optimum department by considering all the variables involved, rather than through piecemeal and sporadic improvement.

2

How to Measure Your Department's Cleaning Needs: The Custodial Survey

The first step in improving your housekeeping program is to carefully analyze the complete operation. Establish performance standards and set up systems and controls to eliminate problems in output, methods of performance, and materials.

In analyzing your housekeeping operations, you should follow the same generally accepted steps for conducting any analysis:

- Define the problem.
- Obtain all relevant facts.
- Critically examine and test these facts.
- Propose methods and courses of action.
- Decide on the proper course of action.
- Act upon your decision.
- Make further improvements periodically where possible.

This chapter shows how to measure your cleaning needs by conducting a housekeeping survey.

HOW TO DETERMINE YOUR HOUSEKEEPING OBJECTIVES

The general and specific improvement of housekeeping requires both long-range and short-range goals. Long-range goals should be in the direction of professionalizing the housekeeping field. Your organization can move more rapidly and on a broader scale by adopting standards for the following:

- Job titles and definitions
- Frequencies of job performance
- Materials specifications (This can include use dilutions, container sizes, and possible color and order coding for identification.)
- Standard times
- Employee selection qualifications, including physical examination and education requirements
- Definition of housekeeping responsibilities

The only possible short-term objective that can yield worthwhile results is a comprehensive approach to housekeeping and cleaning from a managerial and engineering standpoint. Naturally, this will never be done unless you convince management of the value of such an undertaking. Consider obtaining the services of an outside consultant to help "sell" the program. You might also consider forming a special housekeeping committee to investigate the problems and evaluate potential benefits from a program. Such a committee might include personnel from operations, maintenance, quality control, industrial engineering, industrial relations, safety, purchasing, and housekeeping departments. Decide how the work is to be done, whether by internal personnel, through the use of a consultant, or both.

HOW TO TARGET YOUR CUSTODIAL WORKERS' NEEDS

Management has suffered long enough from placing the responsibility for organizing housekeeping in the hands of the workers. Because they lack qualified technical guidance, in their attempt to solve their problems and ease the work load, the custodian and the supervisor have traditionally used a trial-and-error system when choosing cleaning methods, materials, and equipment. Unquestionably, some of the methods developed in this way have become useful and are desirable, but, on the other hand, many of them are inefficient and needlessly expensive.

Unfortunately, the "do-it-yourself" approach is actually recommended by a few sources. One booklet suggests that the *custodian* should list all the duties to be performed daily, weekly, monthly, etc., and should prepare the work schedule! Nothing could be more damaging to the progress of housekeeping and sanitation. Do not allow the untrained custodian to determine the frequencies, schedules, or standards of custodial maintenance. Despite the suggestion that the custodian's recommendations should always be given top priority in the purchase of operational and maintenance supplies, this does not lead to sound purchasing and material-use practices. Housekeeping is everybody's job, but it must begin at the management level.

When you curtail the work force so severely that your workers cannot perform their jobs, you create a situation where they can perform only corrective maintenance; that is, you are "putting out fires." Under such conditions, the housekeeping force will spend most of its time with repetitive daily work and be unable to accomplish such intermittent projects as rewaxing, stripping, sealing, wall washing, light and vent cleaning, general vacuuming, and so forth.

The effects of understaffing in the housekeeping department are not always immediately apparent, but after a time the effects may become drastic and expensive. For example:

- unprotected composition flooring will wear thin and have to be replaced at considerable cost, or hard floors may be stained irreparably.
- soap film on lavatories, scale in toilet bowls and urinals, and excessive dust on air vents and surfaces may all carry and transmit disease.
- floors and baseboards will become darkened due to wax buildup.
- rest rooms will become offensive in appearance and odor.
- woodwork will darken due to excessive dust retention.
- illumination intensity may be reduced by 50% or more because of dust on light fixtures.
- carpeting may be damaged by the fibers breaking or loosening when particles of soil or sand are not removed frequently through vacuuming and dry cleaning.

These declining conditions affect the protection of surfaces in the facility, the appearance of your facility, and whether your facility is a sanitary work place. Moreover, the expense required to live with and later rectify such conditions greatly exceeds the cost of preventing their occurrence in the first place.

In addition, allowing housekeeping workers to determine their own work load may bring only limited results for other reasons. First, housekeeping is not an exact science—it is "loaded" with inexpressible variables. An accurate evaluation can only be provided through extensive research and experience with a variety of housekeeping situations. This point is made positively when you examine the many variables you must consider to determine the time required to maintain a particular floor area. For example:

- The variables of the person doing the work:

Ability to work with others	Manual dexterity
Age	Morale
Attitude toward change	Physical strength
Experience	Quality of supervision

Fatigue factors
Health
Intelligence and literacy
Intensity of supervision
Training
Motivation and incentive

- The variables of the surface to be cleaned:

Condition (smoothness, porosity, damage)
Degree of soiling
Type of soil
Total area involved
Traffic
When and how last cleaned

- The variables of work conditions:

Amount of obstruction
Interference by other employees
Comfort level
Desired quality of result
Distance to custodial facilities
Evaporation rate
Illumination intensity
Method of performing the work
Quality of materials being used
Type, size, and condition of equipment being used

Second, engineered housekeeping is a recent development; therefore reference material is difficult to locate and is very limited and general in content.

Third, where is the time to do this? The greatest pressures come in other fields, such as equipment maintenance or construction, because of the greater total savings possible. The individual in charge of housekeeping often has more persons to supervise than he can possibly handle, and, in addition, he may have been saddled with responsibility for various functions which just do not seem to fit into other departments or are not welcomed by them.

Fourth, internal personnel have limited housekeeping experience. A housekeeping foreman in a typical facility may be justly proud of twenty years' service in his department. But insofar as this service relates to the exchange of ideas and information on housekeeping, and the application of engineering principles to it, such a work history might well be equivalent to six months of service repeated forty times.

Finally, the communications problem, particularly in the larger facility, presents a limiting factor because of housekeeping's permeation into all facets of the operation.

HOW TO ASSESS MAINTENANCE NEEDS IN SMALL FACILITIES

You might ask how large a housekeeping operation must be before a comprehensive study and program become economically feasible. In smaller housekeeping operations (i.e., those having fewer than a dozen workers), you should provide sufficient schedules to inform workers what work to do, when it is to be done, and how to accomplish it. Although such schedules do not usually involve the complexity and comprehensive implementation phases of larger programs, nevertheless the results of good organization and planning can certainly bring worthwhile benefits.

Thus, although the larger operations should receive a comprehensive analysis and programming, the smaller ones can often most benefit by a limited analysis. In addition to the general improvements in organization and scheduling which this can bring, you can sometimes obtain surprising savings on specific aspects of the housekeeping operation.

GUIDELINES FOR CONDUCTING A CUSTODIAL SURVEY

The first step to preparing a housekeeping program is to take a survey. Other names may be given to this activity, but the purpose is to define the housekeeping problem. To do this, some maintenance managers use voluminous forms, notes, plans, drawings, and printed material. The quantity of the material collected appears formidable and confusing to the novice. The operations manual is then prepared based on an analysis of this survey material.

When arrangements are made for a housekeeping survey, it is sometimes after many months of moving in this direction through conversations, correspondence, and meetings. By the time the surveyor begins, both management and the housekeeping department are eager to learn as quickly as possible the nature of the basic problems and the opportunities for improvement. Don't make the mistake of giving premature pronouncements or judgments which you may later find it necessary to correct, on the basis of a more careful analysis of the survey data. Restrict the survey to a gathering of information and to making personal contacts.

It is difficult to determine in advance how much time will be required for a survey. This will depend on the actual size of the physical facility, the detail with which various subjects will be pursued, the support given to the survey by management, and other factors. Since the program will be based on the survey data, its effectiveness will be limited by the completeness and accuracy of the survey. A good survey, therefore, is essential to the preparation of a sound housekeeping program.

You should carefully plan your survey in advance to make certain that you cover all aspects of the inquiry and that the work consumes the least amount of management and personnel time. The survey agenda must be flexible enough to permit adjustments because of unexpected conditions, such as meetings, or sickness. Proper planning of the survey will set the tone for the program to follow and will assure the greatest possible benefits from the time invested.

The surveyor's role must be to obtain as much useful information as possible. He or she must obtain *all* of the information necessary to prepare an effective housekeeping program and must therefore be adept at working with people at all levels of the organizational pyramid in order to evoke a positive response. The program must not be limited by negative feelings or resistance brought out during the survey by expressions of surprise or distaste at unorthodox methods, voiced or implied criticism, impatience, haughtiness, or condescension.

How to Prepare Personnel for the Survey

Certain data can be prepared in advance of the actual survey, thus permitting the time spent during the formal survey to be of the greatest possible value. Such data should be limited to counts, listings, and measurements not subject to interpretation, such as square footage determinations, fixture counts, the collection and arrangement of plans, etc. This approach is of particular value if a consultant is retained, or internal personnel are costed, on a *per diem* basis. The surveyor should furnish a list of such preliminary data required or desirable, together with an indication of the method of arranging, and possibly even collecting, the information.

Announce the forthcoming survey to department heads and to all housekeeping department personnel so that they will understand what is taking place. Have a meeting of the housekeeping department to indicate the nature and extent of the survey and to point out to them that one of the basic purposes of the program will be to benefit them as individuals. A representative of management should introduce the surveyor, explaining the purpose of his work in positive terms. Reassure housekeeping personnel about the beneficial nature of the study, and, if this can be agreed upon with management, the fact that there will be no direct lay-off as a result of the program.

In some plants the security situation will hamper the work of the surveyor. This is true not only of defense industries, but also of other types of organizations which may have secret processes or products under development. As many steps as possible should be taken to simplify the security arrangements in advance of the actual survey.

The following lists the type of information that should be obtained before the survey begins, so as to minimize the total time required for the actual survey:

- Scaled floor plans of buildings included in the survey (reductions of 1/8 scale drawings are preferred).
- A roster of personnel, including supervision, assigned to the housekeeping function to include:
  - — Classification or grade.
  - — Absentee record for past twelve months (an overall percentage for the department is acceptable).
  - — Number of vacation days earned annually.
  - — Wages and salaries, including payroll taxes.
  - — Direct costs of benefits and fringes.
- Union agreement, if applicable.
- Job descriptions for housekeeping workers, supervisors, department managers, and support staff, if available.
- Copies of written procedures, schedules, responsibilities and other documentation available pertaining to the housekeeping function at the plant.
- A list of chemicals being purchased for housekeeping including:
  - — Generic description (i.e., degreaser, floor finish, etc.)
  - — Manufacturer
  - — Container size (55 gallon drum, spray can, etc.)
  - — Purchase price per gallon
- A list of powered equipment assigned to the housekeeping department to include:
  - — Generic description and size of item (i.e., 20″ floor machine, 6 gallon wet/dry tank vacuum, etc.)
  - — Age of equipment
  - — A brief comment about condition (i.e., inoperable, in need of repairs, excellent shape, etc.).

What to Cover in Your Survey

The surveyor, in knowing what facts are needed, and in knowing the emphasis to place on each field of inquiry, is able to conduct an efficient survey which saves the time of all persons involved. Using detailed and meaningful survey forms is important. Omission of a single basic question may lead to the wrong approach in attempting to solve a problem. The survey forms must be general enough to permit use in a wide variety of situations and yet complete enough so that only a limited amount of special data need be recorded apart from these forms. Much time and error can be saved if the forms are provided with boxes to check, selected words to circle or cross out, and other devices to avoid handwriting. Specialized forms may be used for certain situations having unusual problems or arrangements.

Naturally, the forms must relate to the program document to follow. They must provide all the data needed for the preparation of the program, and, for economy of time, should exclude all extraneous material. An outline of some of the subjects that each of these forms should cover follows:

- Management

Objectives
Limitations
Target completion dates
Distribution of information
Attitudes
Contemplated changes and activities
Planned construction or renovation
Basic statistics
Scope of the study
General observations

- Staff Resources

Departmental organization
Size of the housekeeping staff
Work assignments
Relief arrangements
Emergency scheduling
Individual limitations
Job titles
Hiring qualifications
Specific personnel problems

- Money

Budgetary limitations
Custodial wages
Indirect costs
Cost of equipment and supplies
Contract cleaning costs

- Methods

Job performance frequency
Method of performing each cleaning function
Nonhousekeeping jobs performed by housekeeping personnel

Housekeeping jobs performed by nonhousekeeping personnel

When the work is performed

- Materials

Central storage facilities

Custodial closets and utility sink location

Purchasing methods

Distribution methods

Product specifications and uses

- Machines

Application

Size

Location

Condition

Figures 2-1 through 2-6 provide a useful means of obtaining needed data in a housekeeping survey:

- Area Analysis (Figure 2-1)—This form provides current data for how frequently various housekeeping operations are performed in a given building or area, along with other information about surfaces, fixtures, and the like. Try to avoid obtaining information on *desired* frequencies; instead, record *actual* frequency data.

- Area Summary (Figure 2-2)—This form provides a single summation of each floor of each building. (It could also be done for wings in a large building.) This form should help you apply standard task-time estimates, which should help you assign staff.

- Personnel Analysis (Figure 2-3)—This form indicates which titles are in use, the number of personnel in each shift, pay rates, personnel problems, absenteeism, and other data; it also provides the opportunity to sketch in an organization chart, if one is not already in print and can be attached.

- Personnel Schedule (Figure 2-4)—This form indicates the current assignment, working hours, work days, and the like for each cleaning worker. The designation "maid or custodian" might provide a problem with the equal pay law because all personnel should have the same general title. The age might indicate a number of people about to retire. The information on nonhousekeeping duties will show what period of time is not available for cleaning. Finally, the remarks column might show the use of personnel from whom a reasonable day's work might not be obtained.

- Custodial Facilities and Inventory form (Figure 2-5)—This form provides information on storage areas, custodial closets, measuring and dispensing devices, and other equipment and facilities.

Figure 2-1. Sample Area Analysis Evaluation Form

AREA ANALYSIS

FREQUENCY SCHEDULE

Fill in schedule only where special frequencies are requested. Standard frequencies will be used where the schedule is left blank.

Maintenance Jobs	Frequency Data
FLOORS	
Carpets, shampoo	
Carpets, vacuum	
Floors, buff	
Floors, rewax	
Floors, seal	
Floors, damp or wet mop	
Floors, dust mop	
Floors, strip-reseal	
Floors, strip-rewax	
Floors, sweep	
Floors, scrub	
Rubber Mats, wash & treat	
Stairs, mop	
FURNISHINGS	
Curtains & Drapes, vacuum	
Furniture, clean & polish	
Furniture, dust	
Hardware, clean & polish	
Light Fixtures, vacuum	
Light Fixtures, wash	
Radiators & Vents, dust	
Venetian Blinds, vacuum	
Venetian Blinds, wash	
Wst. Rcptl.-Ashtrays, empty	
Waste Rcptl. (edibles), wash	
Wst. Rcptl. (non-edible), wash	
WALLS, CEILINGS & GLASS	
Ceilings, vacuum	
Glass Part. opaque, clean	
Glass Part. transp., clean	
Skylights & Monitors, clean	
Walls & Woodwork, vacuum	
Walls & Woodwork, wash	
Windows, wash	

FIRM ______________________

AREA ______________________

DATE ______________________

RESPONDENT

INSTRUCTIONS

Fill in a form only for each major type of area. For example:

HOSPITALS: Patient Rooms, administrative, treatment, dietary.

INDUSTRIES: Offices, production areas, cafeteria.

COLLEGES: Classrooms, offices, cafeteria, dorms.

AREA INFORMATION

Indicate the typical condition

1. Degree of obstruction

   None Slight Medium Heavy

2. List any special conditions affecting the scheduling of maintenance tasks ______________________

3. Are there any special problems of traffic or soiling to be considered? ______________________

4. Stairway covering material ______________________

5. Any special problem with glass surfaces? ______________________

6. Rest room floor type in this area ______________________

7. Locker room floor type in this area ______________________

8. Describe any special hardware ______________________

9. Ceiling light fixture type ______________________

10. Circle heating methods: Radiators Forced Air Wall Vents Ceiling Vents

11. Any special waste receptacle problem? ______________________

Figure 2-1. (continued)

RESTROOMS	
Restroom, general cleaning	
Restroom Dispensers, refill	
Bowls & Urinals, descale	
SUB- AREA	
Grounds, police	
Parking & Walks, sweep	
Utility Mach. Area, clean	

ABBREVIATIONS

2/D Two times per day
D Daily
2/W Two times per week
W Weekly
2/M Two times per month
M Monthly
2M Every two months
4/Y Four times per year
3/Y Three times per year
2/Y Two times per year

Y Yearly
2Y Every Two years

These are special designations:
S Every Shift
U After each use
* As required
— Not normally performed

12. Any special furnishings? ______________________

____

13. Circle wall type: Paneled Plastered Tile
Brick Concrete Block Stucco Papered Metal

14. Circle ceiling type: Acoustical Plastered
Papered Wood Metal

15. Circle predominant furniture type:
Wood Metal Leather Mixed

16. Indicate types of window coverings:
Venetian blinds ________ Awnings ________
Shades, inside ________ Drapery ________
Screens ________ Curtains ________

17. NOTES: ______________________________________

____

____

____

____

Figure 2-2. Sample Summary Record for Area Evaluation

FOR ______________________

DATE ______________________

RESPONDENT ______________________

AREA SUMMARY

Fill in one line for each floor of every building to be maintained. Be sure accompanying floor plans are complete and properly identified (show building names, scale to which plan has been drawn, and direction of north).

Floor and Building Designation	Name the Departments, Services and Functions Performed on This Floor	Days and Hours of Occupancy	Remarks (Special Problems or Requirements, Conditions, Appearance, Etc.)	Approximate Floor Percentages									
				Carpet	Ceramic	Concrete	Cork	Marble	Resilient (Composition)	Stone	Terrazzo	Wood	
______ Floor ______ Bldg.													
______ Floor ______ Bldg.													
______ Floor ______ Bldg.													
______ Floor ______ Bldg.													
______ Floor ______ Bldg.													
______ Floor ______ Bldg.													
______ Floor ______ Bldg.													

• Materials Inventory and Equipment Inventory (Figure 2-6)—These records provide data on individual products and specific pieces of equipment. It is particularly important that the equipment inventory indicate not only the number on hand, but the size and condition.

Survey Equipment. For a comprehensive survey, the surveyor would require the following equipment:

- Survey forms
- Graph paper
- Analysis paper
- Legal pads for notes
- Slide rule
- Architect's scale
- Photometer
- Litmus paper
- Steel tape
- Colored pencils
- Portable dictating machine, for some cases

Conducting a survey by a team rather than an individual provides advantages in certain situations, such as where the work must be completed within a given time limit. In such cases, the work must be especially well arranged to prevent duplication and confusion, and to insure completeness. When two persons are taking the survey, a logical division of the work requires one person to perform the basic physical survey while the other performs the basic personnel survey.

Personnel from housekeeping, methods, purchasing, safety, and other departments should be encouraged to participate in the survey. This will give them a proprietary attitude concerning the program, which is important because people support what they help to create. In addition, they will be likely to learn a good deal about their own housekeeping operation, and possibly their own general business operation, in the process.

Management determines the need for a housekeeping survey, and the program to follow, on the basis of recognition of this method as a means to achieve specific objectives. This may be done prior to the survey, or as one of the first steps during the survey.

Gathering Survey Data. In the survey you begin to encounter problems in communications which will confront you throughout the development and implementation of the housekeeping program. The information must be collected carefully from well-selected sources. On a given question, two different persons within an organization will often provide conflicting information,

Figure 2-3. Sample Personnel Analysis Form

FOR ______
DATE ______
RESPONDENT ______

PERSONNEL ANALYSIS

1. Circle the name used for housekeeping personnel, and indicate average weekly wage rates:

	1st Shift		2nd Shift		3rd Shift	
	No.	Rate	No.	Rate	No.	Rate
a. Exec. Housekeeper Foreman Supervisor						
b. Ass't. Housekeeper Ass't. Foreman Ass't. Supvr.						
c. Group Leader Lead Man Sub-Foreman						
d. Custodian Janitor Porter Orderly Laborer						
e. Maid Matron Janitress						

2. Total number of housekeeping personnel now employed or authorized ______
3. IMPORTANT! Fill in Personnel Schedule.
4. What personnel problems do you have in the housekeeping department? (Turnover, seniority, absenteeism, status, literacy, attitude, etc. Explain.)

______
______
______

5. How many days per month does the average housekeeping employee lose from work because of sick leave, vacation, absenteeism, etc.? ______
6. Is there an organization-wide vacation shut-down? ______
   If so, when? ______
7. Do you have printed directives or regulations available on the following (obtain copy):
   - ☐ Safety rules
   - ☐ Policy on security, visiting hours, etc.
   - ☐ Rules of conduct
   - ☐ Samples of house organ, news letters, etc.
   - ☐ Union contract
   - ☐ Housekeeping operations
8. Are uniforms provided for the housekeeping staff? ______
9. What are your qualifications for hiring housekeeping personnel? ______

______
______

10. Except for very simple situations, an organization chart helps to clarify the personnel situation. The sketch should include supervisory names, titles, lines of authority, and numbers of personnel. Sometimes area responsibilities can be included. PRINT. (See specimen program for examples—use separate sheet if necessary).

Figure 2-4. Sample Personnel Job Schedule

PERSONNEL SCHEDULE

(Use additional copies of this form if extra spaces are required.)

Personnel Name Or Number	Maid	Custodian	*Assignment Area, Duty, Square Footage, Etc.	Working Hours (AM & PM)	Work Days	Age	Years Employed	Non-Housekeeping Duties		Remarks (Limitations, Etc.)
								Type Duty	Hrs./ Week	

* Obtain copies of any schedules now in use.

Figure 2-5. Sample Form to Survey Custodial Facilities and Inventory

CUSTODIAL FACILITIES AND INVENTORY

FOR ____________________
DATE ____________________
RESPONDENT ____________________

1. Describe the central storage area:
   Size ____________________
   Location ____________________
   Notes: ____________________
   ____
2. Is the central storage area properly equipped? (Check off if OK)
   Water __________ Tool Holders __________
   Drain __________ Ventilation __________
   Lighting __________ Shelving __________
   Racks __________ Identification Signs __________
   ____
3. Who is responsible for issuing materials from central storage? __________
   ____
4. Are materials issued in original containers? (Explain) __________
   ____
   ____
5. Where are custodial closets and cabinets located (indicate on plans or describe here, and show which contain utility sinks).
   ____
   ____
   ____
   ____
   ____
   ____
   ____
6. Are closets and cabinets properly equipped? (Check off if OK)
   Water __________ Tool Holders __________
   Drain __________ Ventilation __________
   Lighting __________ Shelving __________
   Racks __________ Identification Signs __________
   ____

Figure 2-5. (continued)

7. Describe the requisitioning and purchasing procedure (see below): ____________________
____
____

8. What records are maintained with reference to:
   a. Inventory Control ____________________
   b. Requisitions ____________________
   c. Purchases ____________________
   d. Stock distribution ____________________

9. What types of measuring and dispensing devices are in use? ____________________
____
____

   What types are needed? ____________________
____
____

10. Other notes: ____________________
____
____
____
____
____

11. IMPORTANT: Complete Materials and Equipment Inventory on the other side of this form.

Figure 2-6. Sample Form for Surveying Materials and Equipment Inventories

MATERIALS INVENTORY

MATERIAL	BRAND NAME	PRICE	USE DILUTION	ANNUAL CONSUMPTION	AMT. ON HAND	MINIMUM	ORDER QUANTITY	USES
Floor Cleaner								
Cleaner - Disinf.								
Wax Remover								
Rug Cleaner								
Floor Wax								
Mop Treatment								
Floor Seal								
Disinfectant								
Furniture Polish								
Glass Cleaner								
Cleanser								
Metal Cleaner								
Hand Soap, Liq.								
Hand Soap, Pdr.								
Deodorant								
Bowl Cleaner								
Insecticide								

Figure 2-6. (continued)

EQUIPMENT INVENTORY

EQUIPMENT	NUMBER ON HAND	MAKE	SIZE	CONDITION	STORAGE LOCATION	USE AREAS	USES
Auto Scrubbers							
Power Sweepers							
Floor Machines							
Wet & Dry Vacs							
Dry Vacs							
Pack Vacs							
Applicators							
Brooms							
Brushes							
Buckets							
Dispensers							
Dust Mops							
Maid Carts							
Matting							
Mopping Outfits							
Receptacles							
Sprayers							
Squeegees							
Waste Carts							
Wet Mops							
Wringers							

which may arise from ignorance, reliance on obsolete data, misinterpretation of data, or clerical error. An individual may even provide self-contradictory survey information on occasion.

Oftentimes wishful thinking enters into responses to survey questions. This is particularly true on questions dealing with frequency of job performance. These answers often indicate what the personnel should be doing, or would like to be doing, rather than what they are actually doing.

Sometimes information is distorted through fear. The person giving the information may feel that his position is in jeopardy because of some failure to perform properly, and therefore he colors all of his answers accordingly. Occasionally fear will prompt an attempt to discredit the surveyor by having him make his recommendations on the basis of faulty information.

In making a survey, it is good practice to identify the source of the information. When people realize that they will be personally identified with data which are requested of them, they are particularly careful to provide accurate and relevant material. Only through personal contact can the surveyor develop instincts for the situation.

Take a rapid familiarization tour early in the survey, so that the data later collected can be related to the physical situation. Very often supervisory personnel are impatient to conduct a tour of their plant. This healthy pride in the company and its facilities is a response which always kindles optimism for the housekeeping program to follow. The quick walk-through tour, naturally, must be followed later by a detailed physical survey.

Using Floor Plans to Survey the Building. The fundamental requirement of the physical survey is an accurate and complete floor plan of the physical plant. This is obviously necessary where a program is being provided for a building still under construction, where the amount of written survey data will be quite limited. The helpfulness of the plans is increased materially when they are properly marked to show information such as:

- Color coding for floor types
- Traffic load
- Rest room fixture counts
- Area names
- Area uses
- Work routes
- Square footages

Of course, most cleaning problems are caused by the people working within the buildings, rather than by the buildings themselves. Therefore, the plans alone cannot constitute the entire survey.

For multibuilding facilities, a plot plan is required to show the spatial relationships of the various structures and the actual distances between them. Insurance drawings are useful for this purpose. Additional information or understanding can be obtained from aerial photographs in some cases.

Figure 2-7 shows a line drawing (not an architectural plan) of the second floor of a sample building. This drawing is marked to indicate the use of

Figure 2-7. Sample Line Drawing Marked to Show Space Use, Square Feet and Floor Type

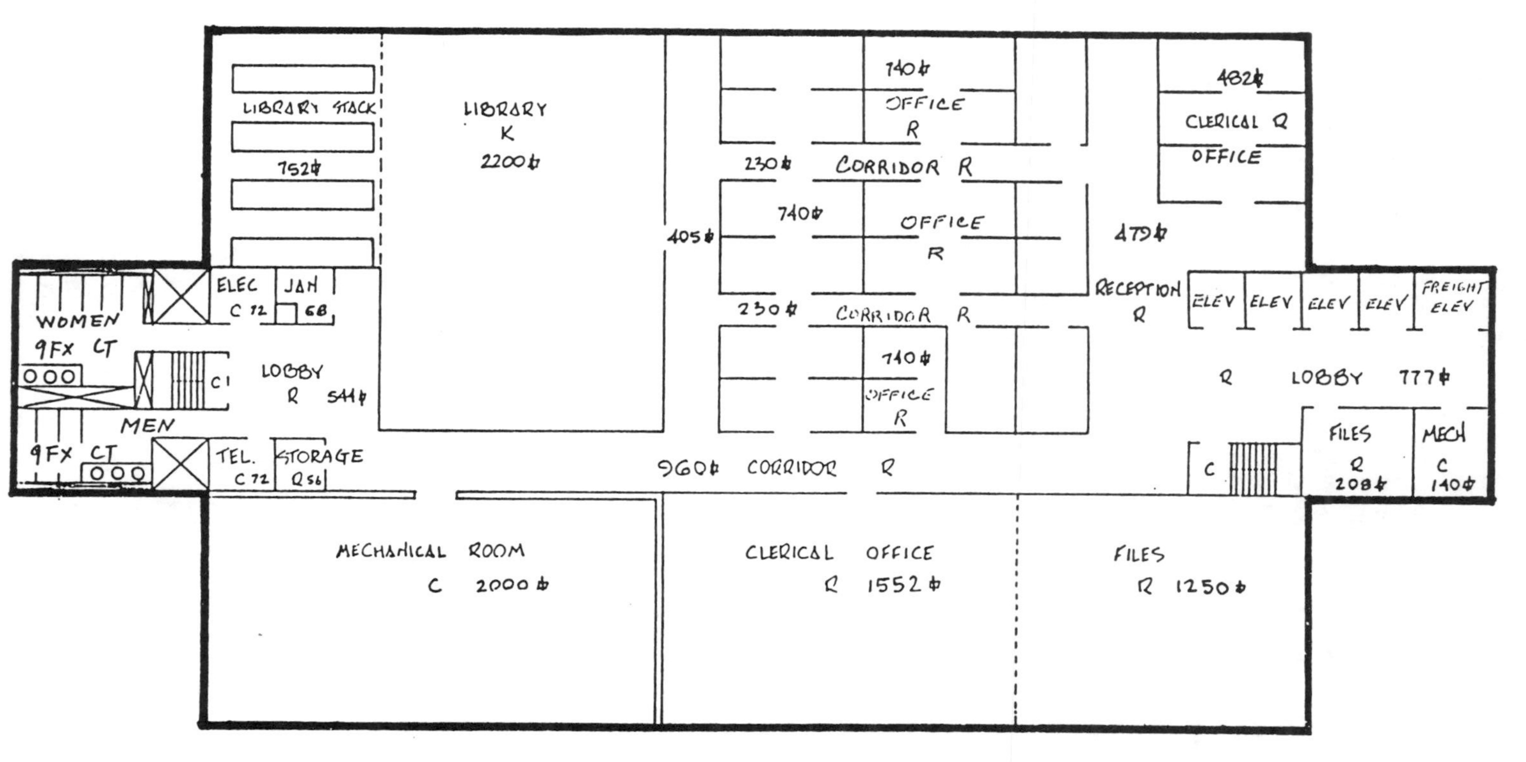

each area, its square footage, and floor type (R = Resilient, C = Concrete, CT = Ceramic Tile).

On another drawing, K might equal Carpet, T = Terrazzo, W = Wood, S = Stone. These plans should also be marked to indicate unusual conditions of soiling, traffic, hours and days of use, inaccessible areas, etc.

Obtaining Other Data to Complete the Custodial Survey. The analysis of the personnel within the housekeeping department, their supervision, and various lines of authority within the housekeeping department and upward to management suggest the need for a carefully prepared organization chart. The organization chart should also show other functions performed by housekeeping department supervisors, staff relationships, cases of overlapping supervision, problems with job titles, etc.

Determining current housekeeping costs and existing or planned budgets is very often an elusive subject, since housekeeping may bear the cost of functions not related to cleaning, while other departments may bear certain housekeeping costs. The figures obtained will not be meaningful unless all costs dealing with cleaning are included, such as overtime, contract costs (for dust mop treatment, laundering, venetian blind cleaning, window washing and rest room service).

Housekeeping materials must be inventoried and surveyed as to type, amount on hand, ordering quantities and minimums, product characteristics (such as pH, use-dilution ratios, solids content, free alkali), cost, and other factors. Equipment must be catalogued as to number, type, location, size, and condition.

Paper work should receive serious attention. This involves communications, reports, schedules, and instructions. It concerns any phase of housekeeping passing between housekeeping, safety, personnel, industrial relations, purchasing, stores, or management personnel. The whole subject of the simplification and provision of paper work is becoming a specialty in itself, but the inclusion of forms in use at the time of the survey will permit basic program recommendations on this subject.

Also, pay special attention to those items which are peculiar to a specific organization, such as chip collection and removal, biological or hazardous wastes, radiologic materials, and special soils created. A number of unusual cleaning problems may arise and must be recorded, such as with the removal of unusual stains, classified waste disposal, bird droppings, and the cleaning of such specialized areas as superclean rooms, gymnasiums, display areas, dispensary, dietary areas, etc.

The surveyor may require other types of information.

- Question department heads about their specific problems, interest, and recommendations concerning housekeeping and sanitation.
- Interview housekeeping workers about their problems, attitudes, and ideas. Such questioning must strictly conform to existing employee relations policies.

- Review the requisitioning, purchasing, and stock control methods for housekeeping materials and equipment.
- Obtain copies of current cleaning contracts and specifications, the union or civil service agreement, personnel rules, safety precautions, and the employee's handbook.
- Survey the personnel department concerning its policies for the recruiting, selection, and training of housekeeping personnel.
- Contact the industrial relations department to determine its ability to encourage greater employee awareness and cooperation about housekeeping.
- Consult the safety department regarding the safety aspects of housekeeping.
- Interview departments regarding problems of hygiene and cross-infection.
- Determine plant-wide employee attitudes about housekeeping and sanitation.
- Study the method of negotiating cleaning contracts.
- Record current methods of work performance for later comparative analysis.

You may find it desirable to question other department heads or personnel about housekeeping to determine company-wide attitudes. The questions should give an opportunity for the persons involved to indicate their opinions concerning the housekeeping services which they like, those which they dislike, where housekeeping might be improved, where they feel their departments could benefit from an improved overall housekeeping effort, and such other information as may seem desirable. Surveys of this type have brought out worthwhile and interesting suggestions, while at the same time focusing attention on the company-wide aspect of good housekeeping.

After the survey is completed, make a recorded inspection to provide a benchmark against which future performance may be measured. Sometimes it will be more desirable to conduct the first inspection during the installation phase of the program.

Reporting the Results of Your Custodial Survey

Document the progress of the program to keep management apprised, to help organize the work ahead, and to keep interest in the project alive. Your report should include:

- Acknowledgment of appreciation for assistance, guidance, and cooperation.

- A review of survey activities, indicating the nature of each activity (meeting, tour, completion of questionnaire), the names of the persons present, and the nature of the information gained, along with the date, time, and place of the occurrence.
- Questions of basic importance should be repeated, together with the answers provided and the identification of the individual supplying the information.
- Any plans, schedules, charts or tables, properly identified together with their source.
- Initial observations on which action can be taken before formal programming, such as involving safety.
- A tentative calendar for future activities.

Where less than a complete study is desired by management, a good deal of useful information can still be obtained with an audit, or overview. This consists of a general review of indicators of performance, followed by a report giving this information:

- The current status of custodial operations
- Opportunities for cost improvement
- Opportunities for quality improvement
- Considerations concerning contract cleaning
- Recommended additional studies

An audit can also provide recommendations on organization and staffing where example portions of the facility are measured and time allowances applied. Naturally, such information cannot be as accurate as a full-scale analysis. Chapter 10 discusses in detail the housekeeping self-audit.

3

Seven Key Areas for Establishing Housekeeping Standards

Standards of quality and cost of custodial maintenance are inevitably interwoven, yet each encompasses a number of different factors which can be considered separately. There are four groups of people who are competing to establish standards: (1) management, (2) occupants, (3) custodial employees responsible for performance of housekeeping work, and (4) those providing funds for all purposes including maintenance.

Basically, the establishment of maintenance standards falls to management; the superintendent of buildings and grounds, plant engineer, or physical plant director is responsible for recommending and justifying to his or her superiors the appropriate standards of housekeeping maintenance. As discussed in Chapter 2, you should not require custodial workers to determine their own work standards. And occupants usually do little to establish the housekeeping standards, but they help to establish the difficulty or ease with which standards are achieved. Finally, those who provide the funds seldom involve themselves directly in questions of maintenance standards.

Housekeeping is one of the last functions to receive standards. This has been due primarily to the perception of housekeeping as unimportant and to the unusually high range of variables.

Standards must be developed for each installation, based on the type of organization, the condition of the physical facility, the type of worker or applicant available, degree of congestion, turnover, management support and employee attitude, the type and amount of waste created, and other factors.

Standards developed for general use lead to misunderstandings and poor results when applied to a specific situation. Standards set up for one operation in a given facility should not be considered appropriate for another facility.

The development of specific standards does not necessarily mean that new time studies should be taken in each case; rather, it generally means correcting factors should be applied for the variables involved. The accuracy of the numerical value of the factors, of course, increases with the experience of the coordinator.

Just as conditions will change, so must standards change to conform to them. Using obsolete standards will limit their effectiveness and cause employee relations problems.

This chapter discusses housekeeping standards for seven main categories:

1. Personnel
2. Frequency
3. Methods
4. Materials
5. Equipment
6. Time
7. Quality

PERSONNEL STANDARDS: HOW TO TELL IF THE WORKER AND THE WORK ARE WELL MATCHED

To establish standards for personnel, you must first identify job descriptions. You are therefore immediately faced with the need to establish and standardize job titles for all housekeeping personnel.

These titles, particularly on the supervisory level, must follow the pattern set for the company as a whole. Thus, in a given company, the head of a department may be called manager, supervisor, superintendent, or some other title. Consider the following titles in a hypothetical organization:

Title	*Job Description*
Housekeeping department manager	In charge of the housekeeping department
Supervisors	Salaried supervisors, one of whom may be designated as "assistant manager."
Group leaders	Working team leaders.
Custodians	The basic housekeeping worker. Custodians may be arranged in several categories or classes, depending upon the type of work performed, such as hazardous duty or shift work.

Standard qualifications for hiring custodians and other housekeeping department personnel will assure that personnel placed within the housekeeping department will be suited for the work.

Developing standard job descriptions may be done in two stages. First, prepare descriptions for categories of workers and then adjust them with use. Following this, set up individual job descriptions. See Chapter 4 for more information on how to write job descriptions.

Both the qualifications and job descriptions will have a bearing on relative wages. The overall wage structure for housekeeping personnel, of course, is involved in management's consideration of relative wages and benefits for all departments.

Chapter 4 also provides details on determining staff requirements.

TASK FREQUENCY STANDARDS: HOW SPECIFIC TIMEFRAMES CAN HELP AVOID COSTLY REPAIRS

Frequency standards should be as specific as possible. Although there are times when designations such as "after each use" are required for the maintenance of a given area or piece of equipment, the great majority of frequencies should be indicated as positive time periods. Designations like "as required" leave too much latitude.

Indefinite frequencies lead to corrective maintenance rather than preventive maintenance. If it is the practice to maintain a surface only if it apparently needs maintenance, there will be several such surfaces needing maintenance at one time, and they will remain in this condition, progressively deteriorating, until the maintenance can be scheduled.

For example, where a floor surface, through observation only, needs rewaxing, then the wax has already worn through and the floor has a bad appearance and is wearing away. You should no more wait until the wax on a floor has been worn away before rewaxing than you should wait until a machine has run out of oil before you lubricate it.

Frequency determination is most useful where each cleaning job done on an intermittent basis is scheduled for each area where the work is required. For example, floor scrubbing might be required daily in one area, weekly in another, and semiannually in yet another. In arriving at frequencies for individual situations, it may be best to begin with a general frequency which is adjusted because of a given situation.

Arrange frequencies on the basis of the type of area involved, such as office and other administrative areas, treatment or service areas, and storage or intermittent-use areas. If you establish a standard frequency for each of these types of areas, you can adjust them depending upon the traffic, soiling, congestion, and other factors found in the various locations within the general area.

Adjust specific frequency standards as conditions change. A program should normally begin with a series of target frequencies, with the housekeeping department operating at those frequencies within a specified number of months. Once these target frequencies are achieved, some of them may be adjusted for performance more regularly. The means to accomplish this additional work will be provided by the increased efficiency of the housekeeping operation.

Set up frequencies in case it is necessary to perform a given job sooner than anticipated. A case in point might be the rewaxing of a lobby floor that had been set up on a monthly frequency. During a snow storm a good deal of water might get in, quickly wearing away the protective coating. In this case, rewaxing might be required after only two or three weeks during this period. If frequencies are set up so that jobs are being done so often as to avoid any change of schedule, the work will be done too often on the average and will therefore be overly expensive.

Although housekeeping operations typically are not performed frequently enough, there are cases where jobs are being done too often. In dusting furniture, for example, only the horizontal surfaces (which catch most of the dust) need be dusted daily; the vertical surfaces require dusting only weekly. Dust mopping of floors need not be done daily in certain types of areas. The time that is saved by extending repetitive frequencies of this type may be used to accomplish intermittent project work which might otherwise be done too rarely or possibly not at all.

When setting frequencies and calculating time requirements for them, remember that some jobs eliminate the need for others. For example, it is not necessary to dust mop a floor which is going to be wet mopped unless the floor is exceedingly dirty, which would cause the mopping water to become ineffective too quickly.

Frequency is also affected by materials. Some polymeric floor coatings take on such a high gloss after initial application that buffing is not required for some time thereafter. Where buffing is desired on a regular basis, it will not improve the appearance of this type of material whatever for the first several days after application.

Figure 3-1 shows a sample listing of tasks for a cafeteria and the frequency with which they should be done.

METHODS STANDARDS: HOW TO DETERMINE THE MOST EFFICIENT APPROACH FOR JOB PERFORMANCE

There is a best way to perform any given job. The best methods are not in use for most jobs.

The value of the method must take into account a number of factors. If a given procedure is extremely efficient, requiring a level of skill that cannot

Figure 3-1. Frequency Standards for a Cafeteria

ROUTINE WORK TASKS AND FREQUENCIES

AREA TYPES: Cafeteria

Frequency	Tasks
D	Empty and spot clean trash receptacles.
D	Empty and spot clean ash receptacles.
D	Replace torn or obviously soiled trash liners.
D	Damp wipe trash receptacles.
D	Police floors to remove litter.
D	Dry clean chalk boards and chalk trays.
D	Spot clean building surfaces.
D	Spot clean furniture surfaces.
W	Dust building surfaces.
2W	Dust furniture surfaces.
D	Rearrange furniture.
W2	Complete dusting.
D	Clean and disinfect drinking fountains.
D	Clean floor mats.
D	Sweep or dust mop non-carpeted floors.
—	Spot mop non-carpeted floors.
D	Damp mop non-carpeted floors.
—	Wet clean non-carpeted floors.
2W	Spray buff.
D	Remove carpet stains.
4W	Partially vacuum carpet.
W	Completely vacuum carpet.
—	Clean and disinfect fixtures.
—	Clean and disinfect building and furniture surfaces.
—	Descale toilets and urinals.
—	Refill dispensers.
—	Clean urinal and floor drains.

D=Daily, D2=every other day, W=Weekly, 2W=Twice weekly, 3W=three times weekly, 4W=four times weekly, W2=every two weeks, M=Monthly be achieved by most workers, it cannot be put into general use. If it requires a timing or coordination that is a mental or physical strain on the workers, your staff may abandon the task. The standard, therefore, must be one which can be met by the general worker who is to be assigned to that job. Workers who cannot meet the methods standard in question should be delegated to perform some job having a standard which they can perform.

Both materials and equipment have a direct effect on methods. Some methods depend directly on some specialized type of equipment or chemical. For example, autoscrubbing is a method of wet-floor cleaning that is primarily a function of the machine itself. In this case the method must be restricted to large uncluttered areas which are suitable for this type machine.

Some materials are designed to perform two or more operations with a single application. Just because a given material will perform a job faster than another is not sufficient justification for its use in a methods standard, however. For example, many custodians use acid-type descalers in the daily cleaning of rest room fixtures. Indeed, this does a very quick cleaning job, but has the disadvantages of presenting a safety hazard, damaging surfaces if improperly used, eroding floors if spilled or dripped, and being rather costly.

Methods also depend on organization. If all the workers in a given housekeeping department are set up on an area-cleaning basis, some of the efficient methods for project work which can be achieved with team cleaning are lost.

The methods specification should primarily answer the question *how*. It should tell the worker how to stand, how to use the equipment, and how to move. It should also indicate *what* equipment and materials are involved and the proper use-dilutions of all chemicals. It should tell what safety precautions should be taken. It is also desirable to point out *why* the job requires performance in a given way, or why it need be done at all, in some cases.

Both time standards and frequencies can be indicated on methods sheets.

Figure 3-2 shows a sample methods sheet for chalkboard reconditioning.

MATERIALS STANDARDS: FOUR BENEFITS OF CONTROLLING PRODUCT USE

It is often easier to establish your own standards for use of housekeeping materials than it is to obtain industry standards.

Housekeeping personnel tend to make unauthorized experiments with housekeeping materials. They may utilize stronger materials than specified for a given job. For example, unless the materials-use situation receives supervisory attention, floor-stripping chemicals may be utilized for general cleaning purposes. These chemicals are, of course, strong cleaners and good emulsifiers, and certainly do clean rapidly, but they also remove wax or

Figure 3-2. Sample Methods Sheet for Chalkboard Reconditioning

MATERIALS

- Lotion cleanser
- Water
- Chalk

EQUIPMENT

- Foam eraser with chamois backing
- Felt eraser
- Clean cloths and sponges

AREAS WHERE APPLICABLE

Chalkboards which have been frequently washed, resulting in a poor writing surface, "ghosts" and excessive glare. Normal, routine cleaning should never involve water.

PREPARATION OF EQUIPMENT

None.

PREPARATION OF MATERIALS

Several gallons of clean water should be placed in a mop bucket or other container.

PREPARATION OF AREA

Remove all chalk and erasers from chalktray.

INSTRUCTIONS

Erase the entire board with the felt eraser.

Clean the board with the rubber or foam side of the foam eraser, then with the chamois side.

Use the lotion cleanser and a damp sponge to go over the entire board. Use a light, circular motion to remove the build-up of the hard chalk binders which have filled the surface irregularities of the board, making it difficult to write on.

Wash the board several times with clear water to remove all evidence of the cleanser. Allow to dry.

"Chalk in" the board lightly using a new piece of chalk held on its side.

Erase the board with the felt eraser, and then with the foam and chamois.

Clean chalktray and any chalk dust that has fallen to the floor.

Supply the board with chalk and replace felt erasers.

Figure 3-2. (continued)

RESTRICTIONS ON USE OF THE AREA

None. Once the reconditioning is complete, the board is ready for immediate use. Boards which do not respond to this method may require complete resurfacing or replacement. Routine care should be restricted to dry cleaning methods only.

CLEAN-UP

Wash out sponges and cloths.

Wipe foam erasers with a clean cloth or soft brush.

Rinse out and dry the mop bucket or other water container used.

other protective coatings from floors, thus leaving them exposed to damage. Experimentation with products also can expose custodians to safety hazards. For these reasons, personnel should use only the materials specified for given jobs.

Even use-dilutions should be standardized. Overuse of a given material may not only cause more work than is necessary, but it also may damage surfaces or constitute a safety hazard.

The determination of material type will be based on such factors as water hardness, the degree and type of soil encountered, the type of floor covering, etc. Operate the housekeeping department with as few products as possible, and utilize dual-purpose products for this purpose. Standardization of products brings these benefits:

- Lower inventory levels because of the fewer number of products involved.
- Purchasing savings and improved storage conditions.
- Simplified training.
- Less danger or error in product selection and use.

If you have standardized both frequencies and materials, you can determine consumption standards by multiplying the amount of material needed to clean a unit surface, for example, times the total surface, times the frequency per year. Under such *controlled* conditions, increased consumption of materials will indicate increased frequency. Under uncontrolled conditions, it may indicate either increased frequency, wastage, pilfering, or misuse. Thus, a successful housekeeping program should increase materials consumption because of improved frequencies and performing some cleaning functions that might never have been done previously.

EQUIPMENT STANDARDS: HOW TO BE SURE THE MACHINE FITS THE JOB

When selecting equipment, the most prevalent error is purchasing undersized machines. In general, the best investment is the largest and the most durable piece of housekeeping equipment that can be fitted to a given job. The additional cost over the life of the equipment, even bearing in mind interest, taxes, and other costs, is usually quite small.

It is also better to buy equipment which can be used for more than one purpose. The flexibility of a piece of equipment, due to its style or availability of various attachments, often determines the amount of time in which it will be in use during the average day.

Of course, you should select equipment that is safe and may be easily repaired or adjusted.

Once you have selected a suitable piece of equipment, you should reorder additional units of the same type rather than going from one supplier to another, even where there is some price advantage. When you standardize on a given type of equipment, you gain these advantages:

- Purchasing savings where the equipment is bought in multiple units.
- Simplified training in the use of the equipment as well as simplified training in its care and minor repair.
- For major repairs, basic parts may be kept on hand where there are a number of pieces of identical equipment in use.
- Suppliers are able to render more service because of the greater sale involved.
- Interchangeability of parts and attachments results in a smaller overall investment.

Figure 3-3 lists sample specifications for an automatic scrubbing machine.

TIME STANDARDS: DETERMINING HOW LONG IT SHOULD TAKE TO COMPLETE A TASK

To many people, "standards" and "time standards" are synonymous. This is probably because so much attention is given to time standards in some operations, and because time standards can lead to controversy.

Time standards do not exist in a vacuum. You can determine time standards successfully only by establishing all the other standards on which they depend.

Figure 3-3. Specifications for a Small, Automatic Scrubbing Machine

AUTOMATIC SCRUBBING MACHINE—SMALL

USES:

To apply detergent or stripping solution to non-carpeted floors, scrub the floor surface and vacuum the solution from the floor.

PURCHASE GUIDELINES:

- ☐ Vacuum motor—¾ h.p. minimum.
- ☐ Brush drive motors—two ½ h.p. minimum.
- ☐ Traction drive motor—½ h.p. minimum.
- ☐ Batteries—three 12 volt 183 ampere within 36 volt system.
- ☐ Solution tank—15 to 20 gallons.
- ☐ Recovery tank—15 to 20 gallons.
- ☐ Brush speed—300 to 350 rpm.
- ☐ Brush force—variable from 0 to 100 pounds.
- ☐ Squeegee width—28 to 32 inches.
- ☐ Forward speed—175 to 200 ft./min.
- ☐ Reverse speed—125 to 175 ft./min.
- ☐ Battery charger—25 amps, 36 VDC output, 12 amps, 115 VAC input.
- ☐ Brushes—two 13 inch diameter with 2″ overlap.
- ☐ Weight with batteries—800 to 900 pounds.

In addition to routine cleaning times, other task times must be provided for, such as:

- Support time (e.g., for supervision, stock control, clerical, group leaders)
- Indivisibility allowance
- Special projects time
- Non-cleaning duties
- Other shift time
- Weekend coverage
- Limited worker allowance
- Absenteeism relief
- Abnormal travel time
- Allowances (for make ready, put away, and personal time).

A specific case is indivisibility allowance, where, for example, if a building is measured to utilize 3.85 people, it would be natural to assign 4 people to that building, since the travel, make ready, and put away time to work somewhere else would consume the rest of the time available.

Standard times can be accurate only when a given job is performed in a standard method by standard personnel using standard materials and equipment under standard conditions. In housekeeping functions, because of the wide variances, time standards are often indicated as ranges. Even these ranges do not comprehend all acceptable times under various combinations of conditions.

Inaccurate standards are one of the fundamental causes of misunderstanding between labor and management. The inaccuracy of standards and of standard times is almost assured when improper or inadequate standards have been established for personnel, methods, materials and equipment, and when specific conditions are not taken into account. Where conditions or the other standard factors differ from the normal, the time standard must be adjusted accordingly, if it is to be used at all.

Without a detailed definition of methods by which the standards were obtained, time standards may mean little. Unless the method can be duplicated in practice, neither can the standard. You must make suitable allowances for fatigue, make-ready, down-time, and personal time. Unless these are built into the time standard, you must apply compensating factors.

There are several methods for determining housekeeping time standards:

- *Actual time study by qualified personnel*—This method will provide the most reliable information, but it is very expensive. Since a large variety of housekeeping jobs would require study, the actual making of a time study may be restricted to only the more important jobs.

• *Internal records*—These are useful if they have been kept carefully and the work was performed under proper supervision, with the use of standard methods, etc. The mere use of historical time rates, where these other standards have not been applied, will tend to perpetuate and officially authorize inefficient work.

• *Published time standards information*—This is available from a number of sources. The use of a synthetic set of housekeeping standards by adopting one of these listings, or by taking the average of several listings, is unsound. Very few serious independent investigations have been made of housekeeping time standards. Much of the information published is a restatement or a duplication of other previously published information which, in itself, may have been estimated in large parts. Standard times developed for other than your own operation should be used only for comparative purposes, or as a checklist, and never applied to the work at hand.

• *Time estimates*—These can be useful in trying to apportion labor to an infrequent or unusual job. Such estimates cannot be used, however, for grading the work.

• *Fixed-time standards for small work units*—These are difficult to use in housekeeping operations. Their value is in establishing larger work units. For example, the amount of time necessary to dust a cigarette urn of a given size is of little practical value, except when it is used to establish a standard for the complete dusting of an office area of unit size. Summary standards are better.

• *Weighted time standards*—An experienced analyst will find it unnecessary to completely develop time standards for every housekeeping job as applied to every surface to be maintained under various conditions. Rather, he or she prepares a standard for a series of given situations and then applies weighing factors against these standards to compensate for the variables which may act on them. Some weighing factors will be positive, others negative, depending on whether the variable would indicate a greater or lesser time standard. The weighing factors must be applied judiciously since they were developed as units, which, however, interact when more than one factor is applied. Experience (and mathematics) provide control figures, maximum and minimum, beyond which the standard cannot be adjusted by weighing factors, since the result would go beyond practical achievability.

Figure 3-4 provides time allowances for various types of areas, indicated by floor type, for a specific facility (these could hardly be used for any facility). Note that the time standard is for number of flights of stairs or number of rest room fixtures, rather than for square feet.

Figure 3-5 indicates standard time allowances for various types of operations, for a typical facility under typical conditions. Note that the

Figure 3-4. Sample Area Time Allowances

AREA TYPE	MINUTES PER 1000 SQUARE FEET OR PER FIXTURE						
	R,Z	K	C	RF	GT	#FX #FLTS	
Building Service Storage			7				
Computer Room				20			
Conference Room		22					
Corridor	18	15	10				
Dining		40					
Elevator - Freight						8	
Elevator - Passenger						12	
Entry			20				
Exam	25						
Executive Office		23					
Files	7						
Kitchen					40		
Library		19					
Library Stack		10					
Lobby	26	22					
Maintenance Shop			13				
Medical Reception	28						
Office	21	18					
Open Office	24						
Parking			5				
Print Shop			35				
Reception	28	24					
Restroom						4	
Security Area	28						
Serving Area					18		
Stairs						10	
Storage	10		7				
Tape Storage	7						
Trash Room			25				
Treatment	27						
Truck Dock			10				

Figure 3-5. Standard Times for Comprehensive Jobs

(These average time estimates include set-up and put-away time, some non-productive time, and assume reasonable mechanization and training level.)

Washrooms, Locker rooms and Related Areas	Work load per 8-hour day
Swing & Locker Rooms (incl. damp mopping)	15,000–18,000 sq. ft.
Swing & Locker Rooms (no damp mopping)	17,000–20,000 sq. ft.
Toilet Rooms in Office Areas	120 fixtures (basins, commodes, urinals)
Toilet Rooms in Plant Areas	107 fixtures
Toilet Rooms (alternate estimate)	3,000–4,000 sq. ft.

General Cleaning Operations	Work load per 8-hour day
Elevators, freight: no damp mopping	24 elevators
Elevators, passenger: including damp mopping	24 elevators
Overhead Office Areas: dusting & vacuuming	7,000–10,000 sq. ft.
Stairs: sweep treads, dust handrails	60, 12-ft. flights
Stairs: mop and rinse	45, 12-ft. flights
Storage & Supply Areas: sweep or dustmop floors	40, 60,000 sq. ft.
Unobstructed Areas: manual sweeping	60, 80,000 sq. ft.
Unobstructed Areas: power sweeping	400, 600,000 sq. ft.

Policing Operations	Work load per 8-hour day
Lobbies and Corridors	200,000 sq. ft.
Stairs	180 flights
Swing & Locker Rooms	45,000 sq. ft.
Toilet Rooms in Office Areas	360 fixtures
Toilet Rooms in Plant Areas	320 fixtures

General Cleaning Operations

Time per operation

Operation	Time	Operation	Time
Chair, cafe, wash	36 sec.	Gen'l ofc. dusting 1,000 sq. ft.	12 min.
Classroom (40 desks) dust	5 min.	Partition, wax or polish, 1 ft.	90 sec.
Desk or table, clean & wax	10 min.	Table, cafe wash	90 sec.
Desk or table, strip & rewax	15 min.	Trash can, wash	4 min.
Desk or table, wash	6 min.	Vacuum & wash drop light	3.3 min.
Desk top, glass, wash	115 sec.	Vacuum & wash floor light	5.9 min.
Dispenser, napkin, refill	90 sec.	Wash receptacle, large, empty	30 sec.
Dispenser, soap, refill	60 sec.	Wash receptacle, small, empty	15 sec.
Dispenser, towel, refill	90 sec.	Windows, washing, typical	40–60 windows
Dust executive office	3 min.		

Figure 3-5. (continued)

Floors

Minutes per 1,000 square feet

Maintenance Operation	Degree of Obstruction None	Slight	Medium	Heavy
Auto scrubber-vac, single pass	16–20	18–24	—	—
Buff, 16″ single-disc machine	23	32	38	42
Buff, 19″ single-disc machine	15	25	30	35
Carpet shampoo, Dry Foam	65	—	80	—
(Same, but before and after)	110	—	125	—
Carpet shampoo, water extraction	400	—	—	—
Damp mop	14	17	20	28
Dust mop	7	9	11	13
Hose and squeegee	20	25	36	43
Rewax (apply 1 coat wax only)	16	19	22	27
Scrub, manual	75	105	120	135
Scrub, 16″ single-disc machine	50	60	85	95
Scrub, 19″ single-disc machine	25	30	40	45
Spray-buff, 19″ machine	30	40	50	—
Strip once and rewax	110	140	—	—
Sweep, administrative areas	9	11	13	16
Sweep, plant areas	11	14	18	22
Vacuum carpets	20	23	29	35
Vacuum, dry	15	19	23	28
Vacuum, wet	29	33	37	45
Wet mop and rinse	32	36	40	48

Furnishings

Dusting, seconds per item of average size.
(Double these figures for damp cleaning.)

Item	Seconds	Item	Seconds	Item	Seconds
Air conditioner, unit	90	Desk trays	8	Radiator, enclosed	30
Ash tray	15	Dictator, covered	8	Sand urn	60
Book case, 36″ × 40″	35	Door, flush, dust	25	Spittoon	180
Cabinet, 3′ × 6′	108	Door, glassed, dust	40	Table, large	40
Calculator, covered	8	File cabinet	25	Table, medium	35
Chair, large	43	Fire extinguisher	16	Table, small	22
Chair, medium	35	Glass part. dust sq. ft.	1.2	Telephone	9
Chair, small	22	Lamp, wall fluor.	8	Tel. switchboard	110
Cigarette stand	25	Lamp, desk fluor.	18	Typewriter, covered	7
Clock, desk	8	Lamp, with shade	35	Vending machine	60
Clock, wall	20	Mural, 3′ × 5′	45	Venetian blind	210
Coat tree	15	Pencil sharpener	15	Sofa or divan	150
Desk, large	28	Pictures, framed	15	Waste basket	16
Desk, medium	23	Rack, 6′ coat and hat	90	Wall, dust per sq. ft.	2.1
Desk, small	18	Radiator, open	180	Window ledge per ft.	2

Figure 3-5. (continued)

Walls, Ceiling and Glass

Cleaning, seconds per item.

Door, wash both sides	150	Wall, marble, wash, per sq. ft.	5.5
Glass part., clear, wash, per sq. ft.	8	Wall, tile wash per sq. ft.	9
Glass part., opaque, wash, per sq. ft.	3	Wall, vacuum, per sq. ft.	4.7
Wall, painted, wash, per sq. ft.	9	Windows, wash, per sq. ft.	7.5

Rest Rooms

Cleaning, seconds per item.

Basin, incl. soap disp.	120	Door, spot clean	50	Receptacle, paper towel	10
Bradley basins, semicircle	180	Drinking fountain	110	Shelving, per sq. ft.	12
Bradley basins, circular	300	Fixtures, destain	180	Toilet, incl. partition	180
Dispenser, napkin	13	Mirror, average	30	Urinal	120
Dispenser, paper towel	7	Mirror, large	60	Wainscot, per 10″	3

comprehensive time in this list provides set-up and put-away time, and some non-productive time in the work area, but does not provide non-productive allowances for travel time and the like.

Figure 3-6 is an excerpt of time allowances developed for a specific facility (in this case a college). The percentage marks in parenthesis indicate the proportion of the area that is likely to be cleaned on each occasion.

QUALITY STANDARDS: FOUR METHODS FOR DETERMINING CLEANLINESS

A basic problem in housekeeping is determining a standard of cleanliness for a facility and measuring the degree of approach to this standard. Certain specific variables are available:

- Dust count
- Reflectivity of surfaces
- Bacteria count
- Color intensity
- Photometric measurement (light transmission through glass, foot-candle illumination level of lighting fixtures)
- Thickness of protective coatings
- Rate of abrasion
- Coefficient of friction

Figure 3-6. Sample Time Allowance for Cleaning a College Facility (Selected Areas)

Area Type: Medical laboratories, examining rooms, clinics and locker rooms with resilient, concrete or terrazzo floors.

Task		
Daily above-floor cleaning*		4
Dust mop daily	(60%)	7
Damp mop with cleaner-disinfectant solution daily	(60%)	12
Spray-buff once per month	(80%)	4
TOTAL ALLOWANCE		27

*Includes daily refuse removal and occasional spot cleaning; however, does not include cleaning of sinks or laboratory equipment.

Area Type: Resilient, concrete, terrazzo and quarry tile ground level hallways.

Task		
Daily above-floor cleaning		1
Dust mop hallways daily		7
Clean entrance mats daily		2
Auto-scrub traffic patterns daily	(50%)	12
Put down finish in work areas once every two weeks	(50%)	1
Buff floor once every two weeks	(100%)	2
TOTAL ALLOWANCE		25

Area Type: Resilient and concrete upper level hallways.

Task		
Daily above-floor cleaning*		1
Dust mop daily	(75%)	5
Damp mop once per week	(100%)	3
Spray-buff once per week	(66%)	7
TOTAL ALLOWANCE		16

*Includes daily emptying of ash trays and refuse containers, with spot cleaning of walls and doors as time allows.

Area Type: Regular classrooms, resilient and concrete floors.

Task		
Daily above-floor cleaning		4
Dust mop daily	(75%)	10
Spot mop daily	(15%)	3
Apply finish and buff monthly		1
TOTAL ALLOWANCE		18

Usually these are sophisticated determinations and are not practical to use for establishing everyday housekeeping levels. Rather, they are useful for specialized purposes, such as determining conditions in superclean areas or hospital operating rooms, and for product development. For everyday use, an arbitrary standard has to be set at the level desired by management. Performance is compared with this standard on the basis of estimated percentages, or merely on the basis of whether or not the standard has been achieved, which is more convenient.

Various individuals have invoked such instruments of measurement as radioactive tracer elements, fluorescent components, and light reflectance; but no one nor any combination of these technics has served to provide overall quality criteria. Standards are set through words, demonstrations and other non-objective approaches. The most basic level of housekeeping maintenance which could be tolerated would be one providing at least adequate hygienic protection for the inhabitants of the environment. This could scarcely be accepted as sufficient, however, for added to it must be considerations of protecting property—a tremendous and hard-won capital investment—and protecting the desired public image. Finally, the level of housekeeping maintained by any organization cannot fail to affect its basic mission.

Assuming that a given housekeeping job is performed adequately each time, the quality of cleanliness maintained will obviously depend upon the frequency with which the task is repeated.

In any given environment, it would be possible through experimentation to determine the frequency with which the elemental tasks must be repeated in order to produce a given quality of result. It is much simpler, however, if you can draw on the experience of some individual or group to establish a frequency likely to give satisfactory results.

These are the basic methods for determining housekeeping quality:

- Direct Measurement—By measuring of dust count, light transmission, or surface wear, you can determine quality in numerical terms. These techniques are expensive and applicable only to critical situations such as superclean rooms, precision assembly areas, etc.

- Calculation of Task Frequency—This relates directly to housekeeping level, all other factors being relatively constant. For example, a floor polished daily will be at a higher quality level than a floor polished weekly; a wall washed annually will remain cleaner than a wall washed every two years. An accurate recording of frequency of job performance, therefore, will in a sense measure the level of housekeeping. This relationship holds more true, the more nearly all work conforms to methods and other standards.

- Arbitrary Quality Standards—You can also measure quality on the basis of mere acceptability or unacceptability of work against an arbitrary standard. For example, such a standard for window washing might require the absence of filming or streaking, visible soil on the glass, and actual soil on the frame and casing as determined by the use of a clean dusting cloth.

• Ideal Quality Standards—A refinement of the use of arbitrary standards is to attempt to measure the approach to the ideal standard. Thus, if a perfect window cleaning job is classified as 100, ten points each may be deducted for streaking, soil on the glass, soil on the frame, unsafe conditions, failure to use a drop cloth, etc. Or, the work may be judged by an experienced observer as *excellent, good, fair,* etc.

The measurement of quality may develop as the program develops, beginning with the consideration of frequency as a guide and working toward the eventual use of numerical grading for general cleaning. It is simpler, though, to determine merely if a job is being completed or not.

Inspecting to Ensure Conformance to Quality Standards

Inspection is a basic function of the service program, providing a number of benefits:

- It can be combined with a rating system to determine compensation or advancement.
- It points out areas where cooperation from other departments has been weak.
- It is an extension of the training and orientation activities.
- It helps assure the satisfaction of executives, supervisors, and employees, who are making their own private inspections of the facility.
- It uncovers situations requiring correction, which will prevent distracting and annoying complaints.
- As a form of supervision, it will lead toward greater worker efficiency.
- It will help measure the effect of various changes and improvements in the housekeeping program.
- Since it should be used for positive purposes rather than for criticisms or threats, inspection can raise morale through judicious commentary and publication concerning exceptionally good work or method improvement by custodians.

Be careful not to treat the inspection phase of the program too lightly. Its apparent simplicity is deceptive. The inspection form must be carefully prepared and faithfully used. A monthly general inspection is desirable, although more frequent inspections can be made for changing conditions or situations of a critical nature. For example, a superclean or "white" room should be inspected daily. Medical, food service, and personnel areas should be inspected weekly.

There are many possible varieties of the inspection form, but basically it is a checklist and progress report. One form should be developed for each area, building, or other physical entity. The various services and cleaning functions performed within this area are listed with a suitable indication either of acceptance or nonacceptance, or for grading by a method of descriptions or numbers. Where a condition requires attention, space should be provided to indicate needed action. The inspection form, or sometimes a separate letter, can cover the inspection of housekeeping materials, equipment, conformity to regulatory statutes, employee attitudes or comments, etc.

Copies of the inspection form should be retained by the program coordinator and the housekeeping department. Also, you should provide upper management with a less frequent written inspection report covering the trend of the regular monthly inspections and supplemented with report information about general program implementation and standing. Provide a comprehensive annual audit and report.

The following forms can help you select a quality assurance recordkeeping system that fits your own needs:

- Figure 3-7 shows a quality control inspection form where the inspector rates the performance by numerical quality levels and provides a total numerical evaluation for the entire area.
- Figure 3-8 provides a "go/no-go" type of inspection, where only deficiencies are indicated with an X mark. This type of inspection usually causes less confusion than the previous form.
- Figure 3-9 is a form that provides the positive title of "progress," which is much better received than "inspection." It also makes it possible to provide the "go/no-go" type of inspection for five given areas.
- Figure 3-10 is a custodial inspection report that is just as valuable as the inspection itself.
- Figure 3-11 is an inspection calendar to keep records of which rooms, floors, or buildings have been inspected over a given period of time. Usually, the particular area to be inspected should be selected by random sampling, but it is also desirable to know how long it has been since the previous inspection.
- Figure 3-12 is a complaint log to keep records of who complains about custodial maintenance, and what the complaint is about. This is an important part of a quality assurance program.

Procedure for Handling Complaints

1. Appoint an occupant of each building and/or each department as monitor.

Figure 3-7. Sample Quality Control Inspection Form

Quality Evaluation Form

Assignment:
Type:
Shift:
Supervisor:
Date:

RATING
2 = At Standard
1 = Incomplete
0 = Unacceptable
X = Not Applicable

SCORE:

(Minimum Passing Score = 90)

#	ITEM INSPECTED	WEIGHT	RATING	TOTAL	MAX.	COMMENTS
	GENERAL					
1	FLOOR, CARPET, MAT	5			10	
2	BASEBOARD, COVING	2			4	
3	WALLS	4			8	
4	FIXTURES	2			4	
5	RADIATORS	2			4	
6	WINDOW FRAMES AND SILLS	3			6	
7	CURTAINS, DRAPES	2			4	
8	BLINDS, SHADES	2			4	
9	DESK, COUNTER	3			6	
10	ASH TRAYS	1			2	
11	TELEPHONE	1			2	
12	FURNISHINGS	3			6	
13	WASTE RECEPTACLES	2			4	
14	CEILING	2			4	
15	LIGHT FIXTURES	2			4	
16	VENTS	3			6	
17	LOCKERS	2			4	
18	DOORS AND FRAMES	2			4	

	WASH, WATER, AND WASTE FACILITIES					
19	SINKS	4			8	
20	PIPES	2			4	
21	MIRRORS	2			4	
22	DISPENSERS	2			4	
23	URINAL AND STALL	4			8	
24	TOILET AND STALL	4			8	
25	SHOWER	4			8	
26	TUB	4			8	
	PATIENT AND ON-CALL ROOMS					
27	BED	5			10	
28	BEDSIDE STAND	4			8	
29	OVERBED TABLE	4			8	
30	LINEN HAMPERS	2			4	
	PUBLIC AREAS, STAIRS, ELEVATORS					
31	FIRE CABINET	1			2	
32	STAIRWELLS AND LANDINGS	5			10	
33	ELEVATOR TRACKS	2			4	
34	DRINKING FOUNTAINS	4			8	
	SERVICE CLOSETS AND EQUIPMENT					
35	EQUIPMENT	3			6	
36	SUPPLIES	3			6	
	EMPLOYEE KNOWLEDGE					
37	QUESTIONS – POLICIES AND PROCEDURES	3			6	
			TOTALS			
			TOTALS / MAXIMUM TOTAL x 100 = SCORE			

Figure 3-8. Sample Quick-Inspection Form

APPEARANCE CHECK

X = Not Acceptable	DUST LITTER AND SOIL REMOVED	SPOTS AND STAINS REMOVED	VACUUMED	MOPPED	BUFFED	EMPTIED	SUPPLIED	DESCALED
CEILING:								
Surface								
Vents & Diffusers								
Light Fixtures								
WALLS:								
Surfaces								
Molding								
Trim								
Ledges								
Fixtures								
Doors								
Push & Kick Plates								
Rails								
Glass								
FLOORS:								
Surfaces								
Baseboards								

REMARKS:

X = Not Acceptable	DUST LITTER AND SOIL REMOVED	SPOTS AND STAINS REMOVED	VACUUMED	MOPPED	BUFFED	EMPTIED	SUPPLIED	DESCALED
FLOORS: (continued)								
Corners								
Steps								
Elevator Tracks								
Mats								
MISCELLANEOUS:								
Drinking Fountains								
Trash Receptacles								
Furniture								
Dispensers								
Fixtures								
Mirrors								
Blinds								
Drapes								

REMARKS:

2. Provide each monitor with a written definition of the custodial tasks to be performed in the various types of areas and how frequently each task is to be performed.
3. Report all complaints by occupants or users to the appropriate monitor.
4. Report all complaints which a monitor believes to be valid and significant to the Custodial Department via telephone.
5. Designate an individual to receive and record complaints from monitors.
6. Enter all complaints in the Complaint Log.

The Complaint Log should record:

- Name and time of complaint
- Monitor's name and telephone number
- Nature of the complaint (being as specific as possible)
- Exact location of the complaint (building number, room, etc.)
- Supervisor responsiblc for the area
- Appropriate area assignment number and name of worker
- Follow up to be performed
- Date and time of resolution or correction of complaint

The frequency, nature, and location of complaints should be summarized in a formal report on a monthly basis by the Custodial Manager (utilizing clerical assistance). The report should be distributed to all supervisors and to the Custodial Manager.

How Quality Translates to Cost

In life you pay for what you get. If you expect reasonable quality standards for housekeeping maintenance, you must expect to pay a reasonable price for the goal. In determining whether you have received your money's worth, you may make the mistake of attempting to compare directly your own housekeeping costs with those experienced by someone else.

It is futile to attempt a direct comparison of cost per square foot between two different environments when it is impossible to standardize all other factors.

In the first place, the quality levels sought or actually achieved by two different facilities are never identical. In the second place, there are seldom enough factors taken into consideration to make possible a valid comparison. A comparison could be made of the annual labor-hours required to clean one square foot of a specific type of area; but what about the quality of the work?

Figure 3-9. Sample Custodial Progress Report

PREMISES OF ______________________

ADDRESS ______________________

DATE ______________________

PROGRESS REPORT

Area A ______________________

Area B ______________________

Area C ______________________

Area D ______________________

Area E ______________________

Item No.	General	Area A		Area B		Area C		Area D		Area E	
		Approved	Req. Attn.	Approved	Req. Attn.	Approved	Req. Attn.	Approved	Req. Attn.	Approved	Req. Attn.
1	Floors protected (waxed, sealed, etc.)										
2	Floors and mats clean										
3	Baseboards clean										
4	Carpets clean										
5	Walls clean										
6	Windows, sills, drapes and blinds clean										
7	Woodwork clean										
8	Ceilings clean										
9	Furniture and fixtures clean										
10	Doors and door hardware clean										
11	Glass in doors and partitions clean										
12	Vents, radiators and pipes clean										
13	Light fixtures clean and operating properly										
14	Pictures and ornaments clean										
15	Drinking fountains clean										
16	Receptacles clean and properly arranged										

17	Grounds, walks and parking area policed										
18	Fire extinguishers, hoses and cases clean										
19	Elevators, stairs and handrails clean										
20	Rooms free from obnoxious odors										
	Rest Rooms										
21	Floors protected (waxed, sealed, etc.)										
22	Floors clean and destained										
23	Light fixtures clean and operating properly										
24	Walls and ceilings clean										
25	Doors and door hardware clean										
26	Windows clean										
27	Mirrors clean										
28	Washbowls and washbowl hardware clean										
29	All dispensers clean, filled and working										
30	Sufficient number of dispensers										
31	Receptacles clean										
32	Stools and urinals clean										
33	Rooms free from obnoxious odors										
	Custodial & Utility Rooms										
34	Rooms clean										
35	Lockers clean										
36	Eqpt. and tools clean & stored properly										
37	Lighting adequate and operating properly										
38	Sinks and drains clean and operating properly										
39	Flammable materials safely stored										

Figure 3-10. Sample Custodial Inspection Report

QUALITY ASSURANCE

CUSTODIAL INSPECTION REPORT

Area/Assignment No. ______________ Inspected By ___________

Worker's Name _________________________ Date _____________

ITEM NO.	OBSERVATIONS	YES	NO
1.	Have the discrepancies from the previous inspection of this assignment been corrected?		
2.	Is the area adequately protected by entrance mats?		
3.	Does the area have an adequate supply of properly located waste and ash containers?		
4.	Have all things which can be economically changed to improve the maintainability of the area been implemented?		
5.	Do tools and equipment have identification numbers?		
6.	Is it obvious from observing and questioning the worker that he has received adequate training?		
7.	Is the worker following the best sequence in performing his assigned tasks?		
8.	Is the worker able to perform without unnecessary interference or delay?		
9.	Is the worker informed as to which tasks must be performed in the various areas and how frequently?		
10.	Is it obvious from the appearance of the areas that the specified tasks are being performed? (See written procedures for appropriate area types.)		
11.	Are the previously approved area procedures still adequate for the areas involved?		

COMMENTS:

Figure 3-11. Sample Custodial Inspection Calendar

INSPECTION CALENDAR

ASSIGNMENT NUMBER						SUPERVISOR						
AREA						EMPLOYEE						
ROOM NUMBER	JAN	FEB	MAR	APR	MAY	JUN	JUL	AUG	SEP	OCT	NOV	DEC

Figure 3-12. Sample Complaint Log for Custodial Maintenance Problems

DATE	TIME	NAME:	EXT.	COMPLAINT	AREA ASSIGN#	SUPER-VISOR	DATE OF ACTION OR CORRECTION

There are important factors other than square footage that will affect the actual maintenance job. These include the amount of congestion, the amount of traffic likely to bring about soiling or to interfere with the cleaning operation, the clean-ability of surfaces, and the number of fixtures or the amount of furniture present. The overall job will be affected by the peculiarities of the situation: the distance of custodial closets and sink facilities from the area of work performance, the convenience of transporting tools and equipment, the existence of soil factors such as boiler and incinerator chimneys, or limitations imposed by the labor market.

The acceptance of so-called average per-square-foot costs as a standard makes no more sense than the acceptance of the average shoe size as being the proper one to buy.

If you accept the formula that quality equals essential tasks multiplied by the frequency with which these tasks are performed, you can see that the addition of one more factor will give you work cost. In other words, cost equals task performed times frequency, times cost per performance. Of course, cost per performance can further be broken down into time consumed, multiplied by wages per unit time.

This is a beginning point for the formulation of cost standards. You are ready for job measurements or the engineering approach to cost control in housekeeping. The trouble is that you have now also reached the point where the unwary frequently are lured into the pitfalls of attempting shortcuts and over-simplification.

4

How to Organize Your Housekeeping Department

The main objective of your housekeeping program is to extend the effectiveness of each person performing cleaning operations. This provides a higher housekeeping level, requires fewer workers and helps save money.

You won't achieve instant improvement with a new brand of wax, a new "miracle" type of equipment, by hiring an assistant supervisor, or any other one-shot approach. There is no panacea to solve the typical set of housekeeping problems. You need a comprehensive initial approach to the problem, followed by a continuous effort to put that approach into practice.

Organization can be defined as the optimum arrangement of all variables which will provide maximum results. The program document provides that organization. Control is gained by careful planning. This involves: (1) developing a written plan, (2) measuring performance against the plan, (3) evaluating results, and (4) applying corrective action.

This chapter will help you organize the structure of your housekeeping department by preparing a carefully organized and comprehensive written operations manual.

HOW TO CHOOSE THE BEST HOUSEKEEPING PROGRAM FOR YOUR COMPANY'S NEEDS

The written program is often referred to as the "formal" document. The word *formal* is good as long as it refers to the comprehensive and detailed nature of the program, rather than to a program that is obscure or stuffy.

Figure 4-1 is a sample table of contents for a housekeeping department operations manual. The main sections in this sample manual include:

- organizing and scheduling
- soil prevention and maintainability
- motivation and employee morale
- supervision
- training
- quality assurance
- routine work standards
- projects work
- chemicals
- tools and equipment
- methods and procedures
- work assignments

Of course, the format shown in Figure 4-1 is only a guideline; you may prefer to choose different names for the forms or sections to be used in your program document. Your goal should be to organize the material as simply and as logically as possible. Some of the basic purposes of the operations manual should be to:

- indicate lines of authority
- provide job descriptions
- state the recommended method of job performance—Figure 4-2 provides an example of special project instructions, in this case, for sealing concrete floors.
- indicate frequencies of job performance
- assign workers to the areas for which they will be responsible
- provide standards for materials, equipment, and methods
- offer recommendations to management (such as on training and quality assurance)
- provide reference data on various aspects of housekeeping
- record time standards
- outline the implementation phases to follow

How to Arrange the Material in Your Housekeeping Program or Operations Manual

One logical arrangement is to base the program document on the six areas covered in the custodial survey: management, staff resources, money,

Figure 4-1. Sample Contents of a Program or Operations Manual, for a Housekeeping Department.

TABLE OF CONTENTS

Figure 4-1. (continued)

Figure 4-1. (continued)

Figure 4-1. (continued)

Figure 4-2. Special Project Instructions

SPECIAL PROJECT INSTRUCTIONS

SEALING
CONCRETE FLOORS

SPI 320

MATERIALS REQUIRED:

CONCRETE SEAL

EQUIPMENT REQUIRED:

1 push broom
dust mop
seal pan
lambs wool applicator
clean rags

INSTRUCTIONS:

PREPARATION

Floor seals reflect the condition of the floor beneath because of their transparent quality. For this reason, it is extremely important that the floor be put into good condition before sealing. Where the concrete is stained by oil or other materials, a series of treatments may be required to bring the stain to the surface and remove it. It is difficult to obtain a uniform finish unless the floor is carefully cleaned in advance and carefully dried. See Special Project Instructions for wet cleaning concrete floors.

SEALING

1. Be sure the floor is clean and completely dry. Since drying time for floor seals is critically influenced by humidity, it is suggested that seals not be applied on rainy days or days of high humidity, unless the work is to be done in a heated building. It is best to seal on the day following the floor cleaning.
2. Carefully sweep or vacuum floors to remove all mop strands and brush bristles.
3. Fill seal pan one-half full with seal and immerse applicator into seal. Press pad against side of pan so applicator is drip-free when lifted from seal pan.
4. Apply seal in straight, even strokes 3 to 4 feet long. Do not try to make excessively long strokes. Carefully "feather" overlap strokes to avoid lap marks and to produce an even appearance. It is not necessary to apply pressure on the applicator; the weight of the applicator itself is adequate.
5. Shiny spots will indicate too much seal or spots where the concrete is more tightly bonded than others. Any excess seal should be spread out; all of the seal should penetrate into the floor.
6. After the floor is thoroughly dry, repaint any stripes or lines on the floor. Allow to dry thoroughly.
7. Apply second coat of seal if required.
8. Be certain that the floor is perfectly dry before re-using.
9. As needed from time to time, patch-seal particularly worn areas (such as aisles) using the same process as described above.
10. Clean and put away all equipment. Close containers tightly. Mineral spirits used promptly will normally remove seal from applicator pan and applicator so that they may be re-used.

NOTE: For drying time, refer to the concrete seal specifications sheet.

methods, materials, and machines (see Chapter 2). For example, the section on management would propose the supervisory structure of the housekeeping organization and the climate for its operation.

You may find it helpful to prepare the program document in two drafts. The first draft should be designed for discussion purposes only. This is especially useful where a number of basic changes or departures for your organization are recommended and require a great deal of study and consideration. The program can then be redrafted to include only those items which have been accepted, modified, or will be considered further. Those items which the organization finds impractical should be deleted. Another approach is to submit an interim report rather than an entire program document, containing only the more fundamental recommendations.

It must be made clear that in authorizing a housekeeping study, management is not agreeing to use every recommendation which the program specifies. The program is, in a sense, a book of ideas. Those ideas which management feels are not applicable or applicable can be modified. The program should be targeted for a measured period of time, rather than rushing into something that may disrupt your entire operation.

Unquestionably you must avoid unnecessary paper work. Even so, the program should indicate the paper work necessary to its successful prosecution and, where needed, should actually propose the forms to be used. Not all of these forms will be needed in every case, even in the larger operations. Their use should depend on need and the benefit derived from them. The determination of forms is not complete without an indication of the number of copies necessary, their routing, and eventual disposition. Here are some possible examples:

Progress reports to management (such as on absenteeism, turnover, and accidents)
Inspection reports
Grading reports
Materials-consumption records
Equipment records
Special project frequency records
Complaint records
Accident records
Work orders

Other items may also be required or desirable. For example:

- Statistical reference material (fixture counts, square footages, personnel data, etc.).
- Schematic layouts indicating floor types by color coding, areas of responsibility, traffic conditions, etc.
- Comparison charts relating existing personnel and their cost to that which may be proposed.

HOW TO WRITE JOB DESCRIPTIONS

Job descriptions are often considered to be unnecessary red tape—a waste of time and good paper. Yet, when properly prepared and properly used, they can be valuable working tools.

Job descriptions for cleaning and maintenance workers, like all job descriptions, serve one or more of these purposes by defining the nature of the work to be done:

- Provide information and instruction for workers.
- Serve as a reference for supervision.
- Are required by a union or Civil Service Agreement.
- Provide a basis for compensation.
- Provide for continuity in case of turnover.
- Fulfill a requirement by some governing or accrediting body.

There are other advantages of having job descriptions:

- They assist in developing compliance with government regulations, such as concerning equal pay and equal opportunity.
- There may be a benefit in fire prevention and safety, as the job description can provide information that indicates possible hazards.
- Job descriptions indicate to workers that management is interested in their position, and a display of management interest is always of great importance to service workers who sometimes consider their status low level.
- Job descriptions are useful in planning training programs, as they indicate the type of information that needs to be relayed to the worker.
- For the industrial engineer or methods improvement specialist, the job description may trigger ideas for improved efficiency.
- Vocational counselors rely on job descriptions to advise workers, such as those who are handicapped, disadvantaged, or inexperienced.
- Finally, and this could be most important, in some cases job descriptions may improve employee relations by providing a basis of understanding concerning job responsibilities, and thus may limit grievances relating to them.

Job descriptions may be confused with other activities, such as job assignments. There may be some overlap, but the assignment describes in general the amount of work to be done and specifically where it is to be done, while the job description tells of the nature of the work.

Job descriptions cannot be used to circumvent the law. One good example is the equal pay law—some organizations have tried to defend a discrepancy in pay between men and women doing essentially the same work by referring to job descriptions which separate these—but the equal pay law not only supersedes job descriptions, it supersedes stronger instruments as well, such as union contracts, Civil Service agreements and state laws.

There are many forms of job descriptions, but they typically include the functions of the job, the type of work to be performed, equipment and materials to be used, may also contain a description of job factors and therefore perform functions of a job evaluation as well as description. The factors most often considered are these:

- Educational background required.
- Training requirements.
- Mental effort.
- Visual effort.
- Physical effort (and dexterity).
- Responsibility for the material or product.
- Responsibility for tools and equipment.
- Responsibility for the safety of others.
- Work surroundings or conditions.
- Unavoidable hazards.
- Effect on the activities of others.
- Effect on the conservation of physical facilities.
- Initiative requirements.

It is important to analyze the "job" rather than the "worker." It is a common mistake to keep in mind an especially capable or successful worker in preparing job descriptions, who does not represent the typical worker for the position.

Don't try to make a job description complete and fully detailed. The workable job description is flexible, and should conclude with a statement such as "the worker will perform other duties as assigned by the supervisor from time to time." Some people prefer to have the wording "other *related* duties."

The job description should also instruct the worker to comply with any assignment, if it is in the nature of an emergency as expressed by a supervisor or manager. The emergency might relate to personal safety (such as water on a finished floor), damage to surfaces (such as water on carpet), etc. The rule is that the job should be done first, then any questions asked or appeals made later.

How to Choose Job Titles That Promote Job Satisfaction

The question of job titles is confusing and requires clarification and standardization. A study of the names applied to housekeeping supervisors and to the persons performing the work themselves is eloquent testimony of the lack of coordination in housekeeping planning. For supervisory personnel, we find these titles being used, among others:

Building and Grounds Manager
Building Service Director
Custodian
Director of Buildings
Director of Sanitation
Environmental Service Manager
Head Custodian
Housekeeping Department Manager
Housekeeping Foreman
Housekeeping Superintendent
Housekeeping Supervisor
Janitor Supervisor
Sanitarian
Service Foreman
Service Manager
Utility Foreman

The person performing the work may be called:

Aide
Attendant
Cleaner
Custodian
Housekeeper
Laborer
Worker
Operator
Orderly
Porter
Sanitationist
Serviceperson
Utility person

Consider using one of these positive names for your operation:

Building Service Department
Environmental Services Department

If you used "Building Service Department" as your housekeeping operations title, the following personnel titles might also be used:

- Building Service Manager (or Director)
- Building Service Supervisor
- Building Service Leader
- Building Service Worker (or Aide)

Sample Job Descriptions

Here are some typical job descriptions for custodial workers.

Custodial Supervisor

I. Characteristic routine work

Definition:

This job entails the supervision of the housekeeping function for buildings and facilities. Employees in this job classification will report to a Shift Supervisor or Services Superintendent. This person will direct the daily activities of all housekeeping personnel assigned to a particular "Zone."

Examples of Routine Work:

- Designate shift and area work assignments for custodians and Group Leaders
- Frequently evaluate the performance of all custodial personnel assigned under their jurisdiction
- Inspect all assigned areas frequently to insure that all work is being carried out in a proper and orderly manner
- Develop and maintain a current training program for all zone custodial personnel
- Conduct classroom-type and on-the-job training for all zone custodial personnel
- Establish and maintain inter-departmental relations with personnel, maintenance, and purchasing
- Evaluate and assist in the selection of new cleaning products, materials, and equipment
- Assign relief custodians to fill in for absentee workers
- Maintain all pertinent records, reports, and other paperwork as needed
- Develop and maintain a projects list for the Project and Relief crew
- Provide a consultation service for management when new construction is planned
- Assist in scheduling and maintaining space for conference groups and other special functions
- Perform other related work as required and instructed

II. Qualifications

Required Knowledge, Skills, and Abilities:

- Knowledge of all cleaning methods, materials, and equipment
- Ability to instruct housekeeping employees in the performance of manual tasks requiring moderate strength
- Thorough knowledge of the operation of all mechanical cleaning equipment
- Ability to schedule repairs for all above equipment
- Ability to prepare written reports and other correspondence relating to the job as necessary
- Ability to establish and maintain good working relationships with subordinates, as well as with other supervisors and management personnel

Minimum Training and Experience:

The custodial supervisor must have a high school diploma or equivalent. Any training and experience gained through attendance at special schools or seminars is desirable. Also required is two years experience in custodial or general maintenance work—of which at least one year spent exhibiting a proven ability to supervise and lead relatively large groups of people.

Group Leader

I. Characteristic Routine Work

Definition:

This is routine manual labor associated with the daily cleaning of buildings. Employees in this job classification are under immediate supervision and perform various duties requiring moderate physical strength. Typical routine duties include, but are not restricted to: dust mopping, sweeping, wet mopping, vacuuming, dusting, waxing, cleaning restrooms, removing trash, polishing furniture, and simple operation of basic mechanical cleaning equipment.

In addition to the above listed tasks, this person will be responsible for: conducting on-the-job training for custodians; requisitioning and distributing supplies and materials; relaying instructions from the Custodial Supervisor; assisting the Custodial Supervisor in employee evaluations; determining the need for special project work to be performed; leading a group of custodial workers and participating in non-routine work listed below, scheduling set-ups for special functions.

Examples of Routine Work:

- Police public areas
- Dust mop, sweep, wet mop, spray-buff floors
- Vacuum clean rugs and carpets
- Scrub and clean restroom fixtures and keep supplied with towels, soap and other items
- Dust and clean walls, doors, windows, woodwork, and other above floor surfaces
- Dust, clean, and polish furniture
- Clean elevators and stairwells
- Keep simple records and make simple reports
- Police around outside of buildings
- Notify Supervisor of equipment and facilities needing repair or replacement
- Replace accessible light bulbs
- Perform related work as required and instructed

II. Characteristic Non-Routine Work

Definition:

This is manual labor associated with the recurring, but non-routine cleaning and maintenance of buildings. Employees will be under immediate supervision and will perform various tasks requiring moderate physical strength. This work is normally of the project-type nature.

Examples of Non-Routine Work:

- Stripping and refinishing floors
- Scrubbing floors with machines
- Vacuuming furniture and drapes
- Window washing
- Wall vacuuming and washing
- Wall, vent, and radiator washing
- Light fixture vacuuming and washing
- Waste receptacle washing
- Snow shoveling and removal
- Carpet shampooing
- Related work as required and instructed

- Set up rooms for conferences, lectures, and special events
- Move furniture

III. Qualifications

Required Knowledge, Skills, and Abilities:

- Knowledge of cleaning methods, materials, and equipment
- Ability to perform manual tasks requiring moderate physical strength
- Ability to operate mechanical cleaning equipment
- Ability to follow simple oral and written instructions
- Ability to establish and maintain working relations with Supervisors, and fellow workers
- Ability to conduct simple inventories of cleaning supplies and products
- Ability to fill out requisitions and related forms
- Ability to communicate effectively to relay instructions from the Supervisor.

Minimum Training and Experience:

A group leader should have at least two years experience in performing cleaning work and other related manual labor. Also required is training and experience, as may be gained through attendance of schools and seminars. Finally, a group leader must demonstrate, through actual experience, the willingness and ability to give exceptional job performance.

Custodian

I. Characteristic Routine Work

Definition:

This is routine manual labor associated with the daily cleaning of buildings. Employees in this job classification are under immediate supervision and perform various duties requiring moderate physical strength. Typical routine duties include, but are not restricted to: dust mopping, sweeping, wet mopping, vacuuming, dusting, waxing, cleaning restrooms, removing trash, polishing furniture, and simple operation of basic mechanical cleaning equipment.

Examples of Routine Work:

- Police public areas
- Dust mop, sweep, wet mop, spray-buff floors
- Vacuum clean rugs and carpets
- Scrub and clean restroom fixtures and keep restroom supplied with towels, soap and other items

- Dust and clean walls, doors, windows, woodwork and other above-floor surfaces
- Dust, clean, and polish furniture
- Clean elevators and stairwells
- Police around outside of buildings
- Keep simple records and make simple reports
- Notify Supervisor of equipment and facilities needing repair or replacement
- Replace accessible light bulbs
- Perform related work as required and instructed

II. Characteristic Nonroutine Work

Definition:

This is manual labor associated with the recurring but non-routine cleaning and maintenance of buildings. Employees will be under immediate supervision and will perform various tasks requiring moderate physical strength. This work is normally of the project-type nature.

Examples of Non-Routine Work:

- Stripping and refinishing floors
- Scrubbing floors with machines
- Vacuuming furniture and drapes
- Window washing
- Wall washing and vacuuming
- Wall, vent, and radiator washing
- Light fixture vacuuming and washing
- Waste receptacle washing
- Snow shoveling and removal
- Carpet shampooing
- Related work as required and as instructed
- Set up rooms for conference, lectures, and special events
- Move furniture

III. Qualifications

Required Knowledge, Skills, and Abilities:

- Knowledge of cleaning methods, materials, and equipment
- Ability to perform manual tasks requiring moderate physical strength

- Ability to operate mechanical cleaning equipment
- Ability to follow simple oral and written instructions
- Ability to establish and maintain working relations with Supervisors, and fellow workers

Minimum Training and Experience:

Custodians require such training and experience as may be gained through schools, some experience in performing cleaning work and other manual labor, or any equivalent combination of training and experience.

Figures 4-3 and 4-4 are sample job descriptions for a housekeeping department manager and a foreperson. Figure 4-5 provides a performance standard for routine work.

No organization is likely to function better than the supervision afforded it. Thus, it is desirable to look at the job at hand and to determine how you can best use the supervisory personnel available and what others will be needed.

You are now ready to establish your working organization, remembering that it should extend down all the way from the top point where policy affecting sanitation maintenance is originated. You should define clearly the lines of supervision, and, finally, you should indicate exactly the assignments of all positions in the organization. The individual duties should be spelled out to include the area of responsibility, a full statement of the tasks to be performed, and description of the tools available for use on the job.

An essential part of the job description is an indication of the frequency with which each task must be repeated.

While the work load for the average custodian will be made up of tasks to be performed daily or at frequent intervals, other cleaning operations must be repeated at less frequent intervals to assure maintenance of the selected level of cleanliness. These are what we normally term "special projects." Except in the smallest of organizations, there usually will be a special project team established, and it will cover this list of tasks on a rotating basis. The scheduling, of course, must be provided by the housekeeping supervisor. When developing a program of this type, management or administration undoubtedly will be concerned about a cost comparison between existing or past organization and the new program organization.

How to Determine Personnel Requirements

The heart of the housekeeping program is personnel. For your program to be successful, you must determine how many workers you need, how their work will be organized, how to schedule their specific maintenance tasks, and many other factors—all of which are interrelated.

To determine the approximate staff needed to perform any function, housekeeping or otherwise, divide the total work load in hours by the

Figure 4-3. Sample Job Description for a Housekeeping Department Manager

JOB DESCRIPTION

Title: Housekeeping

Department Manager

Reports To: Associate Director of Operations

Main Function:	Directly responsible for developing departmental objectives, establishing programs and procedures to meet these objectives, maintaining a high-quality staff of personnel within the department, and controlling and evaluating overall departmental objectives.

Tasks and Responsibilities:

Planning	Formulates plans for improving Housekeeping Service programs.
	Recommends and initiates the purchase of all Housekeeping Services supplies.
	Recommends and initiates the purchase of cleaning equipment pertinent to the Housekeeping Services Department.
	Advises the industrial engineer on standards and procedures.
	Recommends changes of layout and location of equipment to facilitate cleaning of various areas.
Controlling	Formulates the budget for personnel and supplies.
	Establishes procedures and work methods.
	Initiates and maintains the records of all productivity information.
	Maintains a sound interdepartmental working relationship.
Directing	Directs supervision of Forepersons and Lead Maintainers.
	Assists in the delegation of duties for all

Figure 4-3. (continued)

	Housekeeping Services personnel.
	Directs the maintenance and repairs of all Housekeeping Services equipment.
Organizing	Initiates and directs training programs.
	Demonstrates new equipment and methods.
Staffing	Establishes, reassigns or eliminates personnel positions to conform with area changes.
	Interviews, selects and processes all personnel.
	Disciplines and/or terminates unsatisfactory personnel.
	Evaluates supervisor staff.

Entry-Level Requirements and Job Factors:

Education	Some college and/or training in Housekeeping Services supervision and/or management.
Experience	Must have previous experience in institutional cleaning supervision and/or management.
Mental Demands	Moderate in scope; however, instant, frequent, and varied.
Physical Demands	Prolonged standing and walking.
Personal Contacts With	Regular employees, visitors, and staff of other departments.

Figure 4-4. Sample Job Description for a Custodial Foreperson

JOB DESCRIPTION

Title: Foreperson

Reports To: Housekeeping Department Manager

A. Job Summary—Assists the Manager in supervising the Housekeeping Services staff. Assumes the responsibility of the Manager, as assigned, in his/her absence. Follows outlined instructions for projects as designated by the Manager.

B. Job Relationships

1. Responsible to the Housekeeping Department Manager.
2. Workers Supervised: Housekeeping Workers and Lead Maintainers on the Housekeeping Service Staff.
3. Interrelationships with other employees—must know how to deal tactfully with employees of the Housekeeping Services Staff, Department Heads, other employees, and visitors.

C. Duties

1. Arranges and adjusts work schedule for daily needs.
2. Trains sanitation workers in the use of equipment and materials necessary to perform their duties.
3. Supervises all workers in their respective work areas.
4. Helps formulate plans within the Housekeeping Services Department.
5. Recommends changes in personnel and equipment, as necessary.
6. Sets work priorities.
7. Assists in the orientation and training program for new employees.
8. Attends meetings, lectures and training classes.
9. Helps interview all new employees.
10. Counsels all unsatisfactory personnel.
11. Takes inventory of all supplies.
12. Inspects regularly all equipment for defects, and reports to the Manager if there are any necessary repairs to be made.
13. Inspects his/her assigned area daily.

Figure 4-4. (continued)

D. Qualifications

1. Education—Some training in Housekeeping Services is required. Appropriate training and experience may be substituted for education; however, the pursuit of Housekeeping Services training is encouraged.
2. Training and experience—Should have a previous job-related equivalent. Must know how to supervise personnel and deal with people.
3. Special skills—Ability to verbally relate to others, be self-motivated and possess the desire to complete assigned functions.
4. Job Knowledge—Must understand and be able to communicate to employees the importance of clean and sanitary facilities.
5. Mental Requirements—Have the ability to verbally relate to others, be able to work under stress at certain times.
6. Physical Requirements—Prolonged walking and standing.
7. Working Conditions—May have some exposure to dirt, dust, and obnoxious odors. Has to work in dimly lighted areas at times. Must climb stairs and walk on metal grate flooring at times. Must walk from building to building.

Figure 4-5. Sample Performance Standard for Routine Work

Cleaning Basins:

- Using the pump-up sprayer, apply germicidal detergent solution to all the basins (or in larger rest rooms, to all the basins on one wall) and to the wall area beside and between the basins.
- Clean the tops, sides, insides, and wall areas between the basins with a cloth.
- Wipe the metal surfaces dry with a cloth to prevent spotting.

Cleaning Urinals:

- Use the pump-up sprayer to apply germicidal detergent solution to all the urinals (or in larger rest rooms, to all the basins on one wall) and to the wall area between and below the urinals with a cloth.
- This cloth should be of a particular color, used only on commodes and urinals.
- Clean the insides of the urinals with a bowl mop.
- Use the bowl mop to clean the underside of the flushing rim.
- Wipe metal surfaces dry with a clean cloth to prevent spotting.

Cleaning Toilets:

- Use the pump-up sprayer to apply germicidal detergent solution to the insides and outsides of the commodes and to the wall areas beside them.
- Spray the top of the seat first, then lift the seat and spray the remainder of the fixture.
- Clean the seat, outside of the fixture, and wall beside the fixture with the same cloth used to clean the outsides of the urinals.
- Clean the inside of the fixture with a bowl mop.
- Use the bowl mop to clean under the flushing rim.
- Wipe the top of the seat and the metal surfaces dry with a cloth to prevent spotting.

Cleaning Shower Stalls:

- Use the pump-up sprayer to apply germicidal detergent solution to the walls and floors in shower stalls.
- Use a cloth and abrasive pad to damp wipe all surfaces of the shower.

Figure 4-5. (continued)

- **Remove all debris from the shower drain.**
- **In smaller rest rooms, particularly in office areas, using a pump-up sprayer may not be practical. In these areas, use a spray bottle to apply germicidal detergent solution to fixtures.**

Descaling Toilets and Urinals:

- **Use acid-type bowl cleaner and a nylon bowl mop to remove scale, scum, mineral deposits, rust stains, etc. from toilet bowls and urinals.**
- **After descaling, make sure the entire surface is free from streaks, stains, scale, scum, mineral deposits, rust stains, etc.**
- **Use caution to prevent damage to adjacent surfaces caused by spills of the acid-type bowl cleaner.**

number of productive hours per employee. The work load is a product of the number of units to which the work is applied, the frequency of application, and the time required for application.

Frequency of job performance and the desired level of housekeeping go hand in hand. That is, a determination of one dictates the other. To achieve a given level of housekeeping, relate the frequencies of job performance to obtain that level—conversely, given frequencies will dictate a specific level. Often management will ask for a suggested objective for housekeeping. In any case, the first step must be to determine your objective, even if it is tentative.

The total number of work units is a product of the frequency and number of items (or the measure of areas) to be cleaned or otherwise maintained. The information required to develop these figures is taken from the physical survey of the premises. The application of time standards against the work units gives the number of labor-hours per year required for the total housekeeping operation. When you divide this by the number of annual productive hours per custodian, you obtain the theoretical number of custodians required.

The calculation is based on the formula:

$$\text{Personnel} = \text{frequency} \times \text{areas} \times \text{standard times} \div \text{available productive hours per year.}$$

In using this formula it is extremely important to keep the units straight (hours, seconds, weeks; square feet, square yards, etc.) and to cancel them out properly in the calculation.

At this point you will have arrived at a *theoretical* calculation of the housekeeping staff. *It probably will not be correct.* The figure will require adjusting up or down on the basis of the many considerations and variables involved. It would be difficult to overstress the importance of these variables in determining the number of housekeeping personnel, and therefore in determining cost. The general housekeeping program prepared for a typical facility will be of considerable value for a number of years. But when conditions, such as personnel, physical facility, management objectives, occupancy levels, etc., change, the program will require adjusting if its value is to remain high.

The forms used to determine work load are subject to considerable variation. They may consist of a separate sheet for each type of cleaning operation, for example, or possibly one sheet for each floor or each building with all the various jobs listed. The latter has the advantage of relating work load to a natural physical division, which would otherwise have to be done in a separate step. It has the disadvantage of mixing standards and units. Personal preference will dictate the specific form to be used, the column headings, the size of the form, and the method of extension. In any case, these factors must be recorded:

Area designation
Item or surface to be maintained
Measure or count
Housekeeping job
Annual frequency
Annual measure or count
Time standard
Total time

Figure 4-6 shows a square footage building summary. This page summarized individual records that were used for each floor of the building, and here we see total square footage (in thousands) of each type of area, categorized by floor surface type (for example, K = carpet). Standard times and frequencies can then be applied to these figures for work-hour requirements.

HOW TO DEVELOP PLANS FOR SPECIAL PROJECTS AND RELIEF

Special projects may be defined as those jobs that are done less frequently than once per month. Special projects form the backbone of the preventive maintenance aspect of housekeeping. Some types of special projects are best performed by individuals, either with area or project assignments. Other types can most efficiently be done as a team effort. In certain cases, these teams can be augmented by a custodian on area assignments, where the project work is being done in that area and consumes only a short period of

Figure 4-6. Sample Square Footage Building Summary

AREA TYPE	THOUSANDS OF SQUARE FEET									# fx	#flts	UNIT ALLOW	AREA ALLOW
	R,Z	K	C	GT	RF	W							
Building Service Stor			1.54										
Computer Room					4.39								
Conference Room		.26											
Corridor			1.99										
Corridor	2.77												
Corridor		1.10											
Dining		4.22											
Elevator - Freight											1		
Elevator - Passenger											4		
Entry			.86										
Exam	.11												
Executive Office		1.56											
Files	2.08												
Kitchen					2.40								
Library		2.20											
Library Stack		.75											
Lobby	5.27												
Lobby		2.00											
Maintenance Shop			.93										
Medical Reception	.12												
Office	4.90												
Office		2.30											
Open Office	19.02												
Parking			6.71										
Print Shop			1.77										
Reception	.48												

Reception		.58										
Restroom									103			
Security Area	.21											
Serving Area				.41								
Stairs									9			
Storage			1.40									
Storage	.22											
Tape Storage	.77											
Trash Room			.96									
Treatment	.16											
Truck Dock			.30									
TOTAL	36.11	14.97	16.46	2.81	4.39							

CLIENT ____________ JOB # ________ PREP BY ________ DATE ________

TOTAL SQ FT 75.09

BUILDING ____________ NO. ________ FLOOR Building Summary

time. The special projects workers should always be as highly mechanized as possible.

Devise your program to provide the flexibility required to meet unusual situations. These situations may be caused by shortages in materials or equipment, accidents, emergency cleaning situations, or absence due to vacations, special holidays, family emergencies, sickness, or failure to report to work. Set up a procedure in advance to handle each of these possible occurrences.

Organize a separate relief crew to handle most of these problems. Relief must be provided for each shift during which housekeeping work is performed, unless there are only a very few persons on a given shift performing policing operations, where the other workers can "cover." Relief personnel should always report directly to their supervisor at the beginning of the work shift, and, if not required for relief, they may be assigned to perform special projects work.

There is an inter-relationship between special projects and relief personnel. Thus, it is possible to adjust the size of these two combined groups to conform to a housekeeping staff that may vary. This permits permanent daily assignments and instructions to remain in force even during a changing situation. The housekeeping level will fluctuate with the size of this group, because housekeeping must be performed on a continuing quality level through the provision of relief and special projects.

To eliminate interference between housekeeping personnel and others, as much of the work as possible should be performed after the busiest hours of the company. Thus, a safer operation will result, with more efficient working conditions both for housekeeping and other personnel. The advantages must be weighed against the drawbacks, because evening or night cleaning work sometimes intensifies the problems of supervision, security, the ability to recruit workers because of transportation difficulties and the later hours, etc.

Figure 4-7 shows how a relief and projects team could be used to staff a 10-story building with 10 custodians. Without such an arrangement, there is constant turmoil and workers are regularly reassigned. Customers are unhappy with a fluctuating level of service.

Figures 4-8 through 4-13 provide examples of forms and material useful in developing the "special projects" section of the operation manual:

- Figure 4-8 is a special project record showing what type of work is to be performed in what areas, and the target frequency. A calendar provides an opportunity to indicate completion.

- Figure 4-9 is a frequency schedule, similar to the previous form, but providing columns which break down these activities by frequency. Note that this form discusses weekly and semi-monthly activities, but these are not normally considered to be special projects.

Figure 4-7. Sample Organization with Relief and Projects Team

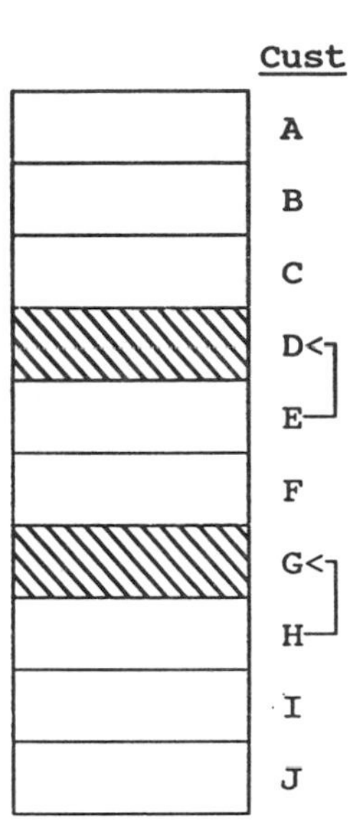

Staffed one custodian per floor

- Workloads not balanced (some floors harder to clean than others).
- If custodian D is on vacation, and custodian G is absent, two other people will "double up"; four floors will be poorly cleaned.
- And all workers are pulled off their repetitive work to do projects.

Staffed for R & P Team

- 8 custodians each have an average of 1 1/4 floors, and perform repetitive work only (frequencies are set accordingly).
- 2 custodians are used for relief and projects.

Figure 4-8. Sample Record of Special Projects

SPECIAL PROJECT RECORD

Building #26

Record the performance of each Special Project on the day it is completed. This record must be maintained accurately and up-to-date to assure regular performance of all Special Projects.

SPECIAL PROJECT	AREA	FREQUENCY	JAN.	FEB.	MAR.	APR.	MAY	JUNE	JULY	AUG.	SEPT.	OCT.	NOV	DEC
Wash Mats & Runners	At Entrances	Monthly												
Clean & Polish Hardware	At Entrances	Monthly												
Clean Utility & Storage Rooms	All Areas	Monthly												
Vacuum Radiators & Vents	All Areas	Monthly												
Descale Fixtures (SPI 400)	Rest Rooms	Monthly												
Machine Scrub Unwaxed Floors	All Areas	Monthly												
(SPI 203)														
Clean & Rewax Composition Floors	Rest Rooms	Monthly												
(SPI 200, 350)														
Clean & Rewax Composition Floors	Foremen's Offices	Every 6 Wks.												
(SPI 200, 350)														
Clean & Polish Furniture (SPI 410)	Foremen's Offices	Every 2 Mos.												
Wash Windows Inside (SPI 110)	All Areas	Every 2 Mos.												
Wash Waste Receptacles	All Areas	Quarterly												
Vacuum Light Fixtures	All Areas	Quarterly												
Vacuum Walls & Woodwork	All Areas	Quarterly												
Wash Windows & Screens (Contract)	All Areas	Semi-Annually												
Strip & Rewax Composition Floors	Foremen's Offices	Semi-Annually												
(SPI 250, 350)														
Strip & Rewax Floors (SPI 250, 350)	Rest Rooms	Semi-Annually												
Seal Floors (SPI 320)	Concrete Floors	Annually												
Damp Clean Light Fixtures (Engrs.)	All Areas	Annually												
Damp Clean Radiators & Vents	All Areas	Annually												
Vacuum Overhead	All Areas	Annually												
Wash Walls & Woodwork (SPI 120)	All Areas	Annually												

Figure 4-9. Sample Work Frequency Schedule

FREQUENCY SCHEDULE

Building ____________________

Date and Initial of Supervisor ____________________

Function Waxing, Stripping, etc.	Areas	Wkly	Semi-Mthly	Mthly	Bi-Mthly	Qtly	Jan	Feb	Mar	Apr	May	Jun	Jul	Aug	Sep	Oct	Nov	Dec

• Figure 4-10 shows a performance standard for two types of project work, in this case dry foam carpet cleaning and dry powder carpet cleaning.

• Figure 4-11 provides a task time per thousand square feet for seven example projects. By multiplying these times by annual frequency and numbers of thousands of square feet, you can develop the total annual project time for each activity.

• Figure 4-12 can be used by a first-line supervisor to schedule projects in a specific floor of a given building.

• Figure 4-13 is a report form to help group leaders or foremen describe the projects completed during a specific month for a designated area.

POLICING ACTIVITIES: HOW TO BE SURE THE WORK GETS DONE

When the basic cleaning work is done after normal working hours, it should be followed during the day by a policing operation. This is not primarily a cleaning operation, but one of waste collection and removal, spillage removal, refilling dispensers, etc. A large percentage of policing labor is consumed in the personnel areas such as rest rooms, locker rooms, etc. The policing operation is one of the more important, yet usually more disorganized, phases of the housekeeping operation, because the very nature of the work makes supervision difficult. This should be overcome by a fixed policing schedule.

The attitude of the public and workers alike concerning the appearance and level of housekeeping of a facility is often based on how well the policing function has been performed.

In some larger operations, the various housekeeping functions are categorized and treated separately, even though certain jobs may be combined to be performed by a single worker or a team of workers. These divisions may include, for example:

Rest-room cleaning
Office cleaning
Dietary area cleaning
Machine cleaning
Waste removal
Chip collection
Power sweeping
Special projects
Relief
Nonhousekeeping responsibilities

In the smaller housekeeping departments we must work with indivisibles. For example, it may require 1.4 people to clean machinery, 2.7 people to clean the offices, etc. In the smaller facility, for example, security guards are sometimes given housekeeping responsibilities. These should be of such a

Figure 4-10. Sample Performance Standards for Project Work

PERFORMANCE STANDARDS FOR PROJECT WORK

Carpet Cleaning, Dry Foam Method

- Dry foam shampooing of carpet is defined as the spot cleaning, vacuuming, shampooing, and revacuuming of all carpet in an area.
- All vacuuming, both before and after shampooing should be done with a medium-duty, pile-lifter type vacuum.
- All stained areas should be treated with spot cleaning solution, following the directions of the manufacturer.
- Spot cleaning should be continued until as much of the stain as possible has been removed.
- The shampooing should be performed using equipment and materials specifically designed for dry foam shampooing and meeting the specifications for such equipment and materials given in this manual.
- The instructions provided by the manufacturers of the equipment and materials should be followed during its use.
- Areas, such as corners, which are inaccessible to the machine, should be shampooed with foam from the machine and manual scrubbing devices.
- After shampooing and allowing sufficient drying time, the carpet should be vacuumed following a pattern which will give the carpet pile a uniform appearance.

Carpet Cleaning, Dry Powder Method

- Dry cleaning carpet is defined as the spot cleaning, vacuuming, application and scrubbing of dry cleaning compound followed by another complete vacuuming.
- All vacuuming shall be done with an upright vacuum.
- All stained areas shall be treated with spot cleaning solution following the directions of the manufacturer.
- Spot cleaning shall be continued until as much of the stain as possible has been removed.
- The dry cleaning shall be performed by applying the "dry" chemical and machine scrubbing the area according to the manufacturer's instructions.
- After sufficient absorption time, the remaining dry cleaning residue and soil shall be vacuumed completely, following a pattern which will give the carpet pile a uniform appearance.

Figure 4-11. Sample Guide to Project Work Times

ESTIMATE OF PROJECTS

Project	Annual Freq.	Task Time Per 1000 Sq.Ft.	No. of 1000 Sq. Ft.	Annual Time
1. Strip & refinish resilient tile floors in public areas		130		
2. Strip & refinish resilient tile floors in non-public areas		175		
3. Surface brighten carpet in offices, non-public areas		90		
4. Surface brighten carpet in non-public areas.		60		
5. Water extract carpet in non-public areas		160		
6. Water extract carpet in public areas		150		
7. High dusting		45		

Figure 4-12. Sample Schedule of Project Work

PROJECTS WORK

19__ to 19__

PROJECT CALENDAR

Building ______________________

Floor/Area ______________________

/ = Scheduled
X = Completed

Project	Allowance	Quantity (KSF)	Est. Man-Hours	Act. Man-Hours	Freq. Per Yr.	JUL	AUG	SEP	OCT	NOV	DEC	JAN	FEB	MAR	APR	MAY	JUN

Figure 4-13. Sample Report of Completed Projects

PROJECT-TYPE WORK COMPLETED

MONTH OF ______________________ AREA ______________________

AREA OF			DATE	HARD SURFACE		CARPET					
BUILDING	FLR	ROOM	DONE	S&R	SCRUB	B.BUFF	EXTR	WALLS	FURN	GLASS	OTHER

nature that they can be interrupted without hazard or without wasting too much time, so that the guards can perform their rounds. Some types of special projects are suitable for this purpose.

ASSIGNING JOBS: TEAM CLEANING VERSUS INDIVIDUAL AREAS

Whether to perform cleaning on an individual or team basis, or some combination of the systems, must depend on a large number of other factors involved in the total picture. The proper choice is actually in the field of technique rather than technology.

Area Assignments

Area assignments have these advantages:

- They give the custodian a proprietary attitude toward his work area, wherein he develops more pride in the results.
- A very healthy competitive spirit can be developed in the cleaning of similar types of areas. Individual assignments lend themselves more to award or recognition systems, based on quality of performance.
- The individual assignment pinpoints responsibility. Any deficiencies or failure to perform the work can be laid at the feet of the custodian involved.
- Because of the fixing of responsibility, security control may be exercised more easily.
- The custodian should be at his work station at all times, except for personal reasons, and therefore he can be easily and quickly located for general supervision, emergencies, etc.

Team Cleaning

Team cleaning may involve anywhere from two to six or more persons for each group. Some of the general advantages of team cleaning are:

- The group can be highly mechanized and yet the investment in equipment for each custodian can be kept to a minimum. A floor cleaning group, for example, might consist of one custodian equipped with a mopping tank, one with a scrubbing machine, and another with a wet vacuum. If the work were to be performed on an area assignment basis, all of this equipment would have to be provided for the individual.

• If a member of the team is a group leader or assistant foreman, supervision of the work can be intense and continuous. This may result in a work efficiency increase of 25 percent or more for the average worker.

• If the team if performing special projects it assures the completion of this important work. Special projects performed by individuals often are postponed for too great a period of time because of the pressures of the daily repetitive work.

• The team has advantages in cases of turnover because new personnel can be placed in the team and quickly trained by working along with the older more experienced employees, and under the direct guidance of the team leader.

• Training is also simplified because the number of jobs which the team as a whole has to perform is limited, and the work of the individual within the team can be even further specialized. This is especially useful in cases of high turnover, such as with plant-wide seniority.

• This specialization permits a higher degree of efficiency because of the decreased make-ready and put-away time, and for other reasons as well.

• The spirit of comradeship, and the mere fact of being a member of a team, improves morale, heightens interest, and may reduce absenteeism through peer pressure.

• Team cleaning assures completion of the project at hand in the shortest possible period of time. Thus it vacates the area for other types of work that may be performed, and cuts down the hazard exposure time in the case of wet floors.

• Team cleaning may be less hazardous for the person doing the work. In case of accident, another worker can give aid or bring help. The specialization and intensive training further tends to reduce accidents.

It is best to use "area complete" assignments for repetitive work, and to use team cleaning for project work, or where the work must be completed in a certain area in a short time.

Here are some general guidelines:

• One worker, completing an entire area—The worker does all jobs in a given area, including infrequent projects.

• One worker, completing an entire area, with exceptions—Worker does all jobs within a given area, except for periodic projects such as floor stripping or wall washing; also, some jobs are handled by specialists, such as rest-room care or waste collection.

• Two workers complete an entire area—Two people are responsible for all or most jobs in a given area, with one specializing in the heavier duty work (typically floor care) and the other specializing in the lighter duty work (typically non-floor care). This is effective in a high turnover situation, where the new employee is placed with the older employee, or where there is a mixture of men and women, with the women doing the lighter duty work.

- Team cleaning (for projects)—Typically with teams of two or three people, this system is used for infrequent or special project work, such as floor stripping and refinishing, wall washing, or overhead vacuuming.
- Group cleaning—Also known as "route cleaning," this system uses several people, such as from 4 to 6, who move through one area or building at a time, doing all the work there, then going on to the next area or building. Within the group, assignments can be made using any of the above or a combination of the above systems.
- Gang cleaning—This system uses more workers than group cleaning (such as a dozen or more) to rapidly clean an area.

The following figures relate to job assignments:

- Figure 4-14 uses a shaded floor plan to indicate a portion of an assignment by a specific worker.
- Figure 4-15 lists the materials and equipment requirements for a specific job assignment.
- Figure 4-16 shows five alternatives for staffing a seven-day custodial operation, indicating the total number of FTE's (full time equivalent staff) to fit the example perfectly, and also showing the number of workers who would be on hand each day.

Improving Staff Productivity by Rescheduling Employee Breaks

You may also want to consider a change in break times (e.g., coffee breaks) to improve productivity. Figure 4-17 illustrates that typical break periods are often exceeded. Some studies indicate that an official ten or fifteen-minute break generally lasts an average of 23 minutes, and it is not infrequent for them to last a half hour.

An alternating break schedule, on the other hand, puts the two 15-minute breaks together, dividing the day up into three rather than four work periods. Even if the double break is exceeded in time, it would only be once rather than twice. Further, greater productivity should be achieved because there are three make-ready and put-away times in the alternate schedule, rather than four in the typical schedule. Many workers like the alternate schedule, since they have two half-hour periods, one of which is paid, and one not paid.

Figure 4-14. Sample Floor Plan with Job Assignments

<u>ASSIGNMENT:</u>

MONDAY - FRIDAY
8:00 A.M. - 3:30 P.M.

Perform routine work task in the areas designated on the drawings below.

Perform other duties as assigned.

 These areas are routinely cleaned by this assignment.

☐ These areas are <u>not</u> routinely cleaned by this assignment.

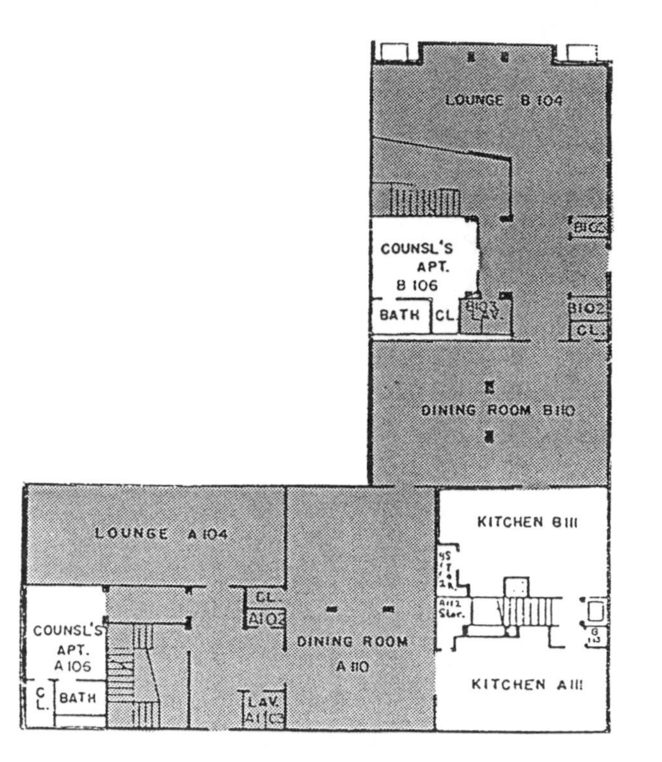

Encouraging End-of-Shift Cleaning by Non-Housekeeping Workers

Many industries provide for the cleaning of production areas up to the aisles by production department personnel during the last ten minutes of each work shift. This practice is consistent with the tendency of production workers to "unwind" near the end of the day. It has the advantage of promoting a greater housekeeping awareness for all such employees by involving them directly in housekeeping and thereby contributing to improved quality control and productivity. Workers will create less waste and soil when it is their responsibility to remove it. Whether this system should be used or not depends upon the type of industry, the method of compensation of the production worker, the difference in wages between production and housekeeping workers, and the type of waste and soil created. The amount of cleaning performed by nonhousekeeping workers directly affects the determination of the size of the housekeeping staff.

Assigning Priorities to Various Maintenance Tasks

Job sequence (i.e., priorities) has an important effect concerning cleaning time and quality of results. Instruction sheets should indicate the work and the proper order of performance to save as many steps as possible and to actually prevent the accumulation of more soil. For example, if the floors are dust mopped before the furniture is dusted, the custodian may complete "cleaning" and leave a floor which has become dusty from furniture dusting. Or, if the wastebasket is emptied before the cleaning of the room is complete, extra trips must be taken to the waste cart.

Using Internal Staff Versus Contract Cleaning

Sometimes it may be more economical for some aspects of the work to be performed on a cleaning contract rather than by internal personnel. In some cases, the entire work might be performed on a contract basis. You should carefully compare contract costs with internal costs, and your decision must involve all of the many other factors already described. No contract cleaning will be successful unless close control is exercised over frequency, materials, methods, and quality of job performance.

Frequency of Performance of Cleaning Tasks

In the usual housekeeping operation, certain frequencies can be extended with no apparent physical change in the overall result. Wax is often applied

Figure 4-15. List of Materials and Equipment for a Specific Job Assignment

DATE: ____________________
LOCATION: ____________________

MATERIALS AND EQUIPMENT REQUIRED FOR ROUTINE ASSIGNMENT #____________________

SECURABLE CLOSET (Location or room #____________________)

0 LIGHT W/BULB PROTECTED
0 TOOL HOLDER RACK
0 SHELVES ADEQUATE FOR SUPPLIES
0 FLOOR SINK W/DRAIN AND SPIGOT
0 HOSE
0 GALLON OF GLASS CLEANER CONCENTRATE W/JUG PUMP
0 GALLON OF GLASS CLEANER SOLUTION W/JUG PUMP
0 GALLON OF NEUTRAL DETERGENT CONCENTRATE W/JUG PUMP
0 GALLON OF NEUTRAL DETERGENT SOLUTION W/JUG PUMP
0 GALLON OF GERMICIDAL DETERGENT CONCENTRATE W/JUG PUMP
0 GALLON OF GERMICIDAL DETERGENT SOLUTION
0 MEASURING CUP
0 FUNNEL
0 PAIL
0 SUPPLY OF SPONGES

0 CONTAINER OF STAINLESS STEEL CLEANER AND POLISH
0 CONTAINER OF LOTION CLEANSER
0 SUPPLY OF LINERS FOR TRASH RECEPTACLES
0 SUPPLY OF LINERS FOR CART
0 SUPPLY OF CLEANING CLOTHS
0 SUPPLY OF WET MOP HEADS
0 SUPPLY OF DUST MOPS
0 CONTAINER OF HAND SOAP
0 SUPPLY OF HAND TOWELS
0 SUPPLY OF TOILET TISSUE

0 LIGHT BULBS
0 FLUORESCENT TUBES

NOTE: ALL MATERIALS SHOULD BE LABELED PROPERLY AS TO CHEMICAL COMPOSITION, MIXING INSTRUCTIONS, SAFETY PRECAUTIONS, ETC.

CART NO____________________

0 SPRAY BOTTLE NEUTRAL DETERGENT
0 SPRAY BOTTLE GERMICIDAL DETERGENT
0 SPRAY BOTTLE GLASS CLEANER
0 LOTION CLEANSER
0 STAINLESS STEEL CLEANER AND POLISH
0 DUSTING TOOL
0 PUTTY SCRAPER

NON-CARPETED AREAS - FLOOR CARE

0 SMALL SWIVEL DUST MOP
0 LARGE DUST MOP, SIZE____________________
0 EXTRA DUST MOP HEADS
0 PUSH BROOM
0 CORNER SCRUB BRUSH
0 "DOODLE BUG" BRUSH

0 CLEANING CLOTHS
0 DUSTPAN
0 RUBBER GLOVES
0 SAFETY GOGGLES
0 SPONGES (DIFFERENT COLORS)
0 SCRUB PADS
0 UTILITY (SCRUB) BRUSH
0 FIREPROOF ASH RECEPTACLE

FOR RESTROOMS
0 **CLOSED FOR CLEANING** SIGNS (2)
0 ACID-TYPE BOWL CLEANER
0 BOWL MOP
0 SOAP FILM REMOVER
0 INSPECTION MIRROR
0 SPONGE (DIFFERENT COLORS)
0 PUMP UP SPRAYER
0 HAND SOAP
0 TOILET TISSUES

0 SPECIAL EQUIPMENT
____
____
____
0 UPRIGHT VACUUM # _______________
0 HOSE AND WAND
0 CREVICE TOOL
0 BRUSH ATTACHMENT
0 STAIN REMOVER
0 GUM REMOVER (PRESSURIZED NITROGEN)
0 BACKPACK VACUUM #_______________

0 FAN-TYPE MOP (YARN SIZE)_______________
0 FAN-TYPE WAXING MOP
0 MOP BUCKET AND WRINGER SIZE _______________
0 "CAUTION - WET FLOOR" SIGNS (2)
0 DECK BRUSH (GROUTED TILE ONLY)
STRIPPING
0 LOW SPEED BUFFER #_______________
0 ACCESS TO LOW SPEED BUFFER #_______________
(150 - 300 RPM)
0 WET/DRY VACUUM #_______________
STRIPPING PADS
POLISHING
0 HIGH SPEED BUFFER #_______________
0 ACCESS TO HIGH SPEED BUFFER #_______________
(950 - 1150 RPM)
0 GOLD PADS
0 WHITE PADS
0 SPRAY BOTTLE SPRAY BUFF
0 METAL LINK POLYMER FINISH - UL LABEL
SCRUBBING
0 20" AUTOSCRUBBER #_______________
0 ACCESS TO 20" AUTOSCRUBBER #_______________
0 RED PADS
0 WHITE PADS
0 32" AUTOSCRUBBER #_______________
0 ACCESS TO 32" AUTOSCRUBBER #_______________
0 RED PADS
0 WHITE PADS
0 FIVE GALLON SCRUBBER CONCENTRATE

0 SPECIAL EQUIPMENT
____
____

NOTE: ALL MATERIALS SHOULD BE LABELED PROPERLY AS TO CHEMICAL COMPOSITION, MIXING INSTRUCTIONS, SAFETY PRECAUTIONS, ETC.

Figure 4-16. Five Examples of Work Schedules

SEVEN-DAY WORK SCHEDULE ALTERNATIVES

ALTERNATIVE I - Stagger off days evenly. (100 FTE)

JOB	M	T	W	T	F	S	S	M	T	W	T	F	S	S	M	T	W	T	F	S	S	M	T	W	T	F	S	S
1						•	•						•	•						•	•						•	•
2					•	•						•	•						•	•						•	•	
3				•	•						•	•						•	•						•	•		
4			•	•						•	•						•	•						•	•			
5		•	•						•	•						•	•						•	•				
6	•	•						•	•						•	•						•	•					
7	•						•	•						•	•						•	•						•
S-1	•	•						•	•						•	•						•	•					
S-2			•	•						•	•						•	•						•	•			
S-3						•	•						•	•						•	•						•	•

ALTERNATIVE II - Employ part-time workers. (98 FTE)

JOB	M	T	W	T	F	S	S	M	T	W	T	F	S	S	M	T	W	T	F	S	S	M	T	W	T	F	S	S
1						•	•						•	•						•	•						•	•

ALTERNATIVE III - One weekend off every three-week period. (105 FTE)

JOB	M	T	W	T	F	S	S	M	T	W	T	F	S	S	M	T	W	T	F	S	S
1	•					•	•				•					•	•				
2				•					•	•					•					•	•
3		•	•					•					•	•				•			

ALTERNATIVE IV - Two weekends off every four-week period. (140 FTE)

JOB	M	T	W	T	F	S	S	M	T	W	T	F	S	S	M	T	W	T	F	S	S	M	T	W	T	F	S	S
1		•	•								•	•								•	•						•	•
2				•	•								•	•						•	•		•	•				
S-1						•	•						•	•		•	•								•	•		
S-2						•	•		•	•								•	•								•	•

ALTERNATIVE V - Two weekends off in four-week period with reduced cleaning level on weekends. (93 FTE)

JOB	M	T	W	T	F	S	S	M	T	W	T	F	S	S	M	T	W	T	F	S	S	M	T	W	T	F	S	S
1		•	•								•	•								•	•						•	•
2				•	•								•	•						•	•			•	•			
3						•	•						•	•		•	•							•	•			
S-1						•	•		•	•								•	•								•	•

Figure 4-17. Alternate Break Schedules

TYPICAL BREAK PERIODS

ALTERNATE BREAK SCHEDULE

Double Break

too frequently; vertical surfaces of furniture require dusting only weekly, rather than daily; acid descaling of rest-room fixtures should be done periodically, rather than daily. Task frequency has an obvious effect on the number of personnel involved.

SUMMARY

The cleaning of a multibuilding facility will require a different approach from a plant consisting of only a single major building. This may also necessitate an entirely different approach in terms of supervisory personnel or arrangement required. The arrangement of the personnel is open to many more possibilities, such as where team cleaning may be done all within a single building or may proceed from one building to the next.

The factors described in this chapter by no means represent a complete list of factors affecting the determination, organization, and scheduling of housekeeping personnel, but merely indicate some of the more usual situations.

5

How to Put Your Housekeeping Plans into Action

Although a great deal of time and effort may go into the custodial survey and program document preparation phases of the program (described in chapters 2, 3, and 4), these activities will not create results in themselves. The housekeeping program only begins when you complete the written program document. Furthermore, your housekeeping program will be just a few ounces of used paper stock if you and your staff do not read and follow up on the program's recommendations and schedules.

This chapter outlines some of the basic steps necessary to convert your program document to the practical objectives sought by management. The purpose of the initial installation phase is to set the machinery in motion. It involves four steps:

1. Reviewing the program document with management.
2. Reviewing the program document with housekeeping supervision.
3. Presenting the program to housekeeping workers.
4. Coordinating the housekeeping program with other departments.

After you've taken these four steps, you and your staff can begin the process of implementing the housekeeping program.

HOW TO PRESENT YOUR PROGRAM TO MANAGEMENT AND STAFF: FOUR KEY STEPS

In attaining your housekeeping objectives, you will be moving from one level of efficiency and performance to another. First, you must establish the proper atmosphere by, in your own mind and in the minds of those persons with whom you have contact, establishing a personal identity with the organization. Every organization has a unique "feel." This indefinable tone can be the result of such diverse causes as the level of business or service activity, the company's advertising campaign, the effectiveness of the personnel relations department, the status of the institution, the personalities of management personnel, the growth potential, the average age and service of the employees, specific grievances, and a number of other items as well. It is not unusual to visit a facility one week where a smile and a friendly greeting is the rule, and then the next week to work in a facility of the same general type where the same approach is met with silence or perhaps a bit of mumbling.

To be of true value to the organization, you must develop an understanding of these feelings and, if possible, the motives behind them. The whole approach must be a positive joining of forces and a sincere offering of assistance and understanding. This should be reflected in references to "our" programs and situations, the steps that "we" must take, and the benefits to "us." Where criticisms are unavoidable, they should be voiced as an expression of the positive benefits to be gained following the correction or reversal of a given situation.

During the installation, as well as in other phases of the program, take advantage of the opportunity to learn. From each contact should come information and experience which will benefit the next contact.

The first part of working with the program document is its presentation to, and careful review with, all involved parties.

This must be done carefully to avoid the resistance that such an imposing document can bring out, as well as the confusion that may occur if the various phases of the program are not explained simply, carefully, and in a logical order.

Step 1: Presenting Your Housekeeping Program to Upper Management

By the time the program is presented to management for consideration, a good part of the psychological aspect of the movement toward better housekeeping has already been assured. A management decision to conduct a housekeeping survey and to program the housekeeping operations is in itself the best possible sign of management's growing interest in this field.

Before discussing the recommendations and concept of the program itself, keep in mind these fundamental guidelines:

- Consider all recommendations in the light of management's objectives and limitations. They must be practical, rather than theoretical.
- Give recommendations by listing specific proposals, indicating frequency or scheduling in chart or table form, and describing methods, materials, equipment, and other aspects of housekeeping in the work schedules.

Because of the comprehensive nature of the program, many of these recommendations are interdependent, and you should consider them in the context of the whole program. Acceptance, rejection, or modification of a given suggestion must involve consideration of other dependent or affected recommendations.

- Consider the program as a target to be worked toward over a reasonable period of time rather than a revolution which would disrupt normal production operations. The program is a flexible, dynamic tool rather than a rigid, static instrument.
- Remember that the final decision of the rate of implementation of the program must rest with management, based on its ability to effect the accepted recommendations. It is your responsibility, based on experience, to suggest the speed of implementation. Typically, in the housekeeping field, improvements must be gradual and cumulative if they are to be lasting.
- Bear in mind that management is neither legally nor morally obligated to accept any given program recommendation.
- Remember that management must set up the machinery for taking action. This is done best through a coordinator assigned to the program, who must naturally be given sufficient authority to perform this work effectively. Agreement on recommendations has no effect without executive authorization.

It will take some time for management to consider the various recommendations and the structure of the program as a whole. Meetings, vacations, priority activities, and other factors have an effect on the amount of time required to consider each basic decision. A number of decisions are often made almost immediately, whereas others are delayed over a period of time, depending on the factors above, as well as the need for liaison and further investigation. Implementation of approved items can begin where interdependence on other items is not a problem.

Revision of various pages or sections of the program may be required in the installation phases of the program, as well as later on perhaps, because of management's decisions on recommendations, reevaluation of objectives, changing activity levels, the appearance on the market of improved equipment, etc. Thus, it is the revised draft which is followed in the actual operation of the program.

Step 2: Presenting the Housekeeping Program to Supervisors

Supervisory personnel within the housekeeping department should be made familiar as quickly as possible with the extent of their responsibilities and the steps necessary for its implementation. The best way to accomplish this is to set up a meeting. Set up a schedule indicating each action that will be necessary to implement accepted recommendations. The schedule should identify the item, name the individual responsible for its completion, and set a target date for its beginning and full implementation. A schedule can go far toward reducing the apparently complex program to a listing of simple and defined objectives.

Express the objectives and limit of authority and responsibility of the housekeeping department in written form, circulating this to the departments which receive housekeeping or custodial services. It should name those general areas for which the housekeeping department is responsible as well as indicate specific areas of responsibility which may not fall to the housekeeping department, such as groundskeeping, changing lights, and so forth.

The installation phase of the program is the proper time to organize the paperwork and record-keeping functions of the housekeeping department. The forms put into use should be standard for the organization as often as possible. There is a tendency for some housekeeping supervisors to seek paperwork, often unconsciously, as a status symbol, thus avoiding the lower-status situation of dealing with custodians. The supervisor must avoid every unnecessary record or piece of paper in order to be able to spend as much time as possible "on the firing line," where the most good can be done. Routine paper work should be assigned to a subordinate or clerk whenever possible.

Set up a procedure so that all matters pertaining to housekeeping, including complaints, suggestions, commendations, and changes in personnel, will be referred to the housekeeping department manager. Keep a permanent record of complaints and review quarterly to discover recurring situations or trends that require a more general rather than specific response. Such a record should indicate the date, source of the complaint, nature of the complaint, the housekeeping personnel involved, corrective action taken, and date of corrective action.

Much of the written program material, especially for personnel assignments and responsibilities, can be reduced to chart form. If such a chart is made up in fairly large size, and posted in the housekeeping department, it provides instant reference information and helps to form an overall picture in the minds of the supervisors and workers. Such a chart, which might be titled "Housekeeping Operation Plan," can be divided into vertical columns, the headings of which could include:

- Custodian name or number
- Housekeeping area designation or number
- Area usage or department name
- Square footage (this column should be totaled)
- Types of floors (approximate percentages can be indicated, an overall percentage)
- Equipment assigned—This can include size designations, serial numbers, location, etc.

- Work schedule—This can include an abbreviated description of the type of work, such as general cleaning, policing, projects, and the days and hours worked
- Total weekly labor-hours (this column should be totaled)
- Custodial closet or facility to be used.

The machinery for the assignment and recording of special projects must be set up. Two basic types of forms are required for this. The first is a register, each page of which should be designated for the building or major area involved and properly identified as to the period of time which it covers, such as for a given month of the year. Column headings might be:

Item number
Area designation
Type of project
Instruction sheet to be used
Individual responsible for completion
Target completion date
Actual completion date

For further control, keep a master record so that you can control facility-wide completion of a general type of project. In large installations, you might use a computer to monitor project completion.

Use standard work order forms to assign the work, or design special forms for use by the housekeeping department. Such forms would indicate the name of the person responsible for completing the work, the name of the person issuing the work order, and special instructions and equipment listings that may be required. These work order forms must be turned in periodically so that the register may be kept up to date through regular postings. Column headings may include: type of project, instruction sheet number, name or number of area, target completion date, actual completion date. Figure 5-1 shows a work order form for projects.

It will be difficult for the housekeeping department to perform the operations assigned to it unless it has itself first "cleaned house." The time of installation of the program is a good opportunity for an actual cleaning of the housekeeping department office areas and storage areas. At this time, you should also organize the central storage area and custodial closets, and establish inventory control procedures. (These issues are discussed in further detail in Chapters 21 and 17.)

Step 3: Reviewing the Housekeeping Program with Custodial Workers

It's normal for workers to feel apprehensive when they learn that their work is being investigated, evaluated, and possibly rescheduled. Therefore, avoid evasiveness. The truth is the most effective tool possible.

Figure 5-1. Sample Project Work Order Form

PROJECT WORK ORDER FORM

This form should be initiated by each Building Services Supervisor for his or her assigned buildings.

CUSTODIAL SERVICES PROJECT WORK ORDER	
PROJECT WORK ORDER #:	
INITIATED BY:	DATE:
PROJECT DESCRIPTION, LOCATION & SPECIAL INSTRUCTIONS:	
ESTIMATED TIME TO PERFORM: ______ × ______ = ______ # 1000 sq. ft. # hours per 1000 sq. ft. estimated # hours	
ACTUAL TIME TO PERFORM:	
INSPECTED AND APPROVED BY: NAME TITLE DATE	
COMMENTS:	

Hold an orientation meeting with all housekeeping department personnel, including their supervisors. Although many housekeeping training meetings will be given to less than the entire department because of working hours or duties, this meeting should, if at all possible, include all housekeeping personnel. A good time to hold such a meeting is at the time of a shift change, holding some workers over while asking others to come in early for this purpose. Representatives of management should be present at this meeting to indicate management's interest and sanction.

Following your introduction, you should address the group, with the basic objective of seeking their cooperation. Keep in mind these considerations:

- To gain a positive immediate reaction, give assurance (if previously approved by management) that the program will not result in any worker losing his or her job. If reductions are to occur, they should, if possible, be handled through normal attrition. Job security is very important to workers and you should reassure them as early as possible.
- Let workers know that any feelings of apprehension or suspicion are perfectly normal and expected in most workers when they are first approached with a program of this type. But experience has shown that this attitude changes to one of interest and enthusiasm since the program invariably works to their benefit.
- Reassure your staff that the program, of which they are such an indispensable part, has been conducted successfully for other similar organizations, and these organizations should actually be named. Show examples from other organizations involving activities for personnel morale improvement, better equipment, employee cooperation, and other improvements.
- Exhibit the attitude that you believe each worker is willing to provide a reasonable day's work. Emphasize that the program is designed to equalize the work load, so that each person is doing no more than his or her share of the work; and improved methods, equipment, and employee cooperation, as well as other aspects of the program, will be steps tending to make the work less burdensome.
- Show that the program represents a target to be approached over a reasonable period of time. Let workers know that they will not be expected to learn new methods overnight, and that they will be given sufficient training and opportunity for practice in learning new techniques. Similarly, benefits will take time, and they should also exercise patience.
- Distribute the actual housekeeping program document. People are more at ease with a tangible object than with an abstract conception. Point out that this document is nothing more than a collection of ideas and suggestions concerning housekeeping. Give a simple physical demonstration of a timesaving method to demonstrate the point.
- Encourage workers to do their share as members of the housekeeping team and of the organization as a whole. Encourage participation through

suggestions and ideas. Remember: your workers will determine the success or failure of the program, because they must perform all of the actual cleaning functions.

Step 4: Coordinating the Housekeeping Program with Other Departments

If the survey phase of the program did not permit an opportunity for a meeting with other department heads in order to familiarize them with the objectives of the program and how they might play an important part in it, certainly the installation phase should accomplish this. As many affected department representatives as possible should attend such a meeting.

Special requests for help from the housekeeping department is a subject suitable for such a meeting. In this case, immediate housekeeping assistance should only be expected in actual emergencies. Requests for additional housekeeping personnel or special assistance should be made at least 24 hours in advance, to permit proper planning and efficient scheduling of personnel. Greater pressure than this to fill special requests can only result in wasted staff time, serious postponements of work (which create their own pressures), and demoralization of workers, none of which is desirable.

Especially important during the installation phases is coordination with the following departments:

- The safety department, to insure the inclusion of good safety practices.
- Operations managers and supervisors, to establish a better understanding between their personnel and housekeeping personnel for waste handling and general housekeeping responsibilities in their work areas.
- Internal methods engineering personnel, for a possible synergism of ideas, if this is approved by management.

Figure 5-2 is a valuable tool in developing the image of the department while securing cooperation from others. Distribute it to all departments to find out how the housekeeping function is perceived by the departments it serves.

At this time, you can take steps to pierce the invisible barrier that is often found separating direct and indirect labor groups. Such barriers tend to soften and dissolve through repeated contacts between the two groups, and particularly through better operating conditions which result from improved housekeeping.

You can improve acceptance of your housekeeping program and enlist the cooperation of others if you become familiar with the basic nature of the organization with which you are dealing. You should be able to correctly use some of their specialized terms and jargon.

Figure 5-2. Sample Housekeeping Evaluation Form

HOUSEKEEPING CUSTOMER EVALUATION

Name ______________________ Date ____________
Title ______________________ Time ____________
Department ______________________ By ____________

A. Describe your contacts with the housekeeping department.

Type of contact:

Frequency:

B. How would you rate the service provided by this department?

Outstanding   Acceptable   Marginal   Poor

Comments: ______________________

C. Are the personnel you deal with:

	All the Time	Most of the Time	Seldom
Professional	☐	☐	☐
Competent	☐	☐	☐
Courteous	☐	☐	☐

Comments: ______________________

D. In your opinion, does this department respond to requests?

Promptly	After a Reasonable Time	After an Unreasonable Time	Requests Are Often Ignored

Comments: ______________________

E. How would you rate the overall condition of this facility in terms of Housekeeping?

Outstanding   Acceptable   Marginal   Poor

Comments: ______________________

You should document the major phases of the program, both as a report to management and as a memory aid to the housekeeping department. The completion of the installation phase is a logical point for such a report. This installation report should include the following sections:

- A review of activities.
- Observations or additional recommendations.
- Recommended agenda.
- Copies of prepared material such as inspection reports, implementation calendars, etc.

Figure 5-3 shows such a report form.

Throughout the program, you must bear in mind the *basic objective* of the organization. Just as in a hospital the care of the patient must be the overriding consideration, and in a school effective pedagogy should be most important, so should the earning of a profit be regarded in industry. There is no better place to begin working with this concept than in the installation phase of the program.

The installation phase of the program may require only a short time, or it may take an extended period. It will be difficult to point out a given time at which the installation phase is completed and the implementation phase is beginning, as certain activities of each phase will tend to overlap. In any case, the installation should be a good beginning.

HOW TO SUCCESSFULLY COORDINATE THE PROGRAM IMPLEMENTATION

How well you implement your housekeeping program depends on how well you have prepared your operations manual. The program document represents a relatively long-range goal rather than an instant achievement. The transition that must take place to convert an existing situation to a proposed situation cannot, and should not, be accomplished overnight. There must be a "phasing in" so that decisions may be properly implemented, sufficient time is allowed for supervisory development, housekeeping personnel are trained in their new duties with the least possible confusion, and housekeeping service is provided for all departments without disruption. Sometimes such a transition takes several steps, requiring interim schedules and assignments until the final proposed material can be implemented.

Guidelines for Scheduling Changes

How long a period of time will implementation require? It is important that a program does not proceed so rapidly that the first gains are lost in the

Figure 5-3. Sample Report of Housekeeping Program Implementation

INSTALLATION REPORT

Housekeeping Program Implementation

Date__________________

By__________________

COMPLETED ACTIVITIES		
Item	Date Completed	Comments

ACTIVITIES IN PROGRESS		
Item	Estimated Completion Date	Comments

ACTIVITIES ON HOLD

Item	Comments

OBSERVATIONS AND RECOMMENDATIONS

ATTACHMENTS:

scramble to make additional gains, thus causing "backsliding." The implementation must proceed at a rate so that gains can be made as quickly as possible while maintaining the improvements already brought about. The first six months will be the period of greatest gains; probably 60 to 70 percent of the program will be implemented during that time. Generally, a year is needed to derive the greatest benefit from a comprehensive housekeeping program, for these reasons:

- To implement long-term projects such as employee awareness campaigns and rating systems.
- To handle revisions and changes in management objectives because of information gained from early operation of the program.
- To further refine the program in methods, materials, standards, and other areas.
- To gain the advantages of a continuing training and morale-building program.
- To develop effective custodial supervision procedures.
- To obtain supplies and equipment as needed.

Sometimes solving one problem creates one or more other problems. Unless the implementation phase of the program is long enough to allow time for problem-solving, the program will not provide the full beneficial effect.

In facilities where housekeeping has been sharply curtailed for fiscal reasons, and where there is understaffing, recommendations are often made to management with the sole objective of raising housekeeping levels. The dollar savings done by this should be returned to the housekeeping department. If the savings result from personnel performing one function of housekeeping, they may be transferred to another function. Where the objective has been to save dollars, any savings in personnel should be achieved through normal attrition.

Figure 5-4 shows a partial example of an implementation schedule. Significant items are the activities to be performed, the responsibility, and the dates. In the long run, the basic commodity is *time*.

It is not necessary for one person to conduct all phases of the program. Specialization may have its advantages here, too. The various phases of the program may be handled by two or three different persons, each specializing in a certain aspect of the work. If the project is handled in this way, it is of extreme importance that each person becomes familiar with all phases of the program, because of their interacting effects.

During implementation, where work assignments are changing, you may be confronted with complaints from workers who compare their workload with other workers' job assignments and conclude that the work has not been distributed equally. For example, suppose you are eliminating special projects

Figure 5-4. Sample Implementation Schedule

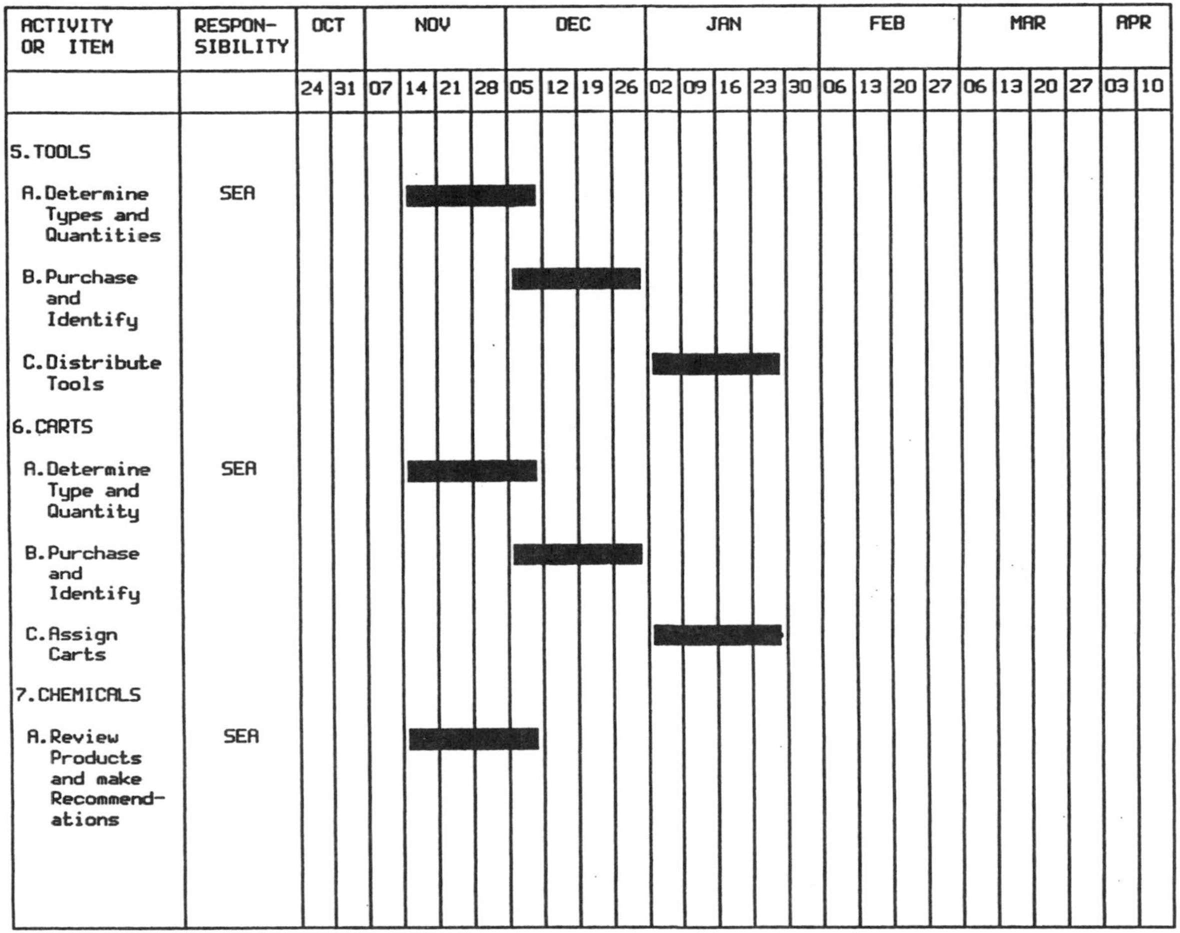

ACTIVITY OR ITEM	RESPON-SIBILITY	OCT		NOV				DEC				JAN					FEB				MAR				APR	
		24	31	07	14	21	28	05	12	19	26	02	09	16	23	30	06	13	20	27	06	13	20	27	03	10
5. TOOLS																										
A. Determine Types and Quantities	SEA																									
B. Purchase and Identify																										
C. Distribute Tools																										
6. CARTS																										
A. Determine Type and Quantity	SEA																									
B. Purchase and Identify																										
C. Assign Carts																										
7. CHEMICALS																										
A. Review Products and make Recommend-ations	SEA																									

work and also providing relief for absence for each area-assigned worker. This would allow you to assign a larger area to some workers, as shown in Figure 5-5. The problem may be that this worker sees that he has a larger area to cover, and therefore feels that he has been overassigned—even if he has not been given more work to do in a day's time. One way to avoid this problem is to move that worker from a given building and put him in another building, such as is shown in Option B in Figure 5-5; then the basis of comparison is removed, and the assignment is more likely to be accepted.

Keep the lines of communication open. Involve the housekeeping department with a number of personnel and departments during the implementation of your housekeeping program. Hold short meetings from time to time to keep all interested persons up to date on imminent and completed changes and to give them an opportunity to ask questions and make suggestions. Every meeting should stress the importance of their personal participation and the resulting benefits to them.

The housekeeping department should receive the full support of the building maintenance department because proper housekeeping results in lower building maintenance costs. For example, wall vacuuming and washing at proper frequencies extends the length of time between paintings. Proper floor care protects the surface from undue wear and avoids replacement of resilient tile or wood, the redensifying of terrazzo, or the patching of concrete. The maintenance trades should understand their responsibilities toward housekeeping in performing their daily work. With very little actual time or effort, maintenance personnel can create much better housekeeping conditions by washing their hands occasionally, spreading drop cloths to catch chips, drippings, and shavings, and by immediately removing any material which may cause damage to floors.

You might also consider forming a housekeeping committee. Such a committee is very helpful in improving liaison through a more complete understanding of the program and the problems affecting it, and a quicker response to suggestions. Although a permanent committee is desirable, the frequency of its meetings should diminish as the housekeeping program becomes properly implemented.

Revising Your Housekeeping Program

New ideas are constantly being born; new equipment is reaching the market; new materials are being developed in laboratories. Your housekeeping program should adapt to accommodate improvements from outside sources and from internal developments. Program revisions will be required where such changes are fundamental or far reaching, or where a number of minor changes have accumulated.

Revisions may be also called for where, at any point during the implementation phase, management reformulates its housekeeping objectives.

Figure 5-5. Avoiding a Basis of Comparison When Assigning Work

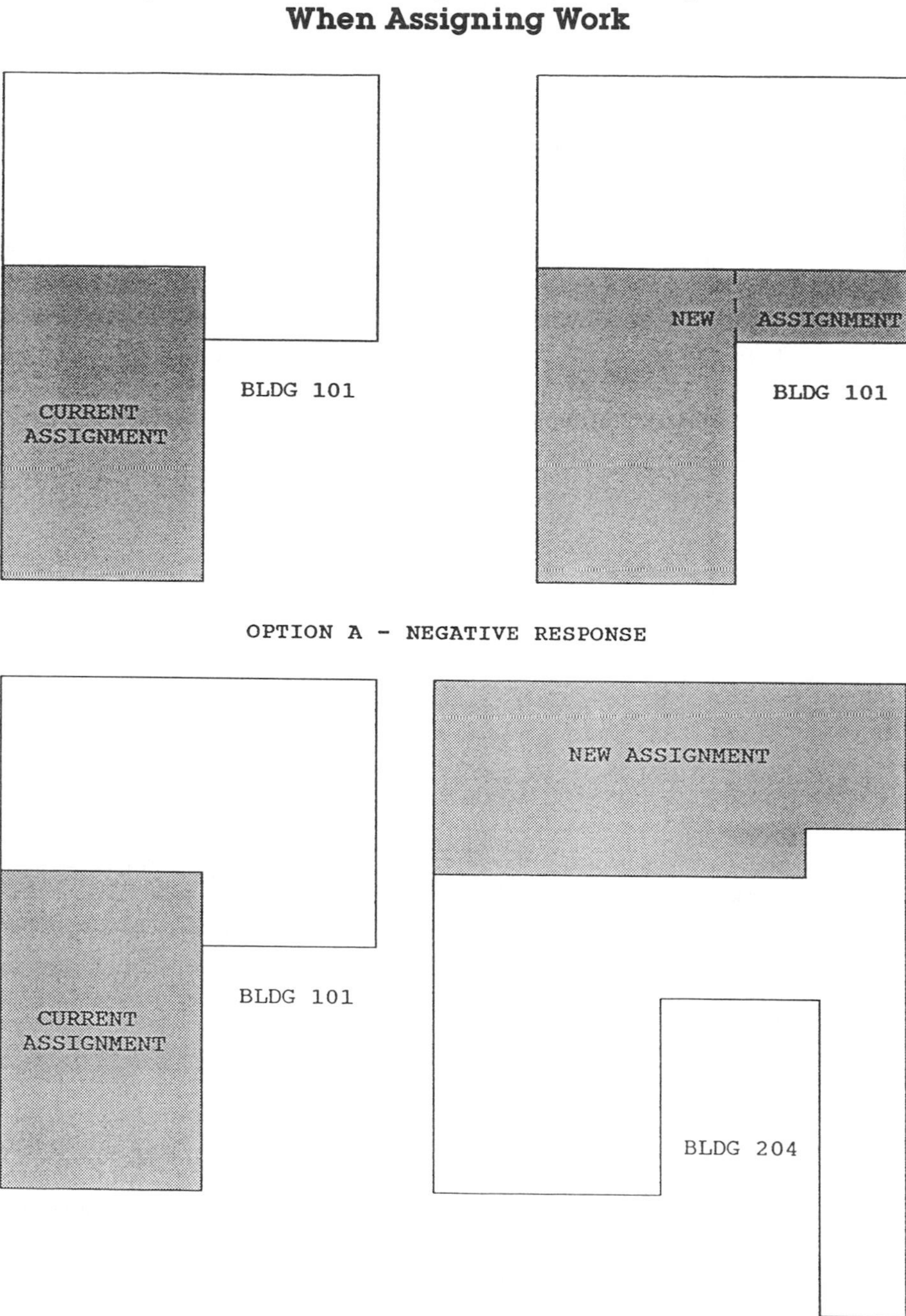

Keep a close check on labor requirements, comparing actual personnel levels to that which was planned and budgeted for. Figure 5-6 is an example of a labor report form.

How to Encourage Employee Awareness and Cooperation: Eight Promotional Techniques

In benefits provided versus cost in dollars, a well-conceived employee cooperation and awareness campaign is unquestionably a superb investment. If this activity is conducted in any way other than as part of a complete housekeeping program, such an arrangement may have a very limited effect. On the other hand, if personnel are able to see improvements which the housekeeping department has made and have seen evidences of management's growing awareness of the importance of this function, they are more likely to provide a positive response. Thus, consider adopting an employee awareness campaign to permit proper planning and preparation—but plan your campaign carefully to coincide with improvements and other activities during the implementation phase of the housekeeping program.

Even housekeeping programs now considered satisfactory can be improved through increased employee cooperation. Consider the effect on the custodial work load of a 10 percent improvement in attitude of a given facility with 1,000 employees. The effect approaches the equivalent of cleaning up after 100 fewer people! Such an improvement might permit an annual saving of several thousand dollars in wages, or the transfer of a custodian to special project work in order to raise the quality level of housekeeping.

Everyone wants a clean, healthy, and safe place in which to work. The important factor in obtaining employee cooperation is to bring this desire to the surface and keep it there. Pride, respect, and conscience in maintaining clean work areas will be the result. This natural employee attitude should be developed carefully, without forcing. A management directive which says "get this place cleaned up" will produce only temporary results. Contrast this to a management policy which states "we recognize your desire for clean, healthy working conditions and are doing everything possible to help you and our housekeeping department obtain them." Certainly everyone can respond positively to such an appeal.

Devise the campaign with the cooperation and guidance of the personnel relations department. They will normally "carry the ball." Literature and methods already put to good use by other organizations will benefit them in their work. Assign this project to one individual and provide him or her with the authority and means for coordinating the work.

It is a waste of money to attempt to awaken and sustain employee interest in cleanliness by a few half-hearted devices on a limited number of occasions. It is better to do nothing than to awaken interest momentarily and then let it die by failing to provide benefits and follow-through. Such an approach will

Figure 5-6. Sample Report for Tracking Labor Requirements

LABOR REPORT

	1st Shift	2nd Shift	3rd Shift	Total
Variance from Budget*				
Staffing	______	______	______	______
Routine Workers	______	______	______	______
Projects & Relief	______	______	______	______
Policing	______	______	______	______
Miscellaneous	______	______	______	______
Supervision	______	______	______	______
TOTALS	______	______	______	______
Overtime Hours	______	______	______	______
Turnover	______	______	______	______
Number Hired	______	______	______	______
Number Fired	______	______	______	______
Number Quit	______	______	______	______

* Based on staffing on the last working day of the month.

only create new cynics and confirm old ones, with respect to housekeeping. Let the campaign continue over a long period of time. There is no reason why such a program should not be a permanent project of the personnel relations department. The intensity of the promotion will decrease, of course, over time, but interest can be sustained through carefully timed applications of a number of different campaign techniques. These include:

1. Orientation booklets.
2. Letters from management to department heads and supervisors.
3. Letters from management to employees.
4. Posters and banners.
5. Company newsletter publicity.
6. Leaflets and stuffers.
7. "Calling cards."
8. Contests.

1. Orientation Booklets. New employees should receive an immediate impression and expression of the organization's attitude toward housekeeping. Ideal for this purpose is a small booklet of ten or twelve pages. The general personnel policy booklet can contain a housekeeping section, rather than having a separate booklet.

Periodic redistribution to existing employees is valuable as well, particularly if the brochure is reprinted in different colors with a modified cover. The booklet would call attention to the steps that management has taken to provide well-maintained surroundings for its workers. It would point out how each employee can contribute to his or her environment, giving examples such as refraining from putting liquids into wastebaskets, snuffing out cigarettes underfoot, leaving desks littered at night, etc. The brochure might also call attention to a rating system, where each department or area competes for a prize or the possession of a good housekeeping cup. In terms as positive as possible, the booklet should list the organization's housekeeping policies and regulations.

To lend authority and dignity to the booklet, it should open with a letter from the chief executive officer. Use drawings, photographs (including the "before-and-after" type), and the judicious use of color and open areas to maintain interest. Typography should be limited, and the type face carefully selected. The booklet is a good place to tie in other programs interdependent with housekeeping, such as quality control, safety, and fire prevention.

Yale University provides an employee booklet called "Pocket Guide for Cleaning Services," which is given to each of its many custodians. Figure 5-7 shows the contents pages from this publication.

Figure 5-7. Sample Contents of an Employee Orientation Booklet

Page		Page	
2	Important Information	67	Cleaning a Student Dormitory Room
3	Introductory Page	70	Cleaning a Typical Office
4	Contents	74	Cleaning Reception Areas
6	A Word About Products	77	Cleaning a Typical Laboratory
7	Important Safety Tips	81	Cleaning Areas with Computer Equipment
11	Employee's Work Schedules		Gymnasium Cleaning:
13	Explanation of Terms in this Manual	86	Swimming Pools
14	Trash Pickup	88	Steam Rooms & Sauna
16	Restroom Cleaning	90	Gym Floors Used for Playing Activity
19	Shower Stall Cleaning	92	Locker Rooms
21	Dusting	95	Fluorescent Tube & Light Bulb Replacement
24	Cleaning & Polishing Furniture	98	Use and Care of a Wet / Dry Vacuum Cleaner
26	Cleaning & Polishing Hardware	102	Miscellaneous Information:
28	Dust Mopping Floors		Sweeping and Removing Snow from Building Entrances
31	Wet Mopping Floors		Changing Shower Curtains
34	Vacuuming Carpets		The Custodian's Utility Closet
37	Rug Cleaning - Water Extraction		Auditoriums & Theaters
40	Rug Cleaning - Dry Extraction Method	105	Care & Storage of Equipment & Supplies
43	Machine Stripping Floors		
46	Re-Emulsion Floor Stripping		
49	Spray Buffing Floors		
52	Putting on Floor Finish (wax)		
55	Cleaning Floors on Stairways		
59	Washing Interior Walls		
61	Cleaning a Typical Classroom		
64	Cleaning All Types of Glass		

From "Pocket Guide for Cleaning Services," by Yale University, New Haven, CT.

2. Letters from Management to Department Heads and Supervisors. Obviously, the attitude of department heads and supervisors toward the housekeeping program can materially influence its success in the areas for which they are responsible. An effective way of obtaining their interest and cooperation is through a personal letter. The letter should bear the signature of an official in top management and should be mailed directly to the individual. Mail supplementary letters periodically for follow-up. Figure 5-8 is an example of a letter to supervisors.

3. Letters from Management to Employees. Employee response to personal letters from management enlisting their support is usually strong and positive. This is particularly true if the letter is "down to earth" and not officious. Such a letter rarely, if ever, brings a negative reaction. At worst, it falls on deaf or passive ears, and even then response may be awakened through repeated efforts. Follow-up letters should come from various management personnel, to stimulate still further interest. It is a compliment to a person to ask for his or her help.

4. Posters and Banners. Posters and banners have the advantage of repeating their message each time they are seen. If such a message is left in place too long, however, it loses its effect and becomes annoying. Posters should never remain in place longer than a month. A good plan is to rotate posters every week. Although a given poster can be placed in a new location immediately, it should not be returned to its original place before one year.

If bulletin boards are provided, they should be well-lighted and colorfully framed and located where most people will see them.

The National Safety Council has a standard series of colorful housekeeping posters available. Special posters and cartoons, which specifically relate to local problems, may be prepared by internal personnel.

5. Company Newsletter Publicity. The company newsletter, magazine, or other periodical is a good vehicle for housekeeping publicity. This type of material provides for dramatization through photographs and case histories, and ties in well with other objectives. Here is a good place to build humor into the employee awareness campaign in such a way that one employee will "kid" another for violating group-imposed rules for housekeeping. Part of the function of these periodicals is to publicize other phases of the awareness and cooperation program.

6. Leaflets and Stuffers. Leaflets may be left on the desks of employees, or stuffers placed in their pay envelope, from time to time. Again, the humorous theme is best applied here through the use of cartoon situations. The watchwords to be used for this type of literature are brevity, color, and humor.

7. "Calling Cards." Good results have been achieved in attempting to obtain employee cooperation on an individual basis by utilizing a personalized

Figure 5-8. Sample Letter to Supervisors Promoting the Housekeeping Program

MODERN RESEARCH COMPANY

Anytown, U.S.A.

(Date)

Dear Tom -

The success of our new custodial program, announced in the meeting a few days ago, is going to depend on a number of important people like you.

Although President Raymond supports this program fully, he knows as well as I do that it is the first-line supervisor -- loyal people like you, Tom -- who is the backbone of this department.

You've heard me say that "without good first-line supervision, you ain't got much!".

Let me review with you the objectives of this program:

- No assignment in excess of a reasonable day's work, with allowance for make-ready, personal, put-away and other times.
- Balanced workloads based on time allowances, reflecting the type of surface, traffic, soiling, etc.
- Optimum, safe methods, equipment, and supplies.
- A regular training program.

I know you will give your custodial manager, E. S. Styles, all the help you can.

Let me thank you in advance -- as I'm sure I will again later -- for your positive support!

Cordially,

Sal

Sal G. Levering
Vice-President for Facilities

calling card. These cards should be used by housekeeping supervisory personnel.

A 3-inch × 5-inch, typical card, of a bright color, is left on the desk or at the work place of a housekeeping offender, pointing out a housekeeping offense and requesting the employee's assistance in creating better working conditions. When reporting to work the following morning, the offender finds this card which personally asks for help. Many employees respond to this individual appeal for help. Where response is not forthcoming, the repeated appearance of a "calling card" on a given desk, for example, may attract the attention of the department head or supervisor. If the employee has been properly motivated by other aspects of the awareness program, undoubtedly he or she will take a personal interest in the problem.

Figure 5-9 depicts both sides of a successful calling card.

8. Contests. Contests offer a wide number of possibilities, limited only by the extent of your imagination. Possible subjects for a contest include:

- Competition for the best-maintained area
- Determination of the best suggestion concerning housekeeping and sanitation
- Phrasing of the most worthwhile slogan
- Design of the best housekeeping poster or banner
- Statement of twenty-five words on the importance of good housekeeping

Awards might be rotating or permanent trophies, pennants, merchandise awards, or merely publication of the winning employee's or department's name.

To improve employee cooperation, particularly as instituted on the department head level, you might consider costing out housekeeping to the various departments. Set up an account numbering system and the forms needed to assign properly dollar costs to specific departments. A simplified method for performing this function is to assign a base, or fixed, amount for each accounting period for the repetitive work performed within the department and then to make specific charges for any project or special request performed in addition.

Figure 5-9. Sample Calling Card to Promote Employee Participation in Custodial Maintenance

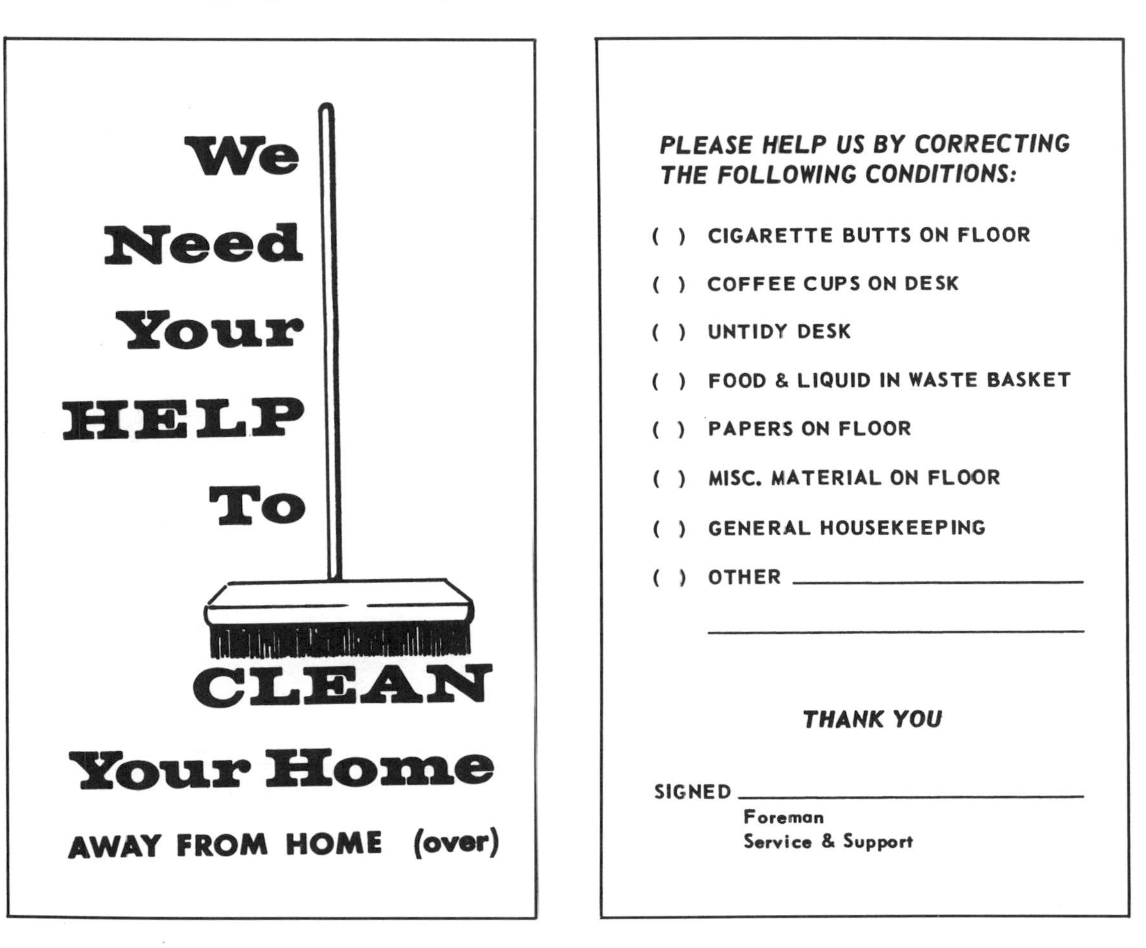

6

How to Make Housekeeping an Integral Part of Building Renovation and Construction

The design of a new building uses the combined talents of many specialists to achieve the best possible results. Don't wait until construction begins—or has been completed—to consider such things as electrical distribution, drainage, sprinkler protection, floor loading, fenestration, heating, or equipment layout.

Is there any reason, therefore, to complete a new structure, or to renovate an old one, without considering many of the aspects of housekeeping and sanitation? Yet, this is usually the case.

How many housekeeping department managers have bitterly considered the serious housekeeping problems which could have been avoided if they had been given an opportunity to advise on a simple point! How many supervisors realize that many cleaning tasks are necessary only because of the lack of housekeeping planning!

Many organizations have constructed new facilities to eliminate the inefficiencies which were unavoidable in an older structure. Unfortunately, the housekeeping problems are carried directly from the old building into the new because of poor planning in this field.

Unquestionably, the greatest possible return on the housekeeping dollar will be realized where the housekeeping program is developed during the planning stages of construction. The housekeeping department manager can work with the architect and design engineer to select materials and surfaces. Together, they can locate and determine the size of custodial storage and operating facilities and provide other conditions that lead to good housekeeping.

The benefits of housekeeping planning can never be realized without management's interest in this direction. The architect, design engineer, and housekeeping department manager all have an obligation to attempt to awaken and sustain this interest.

Some architects are beginning to give housekeeping new thought. They provide greater services by using consultants to improve the maintainability of the structures they design. The provision of a complete housekeeping program may be part of a "package," or it may be an optional service. Through such services, the progressive deterioration that normally follows quickly after building completion can be retarded. Many "new" buildings become "old" in a surprisingly short period of time—there is no reason that the newness cannot be prolonged for many years through proper housekeeping planning and action.

The opportunity to plan for better housekeeping for a new facility comes only once. This chapter will show you how to make the most of that opportunity.

HOW DESIGN IMPROVEMENTS CAN HELP REDUCE CUSTODIAL EXPENSES

Your job is to convince management that your involvement in the design stage can be beneficial for the life of the building. The most direct approach for responsible management is in terms of economics. It is your responsibility to show the economic benefit of making changes that lead to improved maintainability. For example, a better custodial closet would save a certain number of trips per year, at so much time per trip, at so much value per trip, indicating an annual saving; but if the trips are not made, then your custodial staff is not doing its job properly, and the building is not being cleaned well. If the building is not being cleaned properly, there will be resultant problems such as dust in the air (and thus a health problem) or floors wearing out too quickly.

One of the rules of selling is to have as much information as possible related to the subject. Thus, you might consider building a file on maintainability, with information on floors, rest rooms, furniture, entrances, walls, doors, and windows. Use visual materials: take some pictures of poorly designed items and make a slide presentation for management.

Other key elements in salesmanship are enthusiasm and persistence. If you are not enthusiastic about this subject, it will be difficult to develop enthusiasm in others; and don't give up on this subject after just two or three tries, since it may take you six or twelve! Your persistence indicates to management that you are committed to this subject.

If possible, don't wait to get involved until a new building is already half complete—it may be difficult to make changes at that time. Prepare long in advance, even before there is any indication that something is going to be changed.

FOUR CRITICAL MAINTENANCE AREAS TO TARGET FOR DESIGN MODIFICATIONS

There are hundreds of things to look at from the maintainability standpoint, but from a custodial standpoint, the following are listed, in order of importance:

1. Soil Prevention. It is much easier to keep dirt out of a building than to clean it up once it is scattered throughout the premises. Do as many of the following as possible:

- The approach to the building should be a rough-surface walkway to catch gross soils. Further, the walkway should be crowned or tilted slightly to run off water, so as to prevent water and ice from being tracked into the building.
- The vestibule or entry area should contain a grating. The grating should take up the entire vestibule floor, with the rails running perpendicular to the flow of traffic. The pedestrian should walk across at least six, and preferably eight, feet of grating. The rails should be spaced at one tenth of an inch open slot, so as to permit the fall of soil into the pan beneath, but not to catch narrow shoe heels. Where grating cannot be built in, there are surface gratings now available on the market that can be rolled up and carried out for cleaning.
- Once past the grating, the final soils should be removed from the soles of shoes by some type of carpet material, either carpet squares, runners, or sacrificial carpet. This area should be at least 12 and preferably 16 feet long.

All three of these surfaces should also be kept policed, so they do not become a source of soil for trackage within the building. The effect of the matting can be further extended by using carpeting in elevators and on stair landings (where permitted by the fire marshal), to prevent the migration of soils to the upper floors. Chapter 18 describes soil prevention in greater detail.

2. Custodial Closets. Without an adequate number of custodial closets, both size and quantity, as well as the outfitting of the closet, the cleaning worker cannot do his or her job effectively. Too many custodial closets are still being designed and installed in dimensions of about 3½′ × 5′, with the utility sink mounted on the wall where it is difficult if not impossible to use, and the door opening inwards, so one cannot store any equipment or cart within the closet.

Ideally the custodial closet should be about 6′ × 9′ with the utility sink mounted in the floor with a curb around it, a pegboard across one wall so that various equipment can be hung on it (such as vacuum parts, hoses, floor machine pads), and wet mops can be hung to drain into the sink. Two shelves should be in the closet, one at about chest height and the other about 18″ higher, the lower shelf to hold chemicals and other heavy items, and the upper shelf to hold things such as paper goods, fluorescent tubes, floor machine pads, and the like. Underneath the shelving will be the parking area for custodial carts, floor machines, vacuums, and mopping outfits. Chapter 21 discusses custodial closets in more detail.

3. Flooring. The wrong flooring can cost twice the labor in cleaning it as a properly selected floor. Ceramic is the ideal floor for lobbies, rest rooms, kitchens, and other areas where there is a great deal of soil and a lot of cleaning required. Ceramic tiles are available in various shapes and colors, and can be very attractive when properly selected, so as to avoid the "institutional look." It is very important that the grout between this tile be other than white or some light color, since it is so very difficult to keep this grout at its original coloration. Where white is required, for sanitary reasons, such as by a health agency, then the grout should be epoxy, so that it will not pick up stains and will retain its color. Ceramic tiles can easily be auto-scrubbed or wet mopped and will not be damaged by soil; no wax is required, thus avoiding stripping, waxing, buffing, and the cost of the materials in doing this.

Terrazzo floors can also be valuable, but not nearly so much as ceramic, since terrazzo must be sealed and to look its best must be finished and spray-buffed. Both of these materials have a higher first cost, but over the life of the facility, this is more than repaid. Carpet can also be a good choice, especially on upper floors, if due consideration is given to the fiber, the density of weave, the use of looped pile, continuous filament plastic fibers, and glued-down application. Where resilient floors are required, composition vinyl can be used, which may contain a ceramic content of its own, for greater durability. Chapters 25, 26, and 27 discuss floor- and carpet-care guidelines in detail.

4. Coloration. Ideally, floors (and to a certain extent other surfaces as well) should be the color of the soils that will be deposited on that surface. Avoid any solid color of flooring; also, a darker color will show all dust, and a lighter color will show other soils. Solid colors also have the failing that patterns will show up after vacuuming or shampooing. The ideal color carpet is a three- or four-color tweed of intermediate colors, or a floral pattern; resilient floors should also be a mixture of two colors, with a base color having "shots" or streaks of another color within it, such as black or dark streaks to hide rubber heel marks and damage due to scraping.

When it comes to wall surfaces and ceiling surfaces, avoid white—it is very difficult to keep it white! Use an off-white such as cream or ivory. Avoid flat paint; semi-gloss is more practical for washing. Just these few items alone can save untold amounts of money during the life of the building.

WHERE TO SUGGEST IMPROVEMENTS: THE HOUSEKEEPING DESIGN CHECKLIST

A listing follows of *some* of the things that can be done to provide better housekeeping. Your decision to use them should also take into consideration other factors involved in a specific project.

Custodial Facilities

1. Provide an adequate centralized area for the exclusive storage of cleaning materials and equipment.
2. Be sure custodial closets and lockers are adequate in number and size and are properly located.
3. When utility sinks are used, make sure they are low and fitted with a metal lip to prevent the damaging of the enamel.
4. Provide adequate sources of both hot and cold water for custodial use.
5. Locate sufficient electrical outlets for custodial use in corridors and large rooms. Outlets located near doors are helpful in case the outlets inside the room are inaccessible.
6. Where possible, avoid mounting electrical receptacles in floors. They are easily damaged by liquids (such as detergents and waxes), can cause accidents, and are difficult to clean around. They are also rather unsightly.
7. Provide for the laundering and treating of dust mops and cloths in the central storage area. Make provisions for the cleaning of wet mops and venetian blinds.
8. Provide adequate shower and locker facilities for the custodial staff. Make sure that such facilities are conveniently located.
9. Provide adequate storage facilities for all types of waste materials, such as paper, scrap metals, food waste, etc., and provide for recycling.
10. Provide bulletin boards for posting of housekeeping information and instructions.

Floors

1. Do not select a floor that requires hand maintenance or care by other than the average custodian.
2. Try to avoid the meeting of floors at different levels—they should be flush.

3. Where the difference in levels between floors is slight, install a ramp with a nonskid surface. This will be much safer than stairs. It will be easier to clean and will permit easier equipment handling.

4. Design special pedestal floors for data-processing areas so that wiring, ducts, etc., may be run underneath the floor. Removable portions of the floor make the area underneath easily accessible.

5. Specify special floors for superclean rooms or "white rooms."

6. Make subfloors as smooth as possible to avoid depressions and raised areas. These irregularities cause resilient finished floors to wear away quickly and make them extremely difficult to maintain properly. Concrete subfloors should be machine troweled.

7. Cure concrete subfloors properly and correct any irregularities before applying composition tiling.

8. Provide floors which are continuously wet with trench-type drains rather than individual spot drains. The floors should be sloped gently toward these trench drains.

9. For resilient flooring, use vinyl, which provides a serviceable floor at reasonable cost.

10. Prepare for adequate lead time on floor installations so that newly laid resilient tile will not be scrubbed until thoroughly set, which requires from three to six weeks.

11. Select floor patterns carefully, as experiments have shown that floor designs and color combinations can range in effect from therapeutic to annoying.

12. Specify the use of floor sealers in construction contracts for wood, concrete, or terrazzo floors. This permits full protection of the floors before occupancy, as well as a much more efficient application of seal because of the lack of congestion.

13. Make sure corridors have no recesses in the wall or projections into the corridors. Drinking fountains in corridors should be avoided, if possible.

14. Utilize rubber or composition cove bases, rather than wooden baseboards, to provide a scuff-free surface and rounded joints which are easily cleaned.

Furnishings

1. Minimize the amount of furniture to be moved when cleaning by the use of built-in or wall-mounted furniture.

2. Install furniture made of metal, plastics, or composition materials, which are more easily maintained than wooden furniture.

3. Eliminate fabrics when possible; they catch dust. This includes curtains, drapes, upholstered furniture, etc.

4. Avoid ornamental decorations, which catch dust and are normally difficult to clean. On the other hand, decorations such as murals, photographs, or paintings are normally easy to clean if simply framed.

5. Remove signs and calendars from walls, glass partitions, and rest rooms where possible as they are unsightly.

6. Provide bulletin boards for all employees to promote housekeeping, safety, fire prevention, as well as for other purposes.

7. Place waste receptacles immediately adjacent to areas where waste is created.

8. Make sure waste receptacles are of adequate size and provided with covers when used for odor-producing materials, perishable food, or dusty waste.

9. Make sure stairwells, corridors, locker rooms, rest rooms, and custodial closets are well lighted. There is greater loitering and abuse of facilities in poorly lighted areas than in well-lighted areas of similar nature and traffic.

10. Clean or replace air-conditioning filters properly and often. This decreases the cleaning requirements for all surfaces which require washing and dusting. Soil is trapped in the filter rather than being distributed throughout the facility.

11. Where possible, do not place vending machines immediately against a wall, which prevents cleaning behind them.

Openings

1. Make sure walkways leading to doors are of hard material and properly drained to prevent mud and water from being tracked into the building.

2. Use gratings and matting at doors and between areas where it is necessary to catch soil.

3. Use flush doors rather than paneled doors, as they do not collect dust.

4. Avoid curbs or high thresholds at doors, as they prevent the moving of carts and other equipment from one floor area to another. Where curbs must be used because of water on the floor, provide ramps to permit equipment access.

5. Use translucent or tinted glass, which is easier to maintain than transparent glass.

6. Where caulking is used, make sure it is smooth and continuous.

7. Design proper window-washing equipment into the building. This is particularly important in multistory buildings where a railing may be

installed around the roof from which a permanent retractable scaffold is suspended.

8. Avoid blinds, shades, and other window coverings whenever possible.

9. Install marble window sills. Although they are rather expensive, they require very little maintenance and are quite decorative. Their use should be considered for administrative areas.

10. Use aluminum framing rather than wood or painted steel.

Rest Rooms

1. Design rest rooms to permit a circular flow of traffic. Locate wash basins near doors, and mirrors placed on walls opposite the wash basins.

2. Locate rest rooms away from busy areas so as to equalize the traffic load on floors.

3. Install a floor drain for emergency spillages in each rest room.

4. Install adequate lighting in rest rooms. Brightly lit rest rooms stay cleaner, with less abuse of facilities and a reduction in loitering.

5. Use porcelain fixtures.

6. Hang rest-room fixtures and stalls from the wall or ceiling to leave the entire floor area open for ease of sweeping and mopping.

7. Be sure to build in "grab bars" in shower areas or in other places where slipping is possible.

8. Locate ash trays between urinals so as to prevent employees from throwing cigarette or cigar butts into the fixtures.

9. Provide lockers with slanting tops for easy cleaning. Place them on a concrete or ceramic base so that soil will not be trapped underneath, and so that the locker floor area may be mopped or scrubbed freely.

10. Paint walls with light-colored, light-reflective materials. Enamels are suggested.

11. Install large paper towel and tissue dispensers. These require less attention. The tissue dispensers holding one roll in reserve, in addition to the roll being used, are desirable.

12. Make sure dispensers and cabinets are large enough to require a minimum of daily service, if possible.

13. Utilize a type of soap dispenser that may be refilled without spilling.

14. Use dispensers having metal or plastic containers.

15. Do not use bar soap, as it is a medium for cross-infection. It also has a bad appearance. Use liquid, powdered, or paste soap instead.

Walls and Ceilings

1. Select materials which can be properly dusted as well as wet cleaned.

2. Use walls to hang as many items as possible, permitting rapid cleaning and avoiding damage to the fixtures. These items include restroom fixtures, partitions, waste receptacles, water fountains, cigarette urns, etc.

3. Use glazed ceramic tile on the walls of custodial closets, some hallways, stairways, rest rooms, kitchens, etc.

4. Use paints that are durable and washable, or use other washable materials such as plastic coverings.

General

1. Plan the cleaning of a new facility before actual building construction. Most contractors perform a final clean-up, but this usually needs to be augmented by internal cleaning before occupancy.

2. Arrange facilities to facilitate the custodial work load. Office equipment, for example, can often be rearranged to permit a considerably more efficient performance of the housekeeping work without disturbing the efficiency within the office.

3. Whenever possible, seal off inactive areas. Departments with greatly reduced personnel should be moved to smaller areas.

4. Arrange for smooth traffic flow throughout the facility so as to reduce the number of steps that have to be walked. Try to avoid a traffic pattern that will produce overwear in certain areas.

5. Install air conditioning wherever possible. Air-conditioned areas are easier to maintain because of the dust control that is provided.

6. Attempt to eliminate internal window sills, ledges, and all dust-catching surfaces.

7. Provide round corners where possible, so that they will not catch dust.

8. Be sure that plenty of conduits are installed in the building for running all types of wiring, so that wiring need not be exposed.

For a fuller treatment of this important subject, you may wish to refer to *Building Design for Maintainability* by Edwin B. Feldman, P.E.

Here are some examples of poor design for maintainability:

- When restroom partitions are floorstanding they interfere with floor care and must be cleaned themselves; they should be wall and ceiling mounted.

• Partitions between urinals cause a great deal of difficulty; they must be cleaned and disinfected, painted on occasion because of staining and corrosion. When the wall-mounting clips become loose, they must be repaired, and can even damage the ceramic wall tile. A urinal with built-in side panels doesn't need partitions.

• Stairs made of travertine marble ("rotten marble") are not only very difficult to clean, but are safety hazards.

• Carpeted stairs are a very poor investment in view of their short life span. This condition also creates a considerable safety hazard. A much better choice for stairs would be molded rubber with an anti-slip nosing.

• Resilient flooring for an elevator cab is very time consuming to maintain. In essence, it becomes a small room with a floor that has to be waxed, stripped, spray-buffed, and the like. A much better choice would be carpet squares in a tweed or floral pattern (if permitted by the fire marshal); this would also give the effect of trapping soil before it can be taken to another floor. Another alternative, where there are heavy carts and other heavy items to be brought on the elevator, is ceramic tile with dark epoxy grout.

7

Contract Cleaning: How to Decide if It's Right for You

Am I getting my money's worth in housekeeping? More and more managers are asking themselves that question. There are growing pressures to conduct an efficient and productive function from within as well as from without. And contract cleaning is turned to as a possible alternative.

The concern is not only about the dollars and cents of sanitation, but its effects as well. Ineffective housekeeping can lead to deterioration of surfaces, damage to equipment and supplies, an excessive accident rate, fire hazards, an environment not conducive to high productivity, and an increased incidence of cross-infection. Pressures to improve housekeeping can rise from every department head in the organization.

There are plenty of capable men and women operating internal housekeeping departments effectively. Even though any complex job can be improved to a certain extent, in general they are giving their managers a good return for the money they entrust to them in the conduct of their department. But there are also plenty of housekeeping departments that are operated inefficiently and which need considerable improvement.

To compare housekeeping as performed by your own department to that which may be provided by a contractor, you need to take a close look at contract cleaning.

TYPES OF SERVICES: DECIDING WHICH KIND AND HOW MUCH YOU NEED

The typical contract cleaning organization offers a wide range of services. This not only gives you the opportunity of deciding whether or not to use contract cleaning, but you can also consider the questions of "what kind" and "how much." Some of the services include:

- Single daily projects, such as waste collection and removal.
- Single intermittent or periodic projects, such as washing outside window surfaces.
- A contract for a single project, such as cleaning overhead areas.
- Supplementary services to assist the regular housekeeping organization in completing an unusually difficult project, or in meeting some deadline.
- Selected types of services, such as night cleaning only, or the cleaning of administrative areas only, with the balance of the work being performed by internal personnel.
- The complete cleaning of a given portion of the facility, such as outlying buildings.
- Complete cleaning services, with all jobs and responsibilities being done by the contractor.
- Contract management, where the contractor provides the leadership, but the personnel stay on the customer's payroll.

There are many other variations and combinations of the above possibilities.

Advantages of Contract Cleaning

Some organizations find benefits and advantages in contract cleaning. The rapid growth of the industry could not have been sustained without an increase in the number of organizations using contractors and retaining their services from year to year. These organizations profit from one or more of the following:

• Lower custodial maintenance costs—Cost is by far the most important factor. Although contract cleaning can be successful where the wages paid by the contractor to his personnel are the same or even greater than those paid by employers who operate their own housekeeping departments, it flourishes where the contractor is able to pay considerably less.

• Fewer work hazards—The organization's own employees avoid possible hazards. This is especially the case in such jobs as outside window cleaning in multistory buildings, overhead cleaning, cleaning with hazardous chemicals such as the acid-cleaning of new masonry, the descaling and scouring of rest room fixture traps, or equipment cleaning.

• Additional work that can be done without increasing the size of the basic staff. Contract cleaning personnel can be hired for seasonal cleaning.

• No investment in specialized or even standard housekeeping equipment is necessary, thus reducing the total capital investment.

• Emergency cleaning can be done at short notice by contractors.

• Fewer problems with unions.

• Workers may be more qualified because contractors pay *higher* rates, thus providing a more effective program.

• Lower employee benefit costs because contract employees are not on the organization's payroll.

• Use of new equipment supplies and methods because contractors are often more familiar with these than some in-house operations.

• Greater flexibility in staff size. It may be easier to expand and contract a contract staff because of changing requirements (such as production levels, occupancy, etc.) than to hire and fire internal personnel.

Disadvantages of Contract Cleaning

Some of the organizations continuing to use contract cleaning are dissatisfied, to a greater or lesser extent, for one or more of the following reasons:

• Regular deterioration of quality—This is by far the most important disadvantage. Both the client and the contractor contribute to this situation. The client often short-sightedly selects the lowest price bid without evaluating what this means in terms of cleaning quality obtainable at the price. The contractor, on the other hand, may continue to decrease quality to "as much as the traffic will bear" in order to increase his profits.

• A disparity in basic objectives, which may be hard for some managers to reconcile. The contract cleaner is in business and must make a profit. A reputable contractor will make a reasonable profit on every contract, but a disreputable contractor will give as little cleaning as possible to make the largest profit. The objective of the client is just the reverse; namely, to obtain as much cleaning as possible for his dollar.

• Some loss of management control, where work is done in an organization by personnel not belonging to that organization.

• Lack of a proprietary attitude toward the work. An organization is made up of people, and the progress of the organization depends on the attitude of its people.

• Problems in security control, which may become troublesome.

• A loss of flexibility in meeting changing conditions, with some types of contracts.

• More difficult inter-departmental liaison and cooperation.

• Deterioration of service for a particular customer if the contractor transfers a successful manager from one account to another; some contractors recruit managers with this expectation in sight.

One of the greatest problems in this field is the purchase of contract cleaning services on the basis of price. It is similar to the problem involved in furnishing cleaning materials on the basis of price. Specifications in both cases are rather difficult to word and evaluate. Quality deteriorates steadily.

Figure 7-1 indicates how the typical contractor makes use of the money paid by the customer. These figures will change, of course, depending on the factors listed on the bottom of Figure 7-1.

HOW TO INVESTIGATE A POTENTIAL CONTRACT CLEANING SERVICE

How do you decide whether or not to use contract cleaning services, and how do you avoid the pitfalls? The decision to use contract cleaning—or not to use it—should be based on a sound analysis of all the variables involved. The study must include an economic analysis of the competitive benefits and limitations, as well as a discussion of relatively unmeasurable factors such as morale and safety.

You should carefully investigate a potential contractor and require the following information:

- Reference accounts. In addition to being permitted to list the more successful reference accounts, the contractor should also list all accounts if a small concern, or the last twenty accounts if a large concern.
- The contractor should indicate the extent of his or her experience.
- Financial references.
- Insurance coverage, both for cleaning personnel and from a public liability standpoint.
- Qualifications used for hiring, screening and training personnel.
- Methods and extent of supervision to be used.
- Method of payment, along with the period of termination notice required.
- Professional qualifications and background of the organization, and its principals.
- A list of at least three accounts that the contractor is no longer serving. Contact these accounts for further information on the contractor.
- A list of trade organizations of which the contractor is a member, such as BOMI (Building Owners and Managers International), ISSA (International Sanitary Supply Association), IREM (Institute of Real Estate Management), IFMA (International Facilities Management Association), and BSCAI (Building Service Contractors Association International).

Figure 7-1. Example of Receipts Disbursement by a Contractor

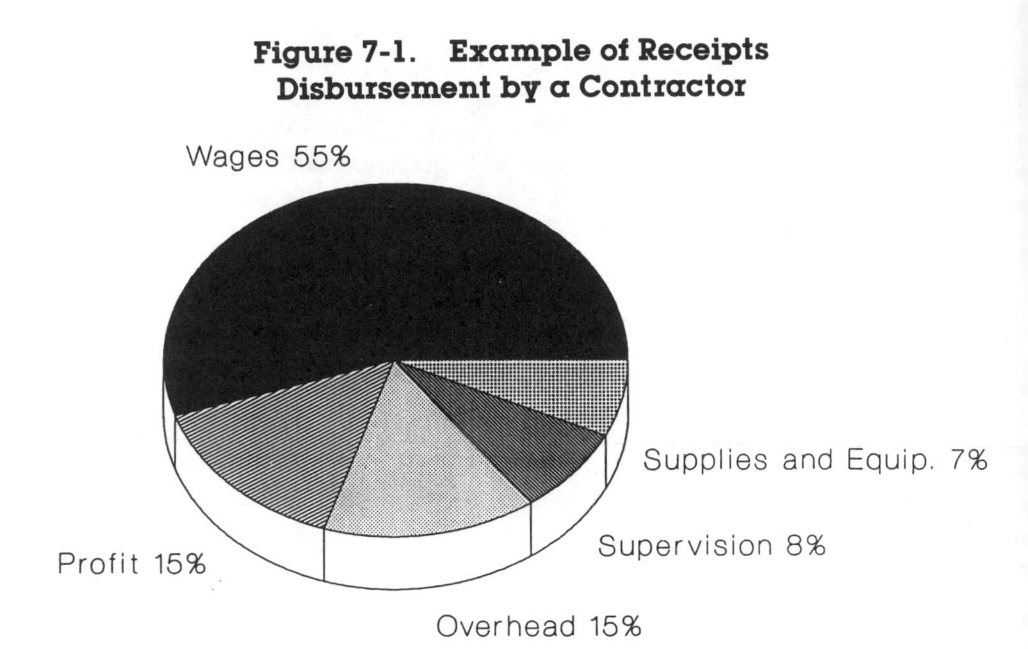

Exact figures would depend on such factors as:

*Size of the job
*Type of job
*Competitive bidding
*Type of contract
*Use of inspector
*System of deductions
*Knowledgeability of the customer
*Location of the job (urban vs. rural)
*Size of the contract company
*Purchasing power
*Compensation and benefits for workers
*Insurance requirements
*Expertise of the contractor
*Type of contractor organization
*Who owns the equipment
*Other factors

BSCA is of particular interest. Its members are contractors, and the dues structure eliminates most newer and smaller firms. Their approach is professional, and you can benefit from their publications. For example, Figure 7-2 shows two sample pages from its publication "How to Select a Building Service Contractor." The address is 10201 Lee Highway, Suite 225, Fairfax, VA 22030.

One of the circumstances that mixes reputable with disreputable contractors is the fact that it is so easy to get into the contract cleaning business, and because there are no regulatory agencies governing the conduct of this type of service. There are even opportunities for entering the field on a franchise basis with a small capital investment and even this can be paid out of profits over a period of time. Thus, poorly qualified organizations may be attracted into this field because of the low capital investment.

AGREEING ON TERMS: FIVE MAIN TYPES OF CONTRACTS

Many of the contracts now being written practically guarantee that no reputable contractor will bid on the job! This may be because the contract is for too short a time; or the specifications are written in an ambiguous or loose way, permitting all types of interpretation; or because there are requirements that can obviously not be met.

It is not unusual for contractors to bid on a one-year job on the basis of "what will it take to get this job?" If the contract is awarded to the bidder, he then subtracts his profit and overhead, and gives his customer the amount of cleaning possible with the funds remaining. The blame lies as much with management as with the contractor in cases of this type, for failure to properly specify the work in all of its aspects, as well as for the naïve entering into a contract that is impossible to perform.

There are five main types of contracts possible for cleaning:

1. A *fixed periodic cost* (either as a grand total or per unit area) based on a quality definition and/or frequency specifications. The periodic rate type of contract is the one most frequently used; it is available in two forms, the most common of which being with *uncontrolled* input. This means that the contractor has agreed to provide a service of a given nature, but is not required to provide any given number of labor-hours, and is usually not required to have on hand any given numbers of types of equipment.

This type of contract is by far the most troublesome and the one least likely to bring a meeting of the minds, and therefore satisfaction to both customer and contractor. But if you use this type, pay attention to these aspects of the specifications:

Figure 7-2. Guidelines for Selecting a Building Service Contractor

6
Monitor the Contractor's Performance

Encourage the idea of a "partnership" with your contractor. Keep open lines of communication. At the supervisory level in particular, set up channels for communication of complaints, communication back from the supervisor of how the complaint was dealt with, and communication to the complainer on the action taken. Lines of communication at branch manager or owner level are also important. These communications should be a regular routine thing, and not just in times of dissatisfaction or crises.

Some formal monitoring of the contractors' performance is advisable, both to determine that you are getting what you are paying for and also to assist you in determining whether or not the contractor should be allowed to bid another of your buildings. Three possible ways of monitoring are as follows:

Regular inspections *with* the contractor of areas selected at random by yourself. Written notes made on these inspections with copies to yourself and contractor.

Monitor the tenants' feelings on the cleaning by informal discussions with them, or by periodic simple questionnaires that can be distributed by the contractor's staff the night before.

Verify the man-hours used by the contractor by method to be worked out with him prior to signing of the contract.

7
Traps the Building Manager Can Fall Into

If the procedures set out in this booklet are not carried out, or not carried out in full, you may fall into some of the following traps:

Price Jacking—Here the contractor bids very low, gains the job, does a first class cleaning performance for initial two or three months (at considerable expense), then asks for a price increase on some pretext such as misunderstanding over the specifications, wanting to pay his staff higher wages, inflation, etc.

Solution:

- Check references carefully for this type of experience. Normally your screening process would eliminate such a contractor.
- If he does slip through, don't give him any pretexts, i.e., have everything spelled out and negotiated beforehand as we have indicated. Instead, insist he performs the contract well at no price increase. It will give you both an inexpensive job for a year, and perhaps cure the contractor of this habit.

Disputes over Pricing of New Works—Disputes may arise over the price to be paid the contractor for new areas to be serviced, or additional services.

Solution:

- Build into the contract some rough guidelines (e.g., price per square foot, etc.) for any new square footage cleaned.

Squeaky Wheel Contractor—This is the contractor who bids low, gains the job. He starts off with a bang and everyone is happy for a month or two. As your attention moves to other things, his work quality gradually slides. Eventually you complain—he oils the squeaky wheel and does a "heroic effort" to get the place cleaned up and back to standard. Once again your attention goes to other things, the quality gradually slides, and the cycle repeats for as many times as your patience will allow.

Solution:

- Throw the rascal out, and select a successor using the selection techniques of this booklet!

From "How to Select a Building Service Contractor," by the Building Service Contractors Association International (BSCAI), Fairfax, VA.

- Define the frequency of job performance, and completely eliminate phrases such as "as needed" or "as required."
- Carefully describe the method of performing the work. For example, if the floor is to be mopped daily, specify whether this should be with a one-bucket, two-bucket, or three-bucket system, and whether or not the floor will be rinsed.
- Specify the nature of the chemicals and materials to be used.
- Control the nature and amount of supervision.
- Describe the type and extent of reports and inspections that should be made.
- Make sure the contractor provides legal and fiscal information. One of the problems inherent in this type of contract is that the contract document is so voluminous that it frightens off many reputable but smaller cleaning contractors and also makes policing the contract very difficult.

2. A *variant* of the *periodic rate* contract has a *controlled* input. This means that the contractor is required to provide a minimum number of labor-hours, in addition to all the other requirements named above. For many jobs, this would be tantamount to having everyone bid on that *minimum* number as the *actual* number of labor-hours to be provided. Where the periodic rate contract is to be used, controlled input is strongly recommended.

3. A *specialized* type of contract is the management fee. In this case, the customer provides his or her own personnel, equipment, and chemicals, while the contractor manages and, perhaps also, supervises the work. This is a hybrid type operation which may be based on a union contract requiring the personnel to remain on the organization's payroll, or perhaps because of management's interest in seeing that all personnel in the organization receive the same benefits and opportunities for transfer to other departments.

4. In an effort to avoid the pitfalls of the periodic rate, a number of organizations have turned to a contract on a *cost plus percent profit basis.* This has the advantages of providing a variable, flexible program, as the customer is free to determine his needs on the basis of changing quality requirements, physical plant, production or occupancy, weather, and the like. On the other hand, this system provides an incentive to the contractor to increase cost in order to earn a larger profit.

5. Finally, a contract may be established on a *cost plus fixed fee* basis, wherein the contractor is reimbursed for all costs but paid an unchanging periodic fee for the management of the program. This provides the advantages of a flexible program along with the elimination of any incentive to spend more money. Since the contractor is guaranteed a reasonable profit as long as he or she continues to perform a meaningful service, this becomes one of the contractor's most valued jobs and he or she is thus anxious to retain

it as long as possible. Because the cost plus percentage profit or cost plus fixed fee contract preferably requires management to be able to develop its own program in detail, the use of a consultant can be of special value here.

Using Controlled Input—Variable Bid Contracts

There may be some problems with controlled input: it tends to "promote" the less qualified firm to the same level as a more qualified firm, since they are proposing on precisely the same thing. It may decrease the incentive to reduce costs by a more reputable firm, since that firm has nothing to gain by such an activity.

On the other hand, if you do not develop a standard control, any bidder can come in with a low bid—it may not be responsive to the needs of the organization at all, even though it cannot be refuted. Then you are subject to the question "why didn't you select the lowest bid?"

A form of contract exists that provides a variable bid around a standard controlled input. The property owner (or his consultant) makes a measurement, develops time standards, prepares labor-hour requirements (based on standard procedures and frequencies), and from this develops a pro-forma budget just as in the standard controlled input system. But the purpose in this case is to be able to evaluate the variable bids, described below.

In this system, the contract bidder is allowed a variation or "spread" on the labor-hour bid—this variation might be 10% or 15% higher or lower than the control figure. This allowed variation on the initial bid is to permit the contractor flexibility in terms of his own system of incentive payments, cost savings ideas, motivation programs, supervisory coverage, new techniques, etc. Once his figure is named, it becomes the new control figure and he must thereafter stick with it. In order to develop these figures, the bid sheet would show the minimum standard and maximum allowable staffing and finally he would enter his proposed figure.

Further, the contractor must furnish in his bid individualized work assignments for each of the positions shown that he has developed for his customer. Each assignment must be described in writing along with a graphic designation of its extent (of course, the property owner provides the drawings). This descriptive material provides the owner a system of monitoring assignments.

The bidder should provide an organization chart, a system of supervisory coverage, training program plans and also his qualifications for performing the work.

For you to determine the bidder's qualifications more easily, ask the contractor to supply information on length of employment for personnel of various categories, turnover rates, absenteeism patterns, training history, supervisory development system, etc.

A final determination of whether or not to use contract cleaning should be made on sound business principles considering the relative cost and quality benefits on either side. Emotional viewpoints should be avoided. Contract cleaning is, simply, a legitimate alternative consideration open to management.

HOW TO GET YOUR MONEY'S WORTH WITH CONTRACT CLEANING

Since there is one best way to clean any given facility, in terms of supervision, staffing, equipment, chemicals, methods, and the like, then that one best sanitation program should be followed whether the work is done by a contractor or by internal staff. Whatever type of contract is used, carefully follow these guidelines:

- Prepare contracts for longer terms, to interest the more reputable contractors. Annual re-letting of bids should be abandoned in favor of a minimum of a three-year contract with one-year options to renew. Contracts up to five years in duration should prove beneficial. A short contract, for example, requires the contractor to cover all costs of equipment and start-up costs immediately, rather than amortizing over a longer period of time. It would be better for the customer in the latter case, of course, where money can be spent on service rather than on covering initial costs.
- Regardless of the length of the contract, include a cancellation clause, either of 30 or 60 days length, depending on the situation. To be fair, this clause should exist for both parties.
- Inspect the work very closely. In larger facilities, this will amount to one or more full-time salaried employees on the customer's payroll who will check method of performance, time cards, quality achievements, utilization of chemicals, safety, and other factors.
- In any type of contract, assess penalties for failure to perform the work or to provide the proper number of labor-hours. Penalties should be especially rigid for failure to provide the proper supervision, since this is so important to the conduct of the job.

The economics of your decision to use a contractor should be based on the same quality service you expect from your internal staff. You can evaluate the contractor only with a measured, controlled input from the contractor. The specification must include the amount of supervision, labor, equipment, supplies, and even training which will be provided by the contractor; this would provide the same service as an in-house basis. If you decide to use contracted services, optimum results require:

- The contractor will fulfill the inputs specified (labor, supervision, equipment, supplies).
- These inputs will permit work to be performed at a given frequency.
- A contract inspector will be on the owner's payroll.
- This inspector will make deductions from payment for failure to conform to the specifications.
- References must be checked.
- Avoid a one-year contract; better results come with a three-year or even longer contract.
- Use a cancellation clause.
- Use a performance bond. The bond would eventually be paid by the customer in the total contract price, so it can be avoided where all of the potential bidders are known reputable firms.

Figure 7-3 lists questions that you should ask of the bidder to determine his capacity to handle an important contract.

Figure 7-4 is a questionnaire to use in prequalifying contractors.

Figure 7-5 is a sample table of contents for a contract cleaning specification.

Figure 7-6 is a sample bid sheet for routine work for such a contract. (Note that only parts of this bid are provided.) Figure 7-7 provides a bid sheet for special project work.

Figure 7-3. Sample Questions to Ask a Potential Contract Cleaner

A. Number

Management	Supervisory	Non-Supervisory	Total

B. Distribution of Years of Employment by Class of Employee

Length of Employment	Management	Supervisory	Non-Supervisory	Total
0–1 year				
1–2 years				
2–3 years				
3–4 years				
5–9 years				
10–19 years				
Over 19 years				

C. Personnel Turnover for ______ (Non-Supervisory Employees Only)

Years Employed	No. of Hourly Employees	
One or less	______	Added to Payroll
	______	Dropped from Payroll
2–4	______	Dropped from Payroll
5–9	______	Dropped from Payroll
10–19	______	Dropped from Payroll
Over 19	______	Dropped from Payroll

Figure 7-3. (continued)

D. Absentee Pattern for the Year (Record a check (✓) in each appropriate box)

Number of Non-Supervisory Employees	Number of Scheduled Work Days Absent								
	0	1	1 4	5 9	10 19	20 49	50 99	100 199	Over 199
1									
2–4									
5–9									
10–19									
20–49									
50–99									
100–199									
200–499									
500–999									
1000–1999									

E. Employee Training — Average Hours of Training Provided/Year/Employee

Length of Employment	Management	Supervisory	Non-Supervisory
New Employee			
1–2 years			
2–4 years			
5–9 years			
10–19 years			
Over 19 years			

Figure 7-4. Sample Contractor's Prequalifying Questionnaire

(Confidential)

1. Name: ____________________
2. Business Address: ____________________

   ____
3. Telephone Number: ____________________
4. Would you be interested in bidding on the janitorial contract for ____________________ at the above address?

   ____

   (If yes, please continue filling out questionnaire; if no, please return questionnaire.)
5. How long have you been in business? ____________________
6. Do you operate as an individual __________ corporation __________ partnership __________
   a. If a corporation, in what state are you incorporated?

      ____
   b. If a partnership, list the names of all partners:

      ____________________ ____________________

      ____________________ ____________________

      ____________________ ____________________
7. Officers authorized to execute contracts:

   ____________________ ____________________

   ____________________ ____________________
8. Is your firm organized so that continuity is maintained in the event of the owner's absence or death? ____________________
9. a. Major area of business: ____________________
   b. History of firm: ____________________

      ____

      ____

      ____

      ____

   Please enclose any literature that describes your company or the scope of its operation.

Figure 7-4. (continued)

c. Number of full-time employees: ______________________

Provide an organization chart indicating the full-time personnel, job titles, locations, telephone numbers, and if the individual is office or field.

10. Average length of employment of personnel:

    Full-Time ______________ Years

    Part-Time ______________ Years

11. Do you normally provide uniforms and/or identification cards if the contract calls for it?

    ________ Yes ________ No

12. Other Services Furnished:

	Normal Work	Subcontracted

13. What is your normal area of operation? ______________________

    ____

    ____

14. Financial References: (Please attach your most current financial statement)

Institution	Contact

15. Are there any judgments, suits or claims pending against your firm? ______________________

16. Are there any liens for labor or material filed on any of your work?

    ____

17. Have you any contingent liability? ______________________

18. Please list five of your largest buildings (including, name, and phone number of manager and square footage) that we may contact to visit.

Figure 7-4. (continued)

Building Name	Manager	Phone	Size

19. a. Previous work for

Company	Contact	Contract Amt.	Period

b. Other projects completed during the last two years.

Company	Contact	Contract Amt.	Period

c. Subcontractors you are using or have used.

Names	Location	Type of Work

20. What officials of your firm may be contacted in emergencies when your office is closed?

a. Name: ________________ Title: ________

Residence Address: ________________

Telephone Number: ________________

b. Name: ________________ Title: ________

Residence Address: ________________

Telephone Number: ________________

Figure 7-4. (continued)

21. According to the following definition, are you a minority business enterprise?

    A minority enterprise means a business, at least 50 percent of which is owned by minority group members; in the case of a publicly owned business, at least 51 percent of which is owned by minority group members. For the purposes of this definition, minority group members are black, Hispanic, Asian American, American Indian, American Eskimo, and American Aleut.

    Yes ________ No ________

Signed: ______________________________
Title: ______________________________
Date: ______________________________
Firm: ______________________________

Figure 7-5. Sample Contents of a Cleaning Contract

SAMPLE

TABLE OF CONTENTS

FOR A CLEANING CONTRACT

Section

INVITATION TO BID

ADVERTISEMENT TO BID

BIDDING INSTRUCTIONS

BID BOND

BID FORM

PERFORMANCE, PAYMENT, AND GUARANTEE BOND

AGREEMENT

GENERAL CONDITIONS

Amendment
Applicable Law
Assignment
Authority of the Contract Administrator
Authority and Duties of the Coordinator
Business Office
Changes in Number of Weekly Routine Cleaning Hours
Changes, Additions and Deletions in the Specifications
Claims
Compensation for Routine Work, Projects and Reimbursable Items
Contractor's Access
Contractor's Freedom to Provide Service
Contractor's Job Manager
Contractor's Responsibility and Insurance
Contractor's Responsibility for Work
Contractor's Supervisors
Custodial Personnel
Daily Report
Deductions for Non-Performance of Routine Work
Default by the Contractor
Definitions
Employment of Aliens
Entrances and Doorways
Escalation of Wages and Prices
Exclusive Performance

Figure 7-5. (continued)

Existing Utilities and Structures
Gratuities
Holidays
Implementation of Termination
Labor Activity
Needed Repairs
Non-Performance of Work
Non-Waiver
Occupational Safety and Health Act
Organization Chart
Other Contracts
Paging Units
Performance of the Work
Performance, Payment and Guarantee Bond
Permits, Laws, Taxes and Regulations
Relief for Absenteeism and Vacation
Review of Records
Royalties
Sanitary Napkin and Tampon Vending Machines
Security and Identification
Soliciting
Specifications
Statements and Invoices
Storage Space
Supplies Furnished by Client
Telephone Service
Termination by the Client
Termination by the Contractor
Time Clocks
Training
Uniforms
Waste Removal
Work Areas

SPECIFICATIONS

Organization
Assignments
Tasks and Frequencies
Routine Work Performance Standards
Project Work Performance Standards
Chemicals
Tools and Equipment

Figure 7-6. Sample Excerpts from a Contractor's Bid Sheet for Routine Work

CONTRACTOR'S BID SHEET FOR ROUTINE WORK

(For performance of routine work only, not special projects.)

BID ITEM R1—WEEKLY CHARGE FOR WAGES AND SALARIES

One full-time, on-site Job Manager $ ________

Day Shift, Monday through Friday:

Supervisors for ______ labor-hours/wk $ ________
Group leaders for ______ labor-hours/wk $ ________
Custodians for ______ labor-hours/wk $ ________

Evening Shift, Monday through Friday:

Supervisors for ______ labor-hours/wk $ ________
Group leaders for ______ labor-hours/wk $ ________
Custodians for ______ labor-hours/wk $ ________

Day shift, Saturday:

Group leaders for ______ labor-hours/wk $ ________
Custodians for ______ labor-hours/wk $ ________

SUB-TOTAL BID ITEM R1 = $

BID ITEM R5—WEEKLY CHARGE FOR TOOLS AND EQUIPMENT

Includes the cost of all tools and equipment used for Routine Work such as vacuum cleaners, carts, spray bottles, and their maintenance, service, replacement costs, etc.

SUB-TOTAL BID ITEM R5 = $

BID ITEM R6—WEEKLY CHARGE FOR OVERHEAD AND PROFIT

Includes all miscellaneous costs such as bonds, bookkeeping, insurance, recruiting, classroom and on-the-job training, transportation of employees to and from the work site, uniforms, telephone service, corporate administration, clerical support, licenses and fees, and all other overhead costs and profit.

SUB-TOTAL BID ITEM R6 = $

Figure 7-6. (continued)

BID ITEM R7—TOTAL MAXIMUM WEEKLY CHARGE FOR ROUTINE WORK	
The total of BID ITEMS R1, R2, R3, R4, R5, AND R6.	
SUB-TOTAL BID ITEM R7	= $

BID ITEM R8—TOTAL MAXIMUM ANNUAL CHARGE FOR ROUTINE WORK	
BID ITEM R7 multiplied by 50	
SUB-TOTAL BID ITEM R8	= $

Figure 7-7. Sample Contractor's Bid Sheet for Special Projects Work

CONTRACTOR'S BID SHEET FOR PROJECTS WORK

(1) PROJECT	(2) ESTIMATED UNIT QUANTITY	(3) NUMBER ANNUAL REPE- TITIONS	(4) UNIT BID PRICE	(5) BID PRICE FOR COMPLETING ANNUALLY (2)X(3)X(4)
P-1 NON-SPECIFIED TASKS	______ Man-hours	1	$______ Per Man-hour	$______ Per Year
P-2 STRIP AND REFINISH FLOORS	______ Sq. Feet		$______ Per Sq. Ft.	$______ Per Year
P-3 CARPET EXTRACTION	______ Sq. Feet		$______ Per Sq. Ft.	$______ Per Year
P-4 CARPET SHAMPOOING BONNET METHOD	______ Sq. Feet		$______ Per Sq. Ft.	$______ Per Year
P-5 MACHINE SCRUB FLOORS	______ Sq. Feet		$______ Per Sq. Ft.	$______ Per Year
P-6 HIGH DUSTING	______ Sq. Feet		$______ Per Sq. Ft.	$______ Per Year
TOTAL ANNUAL PRICE FOR COMPLETING ALL PROJECTS $______				

8

How to Save on Housekeeping Costs by Implementing a Night-Cleaning Program

It is an enticing thought to change cleaning from an evening or night shift to a day shift operation, to save both lighting and other energy costs, such as heating and air-conditioning. Energy conservation has become very important and will become even more important in the years ahead. It affects all aspects of building operation, with specific applications to custodial operations. For example, heat lamps are often installed in rest rooms and other facilities; their creation of heat by electrical resistance is very inefficient and uses much energy. Therefore, avoid use of heat lamps to dry out an area after cleaning—as is often done.

NIGHT VERSUS DAY CLEANING: FIVE FACTORS TO CONSIDER

As with many other subjects, there is no simple answer to the question of whether night cleaning is appropriate for your facility. But here are five factors to consider:

1. If the major consideration is saving on lighting, then you are talking about roughly 15% of the total energy cost for a typical building.

2. For air-conditioning and heating, during custodial activities, the same intensity is not required as for clerical and other administrative functions. Consider the overall operating hours of the building, and the cost of shut down and startup as compared to continuous operation, with perhaps a reduction of heating or air-conditioning during the night hours. The use

of plastic window covering material, blinds and curtains, insulation, and other energy saving systems might do far more than changing cleaning hours.

3. Public opinion is an important factor in the night-time lighting of a building. With all the publicity to save energy, the average citizen may become irritated at the the sight of a building ablaze with light at night-time. Window coverings help to solve this problem, and the future will undoubtedly see many more windowless buildings, which would solve both energy, the cost of window cleaning and re-glazing problems as well. But the basic solution, if night-time cleaning is desired, is in "vertical scheduling," and/or part-time work, both discussed below.

4. The efficiency of building cleaning is considerably reduced when daytime custodial work is performed. Most estimates indicate a loss of efficiency in the range of 20% to 30% for work done during the day. The reasons include:

- Workers often have to do partial jobs because of the traffic of other personnel, such as mopping half of a corridor at a time, which must be chained off, and even then people will track through it, causing additional work.
- Very often a cleaning worker will have to visit an office or area several times before he or she can obtain entry—the worker might have been barred from doing the work by a conference going on, the user of the office making a long distance call, the exposure of confidential documents, or other reasons.
- The custodian is often called away from his work to perform other duties, and may be confused by having "too many bosses."
- Where the work is done on an evening or night shift, a greater security against vandalism and theft is provided, and this is lost if all cleaning work is done during days.
- Even those organizations which attempt to do all cleaning during daylight find that it cannot be done, and certain work is still scheduled for nights, such as shampooing carpets, stripping and refinishing waxed floors.
- A significant proportion of cleaning work is performed by "moonlighting" personnel and some of your best motivated labor will be lost if you avoid evening shift work, since their second job would be a daytime job in most cases.

5. Housekeeping work during the day is a more hazardous activity to the building user, because of wet floors, projecting mop sticks, the use of floor machines, exposure to chemicals and similar items.

HOW TO CONSERVE LIGHTING

Where cleaning is done on the day shift, try to have part of the time available where there is no conflict of activity, such as having the cleaning crew start two hours before the other work force, or end two hours after, so that wet cleaning could be done in corridors, some project work completed and the like.

Figure 8-1 shows two charts developed by the Illuminating Engineering Society of North America, to show the effects of light fixture cleaning on illumination. The top chart indicates fluctuations in illumination based on frequency of cleaning. The bottom chart shows the drop in illumination, for various type of luminaires, based on elapsed time since cleaning. The film of dust on the fixture, which might include the tube, reflector, and cover lens, absorbs light that would otherwise be useful.

You can conserve lighting energy in numerous ways, in addition to changing the cleaning work shift. Some examples:

- Keep light fixtures clean and free from dust. This can provide as much as 25% or more illumination, since particles of dust and other soils tend to absorb light. Also, keep the reflectors not only clean but painted an appropriate reflective color. Clean the light bulbs and fluorescent tubes so that they will last longer. (Caution: Do not clean incandescent bulbs while they are burning, as they may break and injure the worker.) Some fixture cleaning can be done from the floor using a vacuum (such as pack vacuum) with extension wand and bristle brush attachment, while other work has to be done through washing in place or in a fixture cleaning solution tank. (The last case should be reserved only for exceptionally difficult cleaning, such as in an industrial situation with sticky and greasy soils.)

- Use a switching arrangement that permits every other row of lights to remain burning for cleaning work. This is perfectly satisfactory and immediately saves 50% of the lighting energy. The intensity of illumination required for clerical work is not necessary for custodial activities.

- Use lower-wattage bulbs and tubes. Instead of using 40-watt four foot fluorescent tubes, use 35-watt tubes (with necessary fixture adjustments).

- Encourage management to install reflective surfaces in buildings, which provide better general lighting since light is absorbed by dark colors. Examples are the use of semigloss enamel wall paints, light-colored plastic wall coverings, light-colored vinyl asbestos floor tiles and terrazzo floors, and the lighter colors of ceramic floor and wall surfaces.

- Use devices to extinguish lights when there is no occupant in a particular area.

Figure 8-1. How Light Fixture Cleaning Affects Illumination

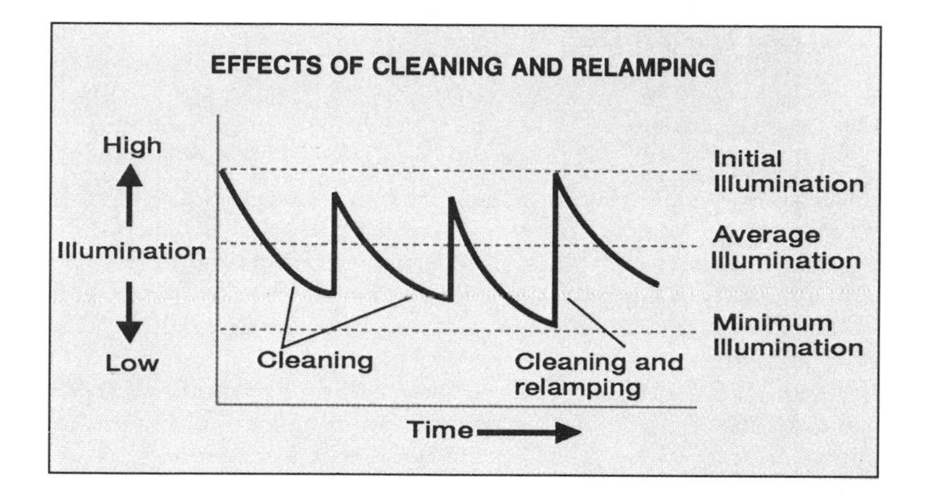

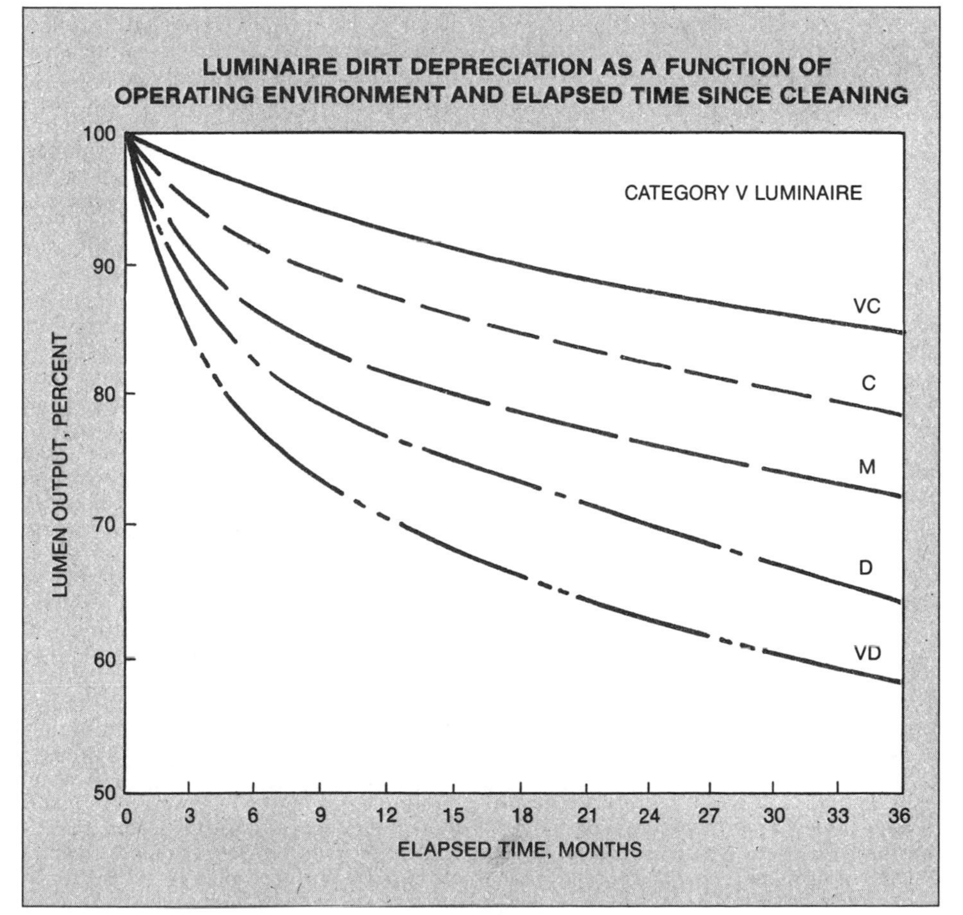

HOW TO CONSERVE ENERGY BY CHANGING CLEANING SCHEDULES

Part-Time Cleaning

Many organizations (including many contract cleaning firms) do cleaning with part-time personnel, often from 6 to 10 P.M. or similar hours. From the energy conservation standpoint, this means that the lights of the building are only on during those hours. In many locations and also at certain times of the year, there is still enough daylight that little interior lighting would be required until the latter part of the work shift.

Where an organization is cleaned with full-time personnel, it is possible to make a transition to part-time personnel through transfers to other departments, and switching over floor by floor, based on normal worker attrition due to retirements and other terminations.

Vertical Scheduling

Custodial work is generally more efficient where the custodian is assigned to a given geographical area and is responsible for all repetitive work in that area. (Project work, such as stripping floors and washing walls, is performed by a team trained and equipped for that purpose, to augment the area-assigned workers' activity.) In a hypothetical case, where a four-story building is maintained by four workers, it is natural to assign one cleaner to each floor. In this case, where the cleaning is done at night, all the lights are always on. This is called *horizontal scheduling*.

An alternate to this is *vertical scheduling*, where the assignments are made from top to bottom of the building rather than from side to side. In the case above, the four workers would each be assigned one fourth of the top floor for his initial activity—they would not be working as a team or gang, but would complete the repetitive work in, let's say, the north wing of the building for one worker. Perhaps at break time, the lights for that floor would be turned out, and after break the lights for the next floor down would be turned on. This tends to help pace the work without getting involved in the distastefulness of short-interval scheduling.

Thus, with vertical scheduling, in the four-story building in question, only one floor is lit at a time. The same principle could apply to a building of any size, with a greater or lesser number of people working in area assignments on a given floor, depending on the size of that floor. It is possible that a twelve-story building might only have two floors lit at a time.

As with everything else, there are pros and cons. A spin-off advantage of vertical scheduling is better personal protection (such as from an accident) since fellow workers are near by—but this very proximity can lead to certain problems of over-socializing.

Other benefits of using vertical scheduling with several personnel on the same floor of a given building at the same time, include:

- Personnel feel much safer from a possible personal attack.
- It is possible to use a working group leader very effectively in this situation, since he would be responsible for all personnel on the floor, under the general direction of a foreman or supervisor. The group leader, in order to have time available for this, would be assigned approximately a 15% lesser time responsibility than the other workers.
- It is easier to supervise people in this fashion, since they are all on the same lighted floor.

The benefits of vertical scheduling far outweigh the shortcomings, and these can be handled by proper supervision, which in itself is simplified by this technique.

The biggest drawback to vertical scheduling is the need and the time needed to take cleaning equipment from floor to floor. Thus, it is more likely to be successful in a low-rise building (involving a smaller number of floors per "team"); where the building is fully carpeted (so no floor machines are involved); and if the elevators are large and not susceptible to damage.

SUMMARY

Where the advantages of several approaches to energy saving can be obtained simultaneously, such as with vertical scheduling, part-time cleaning, and reduced lighting intensity, there is no mandate for performing cleaning work only during the day, and suffering consequently a considerable reduction in performance. The additional pay for shift differentials for evening work have been shown to be an excellent investment in terms of productivity.

9

How to Select and Use Consultants

It is natural to consider every means of improving the cost and quality of sanitation. To consider all possible avenues of improvement, you should consider using a consultant. This chapter discusses how to determine if your organization needs a consultant, describes benefits from using a consultant, offers guidelines for selecting a consulting firm, and tells you how you can make the most of using a consultant.

THE FIRST DECISION: DO YOU NEED A CONSULTANT?

The first and often the biggest question for many managers is whether or not to use a consultant at all. Since your organization continues to operate without the use of a consultant, and perhaps has operated for many years without a consultant, it may appear to be an avoidable expense. But the basic question is, is it an investment that will bring worthwhile dividends far beyond its original cost?

Some people fear that to recommend a consultant to management is a show of weakness. Yet the higher you go in management, the more likely you are to see the use of consultants; attorneys, accountants, psychologists, physicians, and many other dispensers of specialized, experienced advice are consultants.

You may fear that the consultant will criticize your previous performance. A consultant who uses the criticism approach does not know his or her job; the analysis should be creative, should involve and recognize the contributions of the client's staff and should not create barriers to successful implementation through the use of criticism.

The main reason for using a consultant is that there is a reasonable likelihood that the cost of the service will be regained in a reasonably short time, and the benefits obtained will be annually recurring. Sometimes the economic impact can be softened through the spreading of fees over a number of months, or making payments in two fiscal years.

It is possible to make an arrangement with a consultant that the fees are to be balanced with at least an equivalent dollar saving. During austerity periods, this may be unavoidable; but it could lead to a long-term economic loss to the organization because of failure to perform needed services. In custodial operations, you can be arbitrary about the allocation of funds to these services; for example, a 50% cost reduction is possible at any one moment in time, simply by eliminating half the staff, but the results in terms of safety and health, and damage to equipment, surfaces, and the like may be disastrous.

One way to determine whether or not using a consultant would be beneficial is to check with others who have made use of consultants—just be sure they had the same objective in mind as yourself. For example, if your desire is to reduce cost, the firm that used a consultant for the purpose of improving quality within an existing budget would report that they did not save any money (yet, their objective had been fulfilled).

Here are some specific reasons why using a consultant can be a good investment:

- A consultant's recommendations are more apt to be acted on than the same suggestion coming from an internal source, because of the experience and credibility, as well as objectivity, credited to the consultant.

It is frustrating to an internal manager or supervisor to have his or her recommendation ignored regularly, and then have a consultant come in and make the same recommendation which is then accepted and acted on. Management may consider its own personnel to be defensive on the subject, rather than objective, or the internal recommendations may have been made without an economic documentation and justification.

- Dealings with unions, civil service, and political organizations might be facilitated by using a consultant with extensive experience in working with organizations similar to yours.

- Quicker results may be obtained by using a consultant because of the consultant's experience with what will and what will not work, and how quickly people will respond, as well as the best way to elicit that response.

- Some organizations use consultants so that their own image of benevolence and paternalism will not be tarnished where changes are necessary to which the employees might have a reaction.

- Where specialized internal skills are not available for the analysis to be made, it is very often more economical to use a consultant than to augment one's staff, and to spend the time required to research the subject.

• Even if the skills are on hand, the time to utilize them might not be available; the personnel may be needed for daily operations or for other activities.

GUIDELINES FOR SELECTING A CONSULTANT

If you've decided to use a consultant, you may be confused by how to choose the right firm for your needs. Here are some questions to ask a prospective consultant:

• *What is the firm's orientation?* It should specialize in the type of help you are interested in, such as physical facilities consulting, rather than a general management or accounting firm that will perform this service for you as a sideline, perhaps using some junior personnel.

• *What is the company's background?* Is the company under the firm control of a professional, or does it operate as a franchise? Was the company formed because of the failure of some of its members to perform properly somewhere else? Evaluate the personal background of some of the key personnel including their previous employment. (Would you put an important person on your payroll without doing this?)

• *What is the size of the company?* You want an organization that is large enough to give you continuity and backup in case of some turnover within its own group. You can be secure in this regard if the consultant has several professionals able to handle your work. On the other hand, too large a firm might not be able to provide the individual attention you may need.

• *What is the experience of the firm?* You are interested in work done for organizations of similar size and type as yours. The consultant should be required to provide a reference list with at least 100 names on it.

The company should have been in business at least ten years (this not only assures you the proper experience, but also the great likelihood of continuation of the company, since so many firms fail in the first several years of their existence). The consultants in the company should average at least five years experience *in that company.*

• *Is the company a leader in its field?* Acceptance of recommendations, and their implementation, is improved when dealing with a firm with established credentials. For example, has the firm published books on the subject, has its consultants written magazine articles, given presentations before professional groups, conducted conferences and seminars of its own? (Some of these activities are also valuable in perpetuating and updating the initial work of the consultant.)

• *Is the fee reasonable?* It is desirable to deal with a consultant who will name a fixed fee, so that there is avoidance of any "extras." A consulting fee should reflect the time and expense involved in the work; a "share of the savings" arrangement may be extremely attractive to the consultant, but a poor choice for the client. The fee must be related to the nature of the service, the duration of the service, and qualifications of the personnel performing that service.

- *Is an auditing service available?* The consultant should offer an audit service as a relatively rapid and inexpensive means of assisting clients to determine if there is significant opportunity for cost or quality improvement. Where management is already aware of these possibilities, the audit is superfluous.

Figure 9-1 provides a checklist for evaluating several consulting firms.

HOW TO MAKE THE MOST OF YOUR CONSULTANT

Once you have determined which consultant to use, the next question concerns how to best use that consultant's skills. Each firm has its own operating philosophy, which reflects the leader of the organization (remember, you are doing business with three entities: the organization, its president, and its consultants). To get the most out of your consultant, make sure he or she meets the following basic requirements:

- Understands your organization's objectives and limitations, so that activities are not directed toward recommendations that cannot be implemented (of course, limitations must be real and not just imagined).
- Does not use a stop-watch study to monitor work performance and task time. Although work scheduling is an important part of operations, short-interval scheduling—where each minute of the worker's time is controlled, recorded, and reported—is detrimental to the performance of all custodial work and many aspects of general maintenance activities. It tends to cause distrust and antagonism among personnel. There are a number of other ways to determine time requirements or schedule and control performance; Chapter 3 describes some of these methods.
- Obtains input from personnel and supervision, as well as from internal management. This is an important aspect of the study.
- Recognizes that its objective is to build and develop the internal organization, rather than to take it over and attempt to operate it. Where contract services are utilized, the consultant's objective is to develop a specification and contract document that obtains the desired quality at the best cost, while giving the contractor a reasonable profit.
- Conducts seminars and training, not only for personal development, but also to provide the clients with an opportunity for updating their operations. This requirement applies to all the consultants in the consulting firm.

Figure 9-1. Sample Checklist for Evaluating Several Consultants.

	Consultant #1	Consultant #2	Consultant #3
1. Does facilities consulting represent the fundamental service provided by the consultant rather than a sideline?	___	___	___
2. Is the work for each client directed by registered professional engineers?	___	___	___
3. Is the consultant entirely independent of supplier and contract organizations, without interest in such firms?	___	___	___
4. Are fees based only on time involvement by professional consultants?	___	___	___
5. Are services and fees specific, eliminating the need for additional services and fees?	___	___	___
6. Has the consulting firm worked at least ten years in this specialized field?	___	___	___
7. Does the personnel of the consulting firm spend full time in professional consulting activities?	___	___	___
8. Does the consulting firm broaden its fund of experience through producing seminars and other educational activities?	___	___	___
9. Does the consultant provide a comprehensive reference list of work done within the firm?	___	___	___
10. Has the firm contributed to the literature of facilities management?	___	___	___
11. Have the consultants averaged five years or more service in this firm?	___	___	___
12. Is the consulting work performed in professional offices that are open to your visit?	___	___	___
13. Does the firm provide information for D & B reports?	___	___	___

To maximize the value of using a consultant, make sure that both parties have a clear understanding concerning both objectives and limitations. Here is an example of one consultant firm's suggested agreement form:

- OBJECTIVES (check the item that best describes your objective; add to it any qualifying phrase desired; or, write in your objective in the last space provided). You may wish to check more than one box.
  - ☐ Save money without disturbing existing quality.
  - ☐ Save a greater amount of money, even if some quality reduction is required.
  - ☐ Save the greatest amount of money; develop an "austere" program.
  - ☐ Our objective is to save $________________ per year.
  - ☐ Stay within the existing budget and improve quality as much as possible.
  - ☐ Improve quality, but save the amount of the consulting fee.
  - ☐ Improve quality even if additional expense is required.
  - ☐ If additional expenditures are required for equipment, supervision, etc., no more than $________________ per year will be considered.
  - ☐ If additional expenditures for equipment, supervision, etc., are necessary, it must be developed through a reduction in staff.
  - ☐ The benefits of this service should be shared between cost-saving and quality improvement.
  - ☐ No objective has been set; the consultant should recommend one as soon as the analysis permits.
  - ☐ Other: ____

    ____

- LIMITATIONS—Here are some limitations with which we are most frequently charged. Check off those that are applicable to you, and add others if necessary.
  - ☐ Do not consider the use of contract services.
  - ☐ We are using contract service; do not recommend its discontinuance.
  - ☐ We have a union agreement (copy enclosed).
  - ☐ Do not recommend changes in the union agreement for our next negotiations.
  - ☐ We operate under Civil Service regulations (copy enclosed).
  - ☐ Do not recommend changes in Civil Service regulations.
  - ☐ Do not recommend changes in organizational structure.
  - ☐ Do not recommend changes in supervisory staff.
  - ☐ Do not recommend changes in work shifts.

☐ Do not recommend changes in supervisor compensation.

☐ Do not recommend changes in worker compensation.

☐ Do not recommend changes in titles.

☐ Do not recommend changes in materials purchased.

☐ There are no limitations on this study.

☐ Other limitations: ______________________________

HOW TO PRESENT THE CONSULTANT'S FINDINGS TO EMPLOYEES

Very often the degree of success of a consulting job will rest on the way that it is presented to the employees.

Where the organization is unionized, give the union a copy of the presentation to workers so the union can make suggestions. Of course, it should not be necessary to ask the union's permission to engage in a consultation, but it is desirable to seek their cooperation and participation.

The following is a script of a sample presentation to employees about an impending consulting service. If a union is involved, you might wish to consider having a union official make some remarks as well.

Sample Presentation to Employees Concerning Consulting Services

Remarks by First Client Manager

We have asked you to join us in a brief meeting today so that we may introduce you to two people who you are going to see around our buildings over the next months. Their names are ________________.

They are members of the XYZ Consulting Company.

We have retained this company as our housekeeping consultants.

This company has worked for many organizations like ours throughout the country, and we have talked to a number of them to be sure that this was the right firm for us.

They have been hired to make sure that we are using the best type of supervisory organization; the easiest and most efficient cleaning methods; the best equipment and chemicals; balanced, safe and reasonable work assignments. In general, they are here to make observations and recommendations concerning everything that affects the housekeeping situation here.

Now let me tell you some of the things they are *not* here to do:

- They will not criticize or make reports about any individual housekeeping employee.
- They will not eliminate jobs. We want to assure each of you that the study will not cause anyone to lose his or her job.

Remarks by Second Client Manager

You may ask why a consultant firm is needed to review our housekeeping function. We have secured their services not because we have a bad housekeeping function; on the contrary, we have good housekeeping and we want to make it better! We are working hard to improve our service and to protect the great investment that has been made in our property.

We have been reviewing possible approaches to improve our housekeeping. One of the ideas was to secure good expertise in housekeeping and we found that this consulting firm was most qualified to provide that expertise based on its experience in hundreds of organizations. We contacted a number of firms and found that this company's recommendations made a big difference and were well received by all members of the housekeeping team. Recommendations based on broad, successful, and tested experience on cleaning techniques and procedures; training; correct equipment and tools; proper cleaning materials; balanced and reasonable work assignments, will assist us in doing a better, smarter job in the future.

We have a challenge ahead of us as we develop and try to improve our service for all people who use our buildings. As we develop improved methods in other things we naturally want to have our housekeeping operation keep pace and, frankly, we would like to take the lead and make our housekeeping department one of the best in the nation. We can do it if we all work together on this goal. I feel that this firm will assist us in this task.

Comments by First Consultant

It's a privilege for us to be here to work with you. Your organization enjoys a fine reputation and we are certainly pleased to be associated with it.

Our study of your facility will take several weeks and, of course, the first portion of our work will be mostly data collection. As you see us around the complex, please let us know if you have any suggestions or comments that might be of interest to us. You can be assured that any remarks will be held in the strictest confidence. The benefits that we will be able to obtain for you will only be as good as you help make them.

Your management and we know that you are doing important work here. Things can't get done properly around here unless *your* job is done properly. You know that and we know that, but not enough other people in other departments know it, and we think it's one of our jobs to help educate them. Good housekeeping is everyone's business.

In summary, your management has asked us to come here to help find the best way and the fairest way to complete this important task of building sanitation.

Remarks by Second Consultant

Many times, when we begin a project such as this, people are more interested in what we are not going to do rather than what we are going to do. So, at the risk of repeating a few items, let's talk about what we are *not* going to do.

As was mentioned earlier, we are not associated with contract cleaning organizations or supply firms. Therefore, we are not here to try to take over your housekeeping function in any way. That job belongs to your management and they are going to continue to have that job. You see, we can't actually *do* anything here. We are here to make suggestions.

Since we are not affiliated with any chemical supply company or equipment sales company, we are not here to try to load you up with a lot of useless chemicals and equipment that you really don't need, but we'd like to see you get what you *do* need.

As we mentioned earlier, it is not our role here to criticize individual employees; we are not here to spy on you. The relationship that each of you has with each of your supervisors is of no concern to us.

We are not here to eliminate jobs. We can assure each one of you that this study will not cause anyone to lose his or her job.

Over the coming months, we will see a lot of each other, and I am willing to talk to each of you individually. We think that most housekeeping employees know more about housekeeping than they are given credit for. We will be happy to listen to any comments that you would like to make regarding the housekeeping situation here. I cannot guarantee that any action will be taken on any recommendation that you make. But I can guarantee that we will listen to your recommendations and evaluate them on their own merits.

At this time we would be happy to answer any questions you may have.

Final Remarks from Management

10

How to Target Areas for Cost and Quality Improvement: The Housekeeping Self-Audit

A housekeeping self-audit can help you measure how effective your cleaning program is, and serve as a check-list to identify items which need more attention. This chapter provides 75 questions to help you evaluate all aspects of your housekeeping department: budgets and costs; equipment, materials, and facilities; job performance; job assignments; staffing, training, supervision, and worker morale; cleaning procedures; and general organization of the department. To help you consider each point objectively, the questions are arranged in a random order.

To complete the self-audit, follow these steps:

1. Fill in the score with a "0" if your answer is "no"; put the maximum number of points indicated if your answer is "yes"; or enter some figure in between depending on your standing relative to the ideal.
2. Answer each question before going on to the next one.
3. If the question does not relate to your situation, assign it the maximum score.
4. When finished, total your score, and divide by 3.5 to determine the percentage score.

No scoring of this type can tell you exactly how efficient or effective your housekeeping operation is, since so much depends on the objectivity of the answers, as well as on the weighting of the maximum points, which, while established for a typical situation, may require considerable adjustment for your operation. Further, the cost must be related to the size of the staff, and that can only be evaluated through a specific analysis.

Nevertheless, this review can pinpoint opportunities you can use for cost saving or quality improvement—or both. Roughly speaking, a score of 90% is good; over 80% is acceptable; while 50% to 70% is typical. The difference between a low score and 90% gives some indication of the improvement possible.

The Audit

	MAX. POINTS	YOUR SCORE
1. Do you operate your housekeeping department under an annual budget?	3	___
A budget provides a predictable expenditure, while at the same time giving a base against which to measure economic performance. Establishment of the Housekeeping Department budget should always involve participation of the department manager. The budget should be flexible enough so that, as conditions on which the budget was based are changed, the budget may be changed correspondingly.		
2. Are housekeeping services costed out to other departments, both for standard repetitive work as well as for special requests?	2	___
By assigning actual housekeeping costs for both regular and special request work to individual departments, you can better interpret your own operating efficiency and value to the overall organization. It can also help pinpoint unreasonable demands on the time of Housekeeping Department personnel, excessive or insufficient cleaning frequencies, etc.		
3. In calculating the budget, do you use a creative approach, rather than extending past performance?	4	___
Budgets should be designed as a function of a positive, comprehensive program. Extensions of past performance merely tend to perpetuate past practices. The budget should be a target figure, capable of achievement through effort and thought, but not through "coasting."		
4. Have you a way of comparing your housekeeping costs with those of similar organizations?	2	___
While no two organizations have the same requirements in cleaning and sanitation, the same personnel,		

	MAX. POINTS	YOUR SCORE
or the same physical facility, nevertheless cost comparisons can help to turn up areas of potential improvement. Such a comparison might suggest turning away from, or turning toward, contract cleaning for a specific job, for example, or it may suggest the desirability of jobs currently being handled by other departments being pulled into the Housekeeping Department, or vice-versa.		
5. Do you periodically evaluate housekeeping jobs currently performed by outside contractors to see if they could be handled at lower cost by your own Housekeeping Department?	5	___
Because a particular area or job has been handled by a cleaning contractor for a number of years does not mean that this will always continue to be the most economical and efficient way of handling this work. Changes in personnel, organization, mechanization and efficiencies within the internal Housekeeping Department can shift the balance toward performance by your own Housekeeping Department.		
6. Do you periodically evaluate housekeeping jobs currently performed by your own Housekeeping Department to see if they could be handled at a lower cost by outside contractors?	5	___
The contract cleaning industry is dynamic and aggressive; it is moving into broader service fields and geographical areas. As this industry gains knowledge, experience and reputation, it will better be able to handle certain types of work under proper controls and specifications. (Note: if you decide to change from internal cleaning to cleaning by contractors, be very careful—it is often difficult to evaluate the effectiveness of the work done by others, difficult to reinstate an internal department, and may have an adverse effect on the remaining work force.)		
7. Have you studied current housekeeping work procedures to see if savings could be effected?	12	___
A surprising number of outmoded methods are found in the typical housekeeping operation. Where		

	MAX. POINTS	YOUR SCORE
management does not provide efficient work procedures, then workers either develop their own procedures or perpetuate a procedure learned from other workers. The housekeeping field is now moving very fast, and improved methods are appearing regularly.		
8. Have you taken full advantage of the economics to be obtained by mechanizing your housekeeping?	10	___
Housekeeping is one of the last indirect functions where mechanization has received attention. Even in industries which are highly mechanized or, in some cases, automated, a number of the housekeeping functions are still being performed manually where they could be mechanized. Equipment includes such items as auto-scrubbers, power-sweepers, pack vacuums, baseboard scrubbers, wall washers, hand washers, venetian blind cleaners, window cleaners, etc.		
9. Do you have systems or devices for controlling the consumption of cleaning materials?	3	___
Although housekeeping materials account for only 5% to 10% of the total housekeeping budget, nevertheless wasting these materials can consume a considerable amount of money which might have otherwise been usefully employed. Equally important, overuse of product can actually increase the work load on housekeeping employees by increasing the time required for rinsing, or leaving a filmed-over surface which soils easily. Control devices include drum pumps, finger-tip pumps for small containers, controlled packaging, aspirating proportioners.		
10. Have you standardized on materials so that you are using a minimum number of cleaning products?	2	___
Only a half-dozen chemical products are required for normal daily repetitive cleaning, and a few others for intermittent special projects. A larger number of products than this results in intensified training requirements and can cause confusion and even costly error in mis-application or even replacement of damaged surfaces. Finally, standardization permits purchase in larger quantities at lower cost.		

	MAX. POINTS	YOUR SCORE
11. Do you have suitable time standards for housekeeping jobs?	12	___

Time standards permit the scheduling of both repetitive and special project work so that it may be performed efficiently, balancing the individual work loads so that each housekeeping worker is given a reasonable day's work. This contributes not only to overall operating efficiency but to improved personnel morale as well. Preparation of meaningful, weighted time standards requires a comprehensive approach to the entire housekeeping operation, involving standards for methods, materials and equipment; as well as a consideration of the organizational structure, the labor climate, the type of personnel involved, the physical facility, etc. The implementation of a housekeeping program on the basis of time standards alone perpetuates outmoded practices in other phases of the operation.

12. Do you have a basis for comparing your frequencies of housekeeping job performance with those found at other similar organizations?	2	___

There is a strong relationship between the frequency of housekeeping job performance and the level of housekeeping obtained. Thus, a floor which is mopped and buffed daily could be expected to be cleaner and look better than one which is mopped and buffed only twice weekly. Inadequate frequencies cause poor appearance or even damage to surfaces; excessive frequencies may waste effort and money. Frequencies should be balanced by taking time from one function and applying it to another where it may be more needed.

13. Do you know whether the labor-hours devoted to housekeeping are in line with your actual labor-hour requirements?	8	___

As in Parkinson's Law, has the work expanded so as to fill the time allowed for its completion? Many people spend a full day at a job which should require hours less. Work sampling techniques may be used to

		MAX. POINTS	YOUR SCORE
	determine the portion of a worker's time spent in productive effort, although the effort may be more or less efficient. Time, methods, and other standards weighted for use in your own facility are necessary to establish the proper staffing and scheduling.		
14.	Is cleaning performed on a scheduled basis?	8	___
	Where housekeeping services are provided on a corrective basis, or "putting out fires," it can cause considerable confusion and deterioration of morale. Such an inefficient method of corrective maintenance is rarely efficient from a long-term standpoint, because the work is performed only after something unpleasant has occurred. Housekeeping should be performed on a preventive maintenance basis with major projects being performed, insofar as possible, immediately before the need has become apparent. This provides the greatest efficiency and protection of the investment in the physical facility, while also providing the best operating conditions for the entire work force.		
15.	Have specific performance levels been established for housekeeping workers?	6	___
	Custodial workers work best when they know exactly what is expected of them each day in terms of performance, both from a quantity and quality basis. Workers should understand approximately how much time should be involved in the completion of the major segments of their work. The assigning of specific time intervals for the beginning and completion of each job has certain advantages, but these may be counterbalanced by the morale problem involved in having each job timed. The assignment of a reasonable work load, and a description of its method and order, suffices quite well for most cases, where there is a continuing training and inspection program under qualified leadership.		
16.	Do cleaning schedules allow for seasonal variations?	3	___
	A housekeeping organization should be flexible enough to permit special attention being given to		

		MAX. POINTS	YOUR SCORE
	problem areas during inclement weather. Examples are snow removal, cleaning of mats and runners, mopping up water. Scheduling should permit the performance of these duties without creating a crisis in regular daily housekeeping service.		
17.	Do you have a system for rating levels of cleanliness?	3	___
	Describing a level of sanitation and rating a given physical facility against this description requires a good deal of thought and experience. The work must be described both in terms of quantity and quality, and the analysis of results requires a system for measuring or observing both.		
18.	Do you have written instructions for housekeeping work?	7	___
	Written instructions have many advantages: supervisors are properly able to interpret work assignments and methods; training of new employees is simplified; morale is improved because employees know precisely what is expected of them; flexibility is achieved in the simplifying or transferal of people from one area to another.		
19.	Have definite lines of authority, and reporting functions, been established for the housekeeping department?	8	___
	Workers should know precisely to whom to report or ask questions in cases of uncertainty; from whom instructions come; and who will provide training and inspection. Of course, equally as bad as lack of supervision are unclear lines of authority where a worker may be receiving instructions from more than one supervisor.		
20.	Are custodians performing tasks other than those involved in housekeeping and cleaning?	4	___
	When housekeeping workers share their time between cleaning and other duties, the result is often a lack of interest or proprietary attitude concerning any of the jobs. Efficiency is impaired because the "make-ready"		

		MAX. POINTS	YOUR SCORE
	and "put-away" times, which are non-productive, are increased considerably when workers turn from performing one type of job to another.		
21.	Is adequate supervision, and relief for supervision, available?	12	___
	The proper proportion of supervisors to custodians will be based on such factors as turnover, travel time requirements, type of cleaning being done, scheduling (area or group assignment), quality of the worker, etc. In the field of housekeeping, the quality and intensity of supervision is particularly related to the results obtained.		
22.	Has a statistical approach been used with respect to the assignment of housekeeping tasks?	3	___
	Determination of square footages, fixture counts, surface types and condition, etc., contribute to "defining the problem."		
23.	Are persons outside the permanent housekeeping staff permitted to perform cleaning duties only with proper training or supervision?	3	___
	In many cases, cleaning of one type or another will be done by persons not in the Housekeeping Department. Nevertheless, this personnel should attend housekeeping training sessions, and their supervisor should depend on the housekeeping department for recommendations concerning methods, equipment, and materials. In some cases it is also desirable for the housekeeping department to make a periodic inspection of these areas maintained by other personnel.		
24.	Do you have a definite plan for reducing or expanding the cleaning force to compensate for increases or decreases in activity?	5	___
	In many organizations, the number of housekeeping personnel authorized is related directly to the production or occupancy level: this may be an arbitrary management decision over which housekeeping has		

	MAX. POINTS	YOUR SCORE
no control whatever. The housekeeping program should be formulated in such a way that a fluctuating staff will not unduly disrupt housekeeping operations.		
25. If a new building or wing is constructed, do you have an accurate way of pre-determining personnel requirements and housekeeping costs?	4	___
Many new facilities are staffed on a trial-and-error basis, through the use of a small staff which is increased until the work is apparently performed satisfactorily. Or, staffing may be accomplished by applying some average square footage determinations, based on prior history. Pre-determination based on adequate standards, however, assures an efficient work force without danger of damage to the facility, endangering the health of other employees, or affecting the quality level or productive output.		
26. Is an effort being made to improve worker skill and efficiency through a regular training program?	12	___
No one would think of asking a carpenter, for example, to do worthwhile work, using new materials, equipment, and methods, without suitable orientation and training. Yet, cleaning personnel are asked to do this every day. A continuing training program for housekeeping is necessary not only to improve worker skill, avoid accidents, and conserve materials, but also because it provides a vehicle for morale-building in a job where morale is often a serious problem.		
27. Do your workers wear uniforms?	3	___
Uniforms improve morale because of their identification with a specific department, and the impression created that the worker is a specialist. Also, the worker can be identified at a distance by the supervisor. Uniforms are generally a good investment if their use does not create a problem with other departments. A name badge is important too.		

	MAX. POINTS	YOUR SCORE
28. Does each custodian fully understand his or her responsibilities?	5	___

Demoralization results when workers are given nebulous jobs to perform. This may occur because a custodian is used as relief for a worker performing a different type of assignment; when there is considerable turnover and orientation time is not adequate; where supervision is insufficient; or where a well-defined program does not exist.

29. Have you ever tested custodians to determine their understanding of their jobs and their duties?	3	___

It is difficult to evaluate the effect of a training program without periodic testing to provide a history of improvement, and to evaluate the type and extent of additional training required. A complete testing program will include both written tests (or oral if the subject has a language problem) and practical performance tests using normally assigned materials and equipment in a typical work area. Testing can become, if desired, an integral part of an employee rating and qualification system.

30. Have you tried to determine the time custodians actually spend in productive work?	4	___

Industrial engineering techniques may indicate an excessive portion of the total time being devoted to transportation, personal time, absence from the work station, etc., and may indicate the need for action. Carefully done, a work sampling can be taken without disturbing or arousing hostility in employees.

31. Is recognition or reward given to custodians who do exceptionally good work?	3	___

Where good work goes unnoticed or unmentioned, the natural result is for all workers to gravitate toward a barely acceptable level. Recognition, being one of the fundamental human drives, can bring positive and lasting results when well handled.

	MAX. POINTS	YOUR SCORE
32. Are written qualifications used for hiring custodians?	2	___

Where housekeeping is mistakenly considered a job which anyone can perform adequately, employees may be hired who are wholly unfit for the work and who may cause considerable financial losses through supervisory time, retraining, errors and damage to surfaces. In times of stress, the tendency to hire almost anyone who applies for a position in the housekeeping department should be resisted as much as possible. The housekeeping department manager's suggestions concerning qualifications should be given serious consideration. One of the dangers, however, is setting qualifications too high with the risk of hiring employees who may become bored with the work, thus increasing the turnover rate. The housekeeping department will be judged by the performance of its poorest workers.

33. Are rewards given for good suggestions?	2	___

Where an organization-wide suggestion system exists, housekeeping department personnel should be encouraged to participate. Where there is no such overall program, it may be desirable, at least for a period of one year, to set up a departmental suggestion system, particularly if a comprehensive program involving a number of reassignments is in progress.

34. Are physical examinations given to prospective custodians?	3	___

A thorough physical examination should be given to prospective employees in order to eliminate the hiring of disease carriers. Housekeeping personnel work in areas such as rest rooms, locker rooms, food-processing areas, around drinking fountains, etc., where cross-infection can become a serious problem.

35. Do you conduct regular housekeeping inspections?	5	___

A fixed inspection program has the advantage of providing a history of improvement, while providing

MAX. POINTS | YOUR SCORE

a tool for measuring quantity and quality of work performance. It also has the supervisory value of placing the supervisor in the work area periodically, providing contact with the workers where they may ask questions or receive instructions or commendations. Inspection is a necessary aspect of training, while testing, mentioned earlier, also represents a type of inspection.

36. Are custodians hired and trained before new building occupancy? — 3 — ____

Pre-training custodians for the maintenance of a new facility assures efficient work, adequate service, and protection of health and surfaces immediately rather than after a frustrating period of trials and errors.

37. Do you have an effective inventory-control system? — 3 — ____

Good control of housekeeping materials assures early investigation of over- or under-consumption; discourages pilferage, or unauthorized use by other departments; assures proper chargeout to other departments for authorized use; and guarantees a reasonable supply of materials on hand with which the housekeeping personnel will perform their work.

38. Do your custodians feel that management is interested in their work and their problems? — 6 — ____

An active interest in housekeeping by management does much to improve the morale of all housekeeping personnel, while at the same time assuring desirable inter-departmental cooperation and facility-wide personnel awareness.

39. Have you recently reviewed your housekeeping practices from the standpoint of safety? — 4 — ____

A sound safety program without good housekeeping is impossible. Each is a part of the other, and each should be evaluated in view of its inter-relationship with the other. Many claims adjusters estimate that over half of all building accidents involve housekeeping in one way or another.

	MAX. POINTS	YOUR SCORE
40. Are custodial facilities adequate?	5	___

Custodial closets and cabinets, when properly located and equipped, can increase efficiency significantly by cutting the travel time from work station to work area. Note that the custodial cart falls in this category, as it is a mobile work station.

41. Is the housekeeping department consulted when plans are made for new construction?	5	___

The efficiency of cleaning a new facility can be materially affected during the planning stage, generally without adversely affecting the beauty of the structure or its cost. Involved are such things as selection of floor type and color, type of paint, using wall-mounted fixtures, location and size of custodial facilities.

42. Do you have a good materials-distribution system?	2	___

A good system of supplying housekeeping workers with their materials and equipment, on a periodic basis, avoids the need for workers to make repeated trips to the supply room, which also avoids the excuse to be absent from their work area.

43. Are adequate records maintained to insure proper protection of surfaces at proper intervals (such as washing walls)?	3	___

Records of the completion date for special projects, as well as setting up target dates for future projects, are indispensable to a sound preventive maintenance program.

44. Does your organization have a continuing awareness and inter-departmental cooperation program related to housekeeping?	3	___

An improved attitude on the part of all employees concerning personal housekeeping responsibilities can materially decrease the work load on custodians, particularly in policing functions, permitting more time to be used in basic cleaning. This requires a continuing, rather than a one-time, campaign.

	MAX. POINTS	YOUR SCORE
45. Do you require your suppliers to provide demonstrations and other services?	2	___
Most suppliers have services and literature, and provide demonstrations on request to their larger customers or potential accounts.		
46. Are records maintained as to the nature and source of complaints concerning housekeeping?	2	___
The periodic review of such a complaint record can turn up repetitive problems that may indicate corrective action with personnel, materials, management liaison.		
47. Is housekeeping equipment cleaned after each day of use?	2	___
Proper care of equipment, leading to an extension of its life and effectiveness, is a matter of habit. The habit must be formed through guidance by supervision.		
48. Do you regularly read current literature concerning housekeeping?	3	___
Books and magazine articles are regularly appearing in this field. It has been estimated that available information is doubling every ten years—it is easy to fall behind!		
49. Do you have a suitable central supply area for the housekeeping department?	3	___
A good central supply area, conveniently located, provides reserve supplies of materials and expendable equipment, has facilities for minor repairs, lubrication, storage of equipment, and provides for such functions as diluting concentrated chemicals, treating dust mops, etc.		
50. Are regular reports made to management on housekeeping progress?	3	___
Management should be kept informed on budget, personnel, complaints and their rectification, suggestions, overtime, turnover.		

		MAX. POINTS	YOUR SCORE
51.	Are you in contact with other responsible persons in the housekeeping field?	5	___
	Correspondence with persons in similar positions, and attendance at seminars and meetings, help to keep one's technical and managerial abilities sharpened and objective.		
52.	Have you developed specifications for housekeeping materials and equipment?	5	___
	For the larger operation, specifications can be a means to improving quality while significantly cutting costs, particularly where there is not a requirement that the lowest price always be accepted. Actual performance under normal use conditions must be the governing factor.		
53.	Have you posted an organization chart for the housekeeping department?	3	___
	The organization chart can help to clarify lines of authority and responsibility, while helping to build a departmental spirit. Photographs help.		
54.	Has noise control been considered when selecting housekeeping equipment?	2	___
	Often, with very little additional cost, equipment can be selected which reduces much of the noise that is common to many housekeeping duties. Noise-controlled equipment includes such items as mopping outfits, floor machines, vacuums, waste containers, carts.		
55.	Do you have a specific program for improving or sustaining morale in the housekeeping department?	8	___
	Custodial morale and status is one of the fundamental problems in the housekeeping department. Positive steps should be taken to help relieve this problem, on a continuing basis.		
56.	Is good liaison maintained between the housekeeping department and the executive charged with fire prevention?	4	___

		MAX. POINTS	YOUR SCORE
	Many a building has been brought to the ground in ashes because of poor housekeeping. The housekeeping department must play an active role in the organization's fire prevention practices.		
57.	Do you have a housekeeping committee?	2	___
	Such a committee, which may meet quarterly to monthly, provides improved inter-departmental liaison.		
58.	If a labor contract provides "bumping" into the housekeeping department because of seniority, is the worker trained in housekeeping before beginning work?	5	___
	Many organizations mistakenly assume that anyone can perform housekeeping functions, despite the fact that rather sophisticated materials, equipment, and techniques are now being used in cleaning. New persons coming into the housekeeping department must be adequately trained.		
59.	Are the operations of the housekeeping department considered a part of your organization's overall public relations activity?	4	___
	Housekeeping personnel not only personally come in contact with visitors, suppliers, and customers, but their activities have an effect on others' opinion of the organization.		
60.	Is the housekeeping department manager able to communicate with higher levels of management easily?	6	___
	Open lines of communication assure early correction of problems or potential problems, improve morale, and often replace corrective action with preventive action.		
61.	Is your housekeeping primarily performed by a centralized housekeeping department?	6	___
	Where cleaning is done by scattered groups of persons reporting to various department heads, much money and effectiveness are lost through the inability		

		MAX. POINTS	YOUR SCORE
	to standardize equipment and materials, methods, etc. Supervision is often ineffective because of conflict with other responsibilities, and the workers themselves are often indifferent toward housekeeping as it represents a secondary duty.		
62.	Are inactive areas sealed off when not in use?	3	____
	Sealing off inactive areas, and giving them attention only from a policing and fire prevention standpoint, relieves the pressure of the custodial work load, so that more time may be spent elsewhere.		
63.	Have you considered the use of a housekeeping consultant?	5	____
	A qualified consultant is able to bring objectivity and wide, specialized experience to the housekeeping field, which is now changing quite rapidly. A consultant loosens lines of communication and focuses attention on cost and quality opportunities. The consultant should be able to assist in training and implementation.		
64.	Have working hours been carefully selected?	5	____
	Housekeeping work can be much more efficiently performed when done during other than normal business hours (except for policing functions); this also gives the benefits of reducing potential accident hazards.		
65.	Are you confident that the housekeeping department contains the correct number of workers?	12	____
	The optimum staff should contain that number of persons where an increase of one would not be economically justified by the amount of savings from all sources, or a decrease of one would cause a greater loss than the cost of maintaining him on the staff.		
66.	Are sensible job titles in use?	2	____
	Titles should indicate status, responsibility, and should have a positive connotation. Example: Environmental Service Aide or "Custodian" rather than "Janitor" or "Laborer."		

		MAX. POINTS	YOUR SCORE
67.	Have unhardened concrete floors been sealed? Sealed floors provide a surface which permits easier, more effective cleaning, protects the floor from wear, and avoids odors and stains because the porous structure is closed.	2	___
68.	Are multi-purpose chemicals in use? Chemicals are available, for example, that clean, deodorize, and disinfect in one operation; or that combine an insecticide with a floor wax. Multipurpose chemicals eliminate separate duties.	3	___
69.	Are you trapping soil at strategic points? Matting, carpeting, runners, shoe cleaners, mechanical mats and other devices keep soil from being distributed throughout the facility.	4	___
70.	Are adequate litter-control devices provided? An insufficient number of urns and receptacles is an invitation to use the floor as a depository for waste.	4	
71.	Have you considered computerizing your cleaning operations? The computer is helpful in budgeting, changing frequencies and times, and related items.	5	___
72.	Is work assigned on an area complete basis? Each worker should perform repetitive cleaning in a given geographical area (pinning down responsibility), supplemented by projects and policing personnel.	8	___
73.	Are area assignments described by marked floor plans? Book-size floor diagrams can be marked to show area borders, as well as cleaning frequencies.	4	___
74.	Are you using the benefits of participatory management? Workers are more responsive and supportive if they help to select floor machines, uniforms, hand soap, and the like.	6	___

	MAX. POINTS	YOUR SCORE
75. Is each supervisor provided with a relief and projects team?	12	___

Without necessarily increasing staff, but by reassignment, a relief and projects team can fill in for absences; and when not required for that, perform project-type work assuring customer satisfaction through completion of all repetitive work.

	TOTAL	350
Percentage score = your total score ÷ 3.5 =	%	___

SUPERVISION

11

Practical Techniques for Supervising the Cleaning Work Force

The housekeeping supervisor's job is to transform the organization's housekeeping and sanitation objectives into actual physical conditions. Many pressures are exerted on the housekeeping department. For example:

- Other departments want safe and clean working surroundings.
- Administrative personnel want attractive surfaces and wish to be free from interference or interruption.
- Management is concerned with overall cost control.
- The custodians want more status and cooperation.
- Management wants to satisfy as many of the housekeeping objectives as possible but is concerned about the time required to do this.

In order to cope with these pressures and to fulfill its objectives, the housekeeping department must be properly supervised. Providing sound supervision, both in quality and quantity, is one of the best dollar investments that can be made in a housekeeping department. Particularly in the field of housekeeping and sanitation, the intensity and quality of supervision has an almost direct relationship to the results obtained.

The problems in morale, status, relative pay, education, and training make housekeeping the trade requiring the most intense supervision.

The basic objective of housekeeping supervision is to maintain high productivity from each custodian, within the framework of a well-defined program. In accomplishing this, three subordinate functions are performed:

1. It has been said that 95% of workers depend on the other 5% for their planning. Although individual workers can sometimes bring specific improvements to their work, the basic planning and organizing must be done by others.

2. The supervisor is the custodian's most direct representative of management. Thus, the employee identifies and communicates with management largely through the supervisor.
3. Supervision provides control. People do not work together automatically—they need the cohesive force of leadership and purpose.

HOW TO BE A SUPERIOR SUPERVISOR: 17 TRAITS TO STRIVE FOR

Let us look at some of the abilities, characteristics, and qualities of an effective supervisor.

1. The supervisor specializes in obtaining the most efficient effort from others, rather than in providing it himself. Although the hardest worker is able to lead by example he or she may lack the ability to use motivation and control as well as required leadership. There is a certain value in the concept of a good supervisor as an inherently "lazy" person, but the laziness must be expressed as an untiring effort to avoid all *unnecessary* and *inefficient* work.

2. The supervisor is observant and perceptive. Often supervisors (and workers and management as well) have become inured to slovenliness and soil through long and intimate association. The effective supervisor can react to poor housekeeping in only one way: he or she determines to improve the condition as quickly as possible. As simple as this may sound, good eyesight, or properly corrected eyesight, is required for proper observation of housekeeping conditions.

3. The supervisor considers every occurrence involving personnel, no matter how bad it may appear at first sight, as an opportunity for a possible improvement. Every termination and every prolonged illness of a custodian should cause the supervisor to consider what changes this might permit which will benefit the housekeeping operation.

After any controversial or difficult contact with personnel, ask: "What have I learned from this experience?" This deliberate analysis leads to a storehouse of valuable human relations material from even the most apparently commonplace occurrences. Such a fund of *studied* experience is invaluable in everyday human relations.

4. The supervisor is persistent and determined. The supervisor should be prepared, with help, to find a way or make one to accomplish the task at hand. He or she is often working with people who do not consider their work important, in an atmosphere where the "janitor" is mocked, and some or all levels of management may be indifferent.

5. The supervisor must have the ability to think. This requires concentration on important matters and the self-discipline necessary to restrain from becoming overinterested in matters relatively unimportant or trivial.

6. The supervisor must be understanding and tolerant of nationality, race, sex, age, language difficulties, and limited abilities.

The supervisor must be flexible and not so steeped in outmoded attitudes and methods that he or she cannot comprehend and work with new ideas and methods. He or she must not only do a great deal of learning, but also an equal amount of unlearning. One of the limiting factors of the housekeeping program is the ability of the leader of the housekeeping department to work with new tools and concepts.

7. The supervisor is self-confident. This is not the same as pomp or self-esteem based on power or authority.

8. The supervisor is well organized. The simple checklist is one of the easiest and most effective tools available for this purpose. The supervisor should also make use of a pocket-memory aid. Several organizations publish and distribute these monthly. They contain calendars, memo space, suggestions for better supervision, etc. No supervisor should leave the office without a pencil and note pad, or memory-aid booklet, because he or she must be prepared to write down observations and ideas as they occur.

9. The supervisor does not fear manual work or soiling his or her hands or clothing. Custodians involved in soil removal need to be shown that there is nothing distasteful or derogatory about such work in the mind of their supervisor.

10. A good supervisor does not fear criticism.

11. Housekeeping leadership requires a certain amount of scheduling and paper work, which the supervisor must be prepared to handle, and handle well.

12. The supervisor is able to translate thought and planning, based on past information and imagination, into action. This requires not only the initial action but also a determination to "stick with it."

13. The supervisor inspires confidence and cooperation in others through integrity and honesty, humor, and example.

14. The good supervisor can communicate. He or she can speak to personnel in groups to conduct training classes and for the informal exchanges of suggestions and ideas. The supervisor is able to put ideas and instructions into writing that is easily understood.

15. If the supervisor does not want custodians to be clock watchers, the supervisor must not be one. He or she must be prepared to vary work hours according to the given situation when required by emergency, absenteeism, special requests, or when a special training session is needed.

16. The supervisor supports his personnel. He or she does not stand by and permit supervision and instructions to come from outside the housekeeping department.

17. A good supervisor knows the names of his or her workers. No supervisor should ever be guilty of not knowing the names of the persons working directly for him or her. Some supervisors go further by learning some of the personal details concerning their workers, such as nicknames, hobbies and family details. Naturally, the use of this information should be applied carefully and without familiarity. Along the same lines, sending a personal birthday note to each employee can help weld a housekeeping organization into an effective team.

THE FOUR MAIN LEVELS OF SUPERVISION AND HOW THEY WORK

There are four basic levels of supervision in housekeeping operations: (1) housekeeping department manager, (2) assistant managers, (3) supervisors, and (4) group leaders.

The *housekeeping department manager* is in charge of the overall sanitation and cleaning function. Alternate titles can include general supervisor, general foreperson, service manager, etc. The manager should provide the creative effort needed to improve the level of productivity of the custodian. In many ways the housekeeping department can be compared to a business within a business, in which case the manager can be considered its "president."

The *assistant housekeeping department manager* (or other applicable title) can be considered the vice-president of the housekeeping business. This level of supervision is only required in larger organizations where the number of supervisors required is beyond the capacity of one individual to properly supervise and motivate them. A typical example would be a large organization having the department manager on the day shift, with a few supervisors reporting directly to him, while having one or more assistant managers on the later shifts, each with several supervisors subordinate to him.

The *housekeeping supervisor* is directly in charge of the work being performed. This is the basic level of salaried supervision. The extent of responsibility should always be defined as a specific number of personnel, and usually within a given work area, such as a certain number of buildings, or a subdivision of a single large building.

On the day shift where he is supervising policing operations, one assistant supervisor may be in charge of the entire policing operation in all buildings. The effectiveness of the supervisor will be almost directly proportional to the amount of time spent in contact with the custodians.

The *group leader* is a working lead person. Where group leaders actually perform a fair share of the work and lead primarily by example, the advantages that they can provide are considerable. For example:

- If group leaders have accumulated considerable seniority, "bumping" within the housekeeping department can be better tolerated because the group leader is able to train workers at the work scene and "ride herd" on them.
- Working group leaders bring an on-the-spot practical direction to housekeeping work which is difficult to obtain otherwise, particularly where the work is spread over a considerable area. Thus, it is through these persons that some of the problems normally associated with training, limited abilities, motivation, etc., can be alleviated.
- The pyramid of supervision, and especially the group-leader level, provides a method and incentive for advancement for ambitious and deserving workers.
- Some of the advantages of team cleaning can only be realized when the team consists of a leader and several custodians.
- Special projects can be efficiently performed in such a team led by a group leader. It is often desirable to assign the most expensive or complex mechanical equipment used by that team to the group leader for his or her personal care and use.

Just as relief is required for custodians who may be absent for any number of reasons, so must relief be provided for supervisory personnel. If there are more than a very few supervisors, they will be unable to "cover" for each other, and specific relief may have to be provided for this purpose. Where a number of group leaders are used, it is possible for them to handle the situation in the absence of an assistant foreman for a short time.

Just as training of the custodians becomes an important function, which is discussed further, so must subordinate supervisory personnel be trained. Training and general discussion sessions can be combined at times. One supervisor can train the others in some aspect of housekeeping with which he has become particularly proficient.

Although there is a great amount of information available to industrial supervisors in general, actually there is a dearth of material directed to the *cleaning* supervisor. The specific material available to production supervisors, for example, is of only limited use and interest to the housekeeping supervisor—there must be material made available and directed to him.

Supervisors benefit materially from seminars and clinics conducted by professional service engineers and management personnel, either on the premises or at another location. The supervisor should join some national organization in order to receive periodic literature and information concerning methods and techniques. In addition, from a general standpoint, there is a great wealth of literature on this subject, both books and periodicals, which are of real value. Membership in a professional cleaning society is desirable.

DETERMINING THE NUMBER OF SUPERVISORS NEEDED

Determining the make-up and size of the supervisory staff of the housekeeping department depends upon many factors, including the abilities of the workers, the quality of the supervisors, morale levels, interdepartmental cooperation, physical distribution, etc.

There must be one person in whom the responsibility and authority for the whole housekeeping function is ultimately vested. In everything but the smallest operations, housekeeping should be this individual's primary responsibility. In larger operations it should be his sole responsibility and, as the organization grows in size, he will need supervisory assistance.

The ratio of supervisors to custodians varies considerably. Even a range of one salaried supervisor for each ten to twenty workers would not cover every legitimate situation. The majority of cases should fall in the range of one supervisor for each twelve to sixteen custodians. On the wage-earning level, the range is one group leader to each two to six custodians, occasionally a little more. A relatively high supervisor-custodian ratio indicates management's understanding of the scope and depth of the problems that should be handled at the supervisory level. The case where a housekeeping department is staffed with too many *qualified* supervisors would constitute an extreme rarity.

You can determine the number of supervisors by taking the number of minutes per day a supervisor has available to be in direct contact with the worker, and dividing by 15. Effective supervision requires at least 15 minutes per worker per day, on the average (more for newer and difficult workers, less for others). This time should be broken into at least two periods.

The time available to the supervisor for this face-to-face contact with the workers is surprisingly limited. After deducting travel time, paperwork, coffee breaks, planning, meetings, personal time, etc., from the day's time, the typical supervisor has only three hours left!

Three hours divided by fifteen minutes equals twelve. Fewer than twelve workers can be supervised if the supervisor has less than three hours, perhaps because of much travel time around a multi-building facility. More workers can be handled if a clerk can handle the paperwork, if group leaders are used effectively, if a vehicle is available, etc.

HOW TO DISCIPLINE

Housekeeping personnel, like other personnel, should know exactly where they stand with respect to breaking company rules. Some companies believe the punishments for breaking company rules should be posted in a conspicuous place, such as a locker room or near the time clock. A typical company might use the following rules.

The following offenses are usually met with immediate discharge:

- Commission of an important crime on or off the premises.
- Deliberate damage to company property.
- Falsification of documents, such as reports or applications.
- Insubordination to a superior or abuse of authority to an inferior.
- Unauthorized or illegal strike action (counsel should be sought on all labor relations problems).
- Subversion.
- Theft or pilferage.
- Time-card falsification.
- Working for competitors.

The following offenses may be met on the first offense with a lay-off of a suitable period of time without pay, and on the second offense by a discharge:

- Brandishing a weapon.
- Fighting on the premises.
- Gambling on the premises.
- Sleeping while on duty.

The following offenses are usually met first with a warning, a lay-off on second offense, and discharge on third offense:

- Drinking on the premises.
- Garnishment. (In this case, the discharge is sometimes preceded by two warnings rather than a warning and a lay-off.)
- Horseplay.
- Peddling of unauthorized items.
- Flagrant safety violations.
- Smoking in restricted areas.

In the following cases, the first two offenses may be met with warnings, followed by lay-off on the third offense, and discharge on the fourth:

- Unexcused absence from work, or a pattern of absence.
- Carelessness or indifference.
- Chronic absenteeism or tardiness.

It is up to the housekeeping supervisor to maintain discipline within the department. This must be done within the framework of company policy, but this policy alone will not provide all the answers.

Exercising discipline protects the department, and the whole organization as well, from personal abuse. Demonstrating to the workers that each person must conform to given principles and rules and carry his full share of the load insures good discipline. For discipline to function effectively, the workers must clearly understand precisely what is expected of them.

Discipline does not mean trying to keep everybody gloriously happy—nor does it mean brow-beating or threatening the use of extreme measures. Discipline must be impartial. It must also be *predictable,* with no playing favorites. It should not relate to age, length of service, membership in a minority group, friendliness, former association; it must be *fair.*

HOW TO SELECT A SUPERVISOR

When it becomes necessary to select a supervisor for the housekeeping department, because of an expanded operation, retirement, or for other reasons, it must be done very carefully. Just as the best machine operator does not necessarily make the best shop foreman, neither should a supervisor be selected by choosing the best custodian.

One of the following methods can be used for filling the top housekeeping department positions.

- Promote someone from within the department, or someone from another part of the organization, to this position. The person promoted from within has the advantage of knowing the work, the rules, and the facility.
- Hire an older, more experienced person. Here you have the disadvantage of cost if the person is qualified, of attitude if he has done other types of work which he might consider to carry higher prestige, and the problem of "unlearning" systems which may have been satisfactory in other locations and times, but perhaps which should not now be used. Many organizations have achieved excellent results in utilizing retired military personnel for this position.
- Hire a young college graduate or a person having two or three years of college training. A younger person may be attracted to an organization because of its reputation and can be convinced of the importance of housekeeping. The job can become an opportunity for growth and expression. The position may well become a training ground for advancement into other positions within the organization. The supervisor will be receiving excellent experience in dealing with subordinates and superiors, in methodology, and in liaison with most of the other departments.

Custodians need many things in order to give their best efforts. First, though, they need and *want* good supervision. We owe it to ourselves to give it to them.

USING EMPATHY EFFECTIVELY AS A SUPERVISOR

The biggest word in the fields of personnel relations and supervision is *empathy*. It is mentally putting oneself into the other person's shoes. When a specific problem arises in dealing with personnel, attempting to visualize the other person's feelings will often give a clue as to the proper steps to be taken. This approach is particularly valuable when the employee is trying to explain a problem or grievance. At times, the reverse can be used as well. The subordinate may be asked to try to visualize the problems and situations with which the supervisor is faced. Asking the subordinate to sit at the desk of the supervisor brings interesting results.

Empathy enables the supervisor to be a good listener. Under such circumstances, a problem loses its theoretical nature and takes on the aspect of a personal problem of considerable significance. The effective supervisor is much more a good listener than a good talker.

Consider these principles of supervision:

- Always discuss an individual's approach to work in relationship to the situation, rather than in vague general terms.
- Criticize only when clearly deserved, and always privately and constructively.
- Make commendations only when clearly deserved, but make them publicly. Probably the five most effective words in the whole field of human relations are "I am proud of you."
- Show your respect for the individual, and your belief that he can do a better job than perhaps he himself feels. Place your trust in him to do a better job.
- Most people experience confusion and anxiety in a changing situation. Be certain to give all the facts as early and as clearly as possible.
- Don't lose your temper. A display of temper is a certain signal that something irrecoverable has been lost.
- It is a compliment to an individual to ask for his or her help. Many an unusual effort has been made in response to this appeal.

12

How to Select and Evaluate Housekeeping Workers

The success of your housekeeping program will depend on the ability of your custodians to work efficiently and effectively. This chapter offers practical guidelines for selecting custodians and evaluating their performance.

GUIDELINES FOR SELECTING A CUSTODIAN

The housekeeping department can never achieve its most effective performance unless applicants are carefully screened before hiring and only competent workers are employed. The housekeeping department will be judged by some on the basis of its most limited worker. And so may the custodians themselves judge the importance of their work to the company, to themselves, and to the public.

Review with your personnel department your company's practices regarding recruiting, interviewing, hiring, and indoctrination. This will ensure that workers are properly prepared to play their part in a dynamic and progressive housekeeping program. Unfortunately, some housekeeping departments have become hosts to incompetent people who have sifted down through other departments. These people may be mentally limited, suffering from ill health, or may have serious physical limitations. For example, the housekeeping department seems to become the natural resting place for the compensation case arising from an accident in a production department. Such a procedure should be seriously reconsidered because of its catastrophic effect on the morale of the housekeeping department. Injury cases should be retained in the production departments, where work suitable to their limitations are sometimes even more readily found than in the housekeeping department. This would further provide an incentive to reduce accidents.

Five Tips for Hiring an Employee

The method of recruiting and hiring employees for the housekeeping department, or of transferring personnel into this department from others, is generally quite clearly defined in most facilities. Where rules permit, follow these 5 points:

1. Give prospective employees a thorough physical examination to help prevent the hiring of disease carriers. If this cannot be provided for all housekeeping workers, it should at least be given to custodians who will work in rest rooms or medical and dietary facilities.
2. Hire an employee who generally fits the predetermined qualifications. Emergencies do not justify the hiring of an incompetent worker.
3. Do not hire part-time workers or full-time workers who also have other jobs, just because they can be obtained at a low pay rate. The attitude of such persons is usually poor, and they are usually fatigued even before beginning work. The additional amount of pay required to obtain personnel without other jobs will be economical in the long run.
4. Do not allow hiring decisions to be based on a supervisor's prejudice. Prejudices on the part of supervisors can have a serious effect on the quality of personnel, and therefore on the results obtained from the housekeeping effort. If housekeeping department supervisors select personnel, prejudice will result in a diminished field of acceptable applicants. Where the selection is made by another department, prejudice in the housekeeping department will increase turnover and, through lower morale, decrease the effectiveness of remaining workers. In a large housekeeping department, the total absence of workers from a local minority group should be investigated carefully.
5. When preparing labor agreements, encourage both the union and management to consider the effect on all workers of utilizing the housekeeping department as a bumping station for all the other departments. Make sure the housekeeping department has its own seniority "family tree." At the very least, the agreement should protect the group leader position in the housekeeping department by insistence on its being a *working* job, by permitting only persons well trained in housekeeping techniques to occupy such a position, and by preventing their displacement by bumping.

How to Identify Qualified Workers on the Job

Large organizations should use a worker-qualifications system to identify qualified custodial workers. Such a system combines the approaches of training, identification, employee awareness, and may also include a form of rating for compensation. The system identifies the custodian as a trained, qualified specialist:

1. Hire employees as "housekeeping trainees."

2. After six months of service, give the employee the opportunity of taking a written (or oral) examination and practical performance test. If he passes this test, the employee will receive the rating of "Qualified Sanitation Specialist, Third Grade." He should be provided a certificate of achievement (suitable for framing) and a distinctive shoulder patch bearing a single red star.

3. Make advancement into the grades of "Qualified Sanitation Specialist, Second Grade" (two stars) and "Qualified Specialist, First Grade" (three stars) contingent upon the satisfactory completion of tests at one-year intervals. Provide certificates and shoulder patches on each of these occasions.

4. If the employee fails the test, he may repeat it after a period of six months, during which time he will undergo more training.

5. Allow personnel already assigned as group leaders to take these tests without the need for the intermediate waiting periods. All new group leaders must have passed all three tests.

6. Give foremen or supervisors shoulder patches for greater identification with their department; they should be permitted to do so without the need for qualification tests.

7. Tie the program in with wage incentives, if this is not contrary to the organization's operating policy. Each qualification could carry with it a set wage.

The program may be set up as either mandatory or optional.

Welcome each new employee of the housekeeping department in person, as well as with a personal letter, from the housekeeping department manager. Such a letter can accompany the copies of the portions of the program new personnel receive when they begin their training.

How to Use Job Rating Sheets with Job Descriptions

Set up job rating sheets for housekeeping workers. Normally it is not necessary to prepare a separate specification for each worker. A breakdown such as the following will usually suffice: (1) group leader, (2) custodian, (3) window, overhead, or other hazardous type area cleaner.

The specification must consider the local labor market and the specific needs of the organization.

The typical job-rating specification sheet will cover a series of a dozen or so characteristics which may be described either verbally, or by using numbers or letters in a standard code. A typical list of such factors would be:

- Education
- Experience
- Mental ability or aptitude test ranges
- Manual skill or manual dexterity test ranges
- Initiative and ingenuity
- Physical requirements
- Materials and equipment to be used
- Difficulty of the work
- Hazards of the work
- Comfort of the work
- Responsibility for damage or loss
- Responsibility for wastage
- Responsibility for safety

Be careful not to set the standards too high, even in a very good labor market, because of the problem of boredom.

Always prepare job descriptions with the housekeeping duty as a full-time function. Only in the smallest industries should custodians be permitted to perform other functions, such as guard service. Where there is a split in duties, the worker normally lacks interest and fails to develop a proprietary attitude concerning either duty. In addition, the make-ready and put-away times (which are, of course, nonproductive) are increased considerably when custodians turn from performing one type of task to another.

The job descriptions should be based on a personnel assignment that assigns specific jobs to the individuals so that responsibilities do not overlap. This provides a system whereby any deficiencies may be traced to the individual, and whereby workers may take pride in the results of their own personal efforts.

HOW TO DETERMINE A FAIR WAGE RATE

Whenever personnel are surveyed concerning those aspects of their work which they consider to be of the greatest importance, it is a surprise to some that wages are not at the top of the list. This is apparently because the typical worker is generally satisfied with the wages he receives, but not as relatively satisfied with other phases of the work, such as job security or supervision.

Wage rates for custodial workers are normally prepared for the housekeeping department and stated as figures which may not be exceeded for starting rates, raise increments, or maximum wages.

The group or individual responsible for determining these wage rates should consider such factors as:

- Labor contracts
- The current minimum wage
- Quality and intensity of supervision
- Cost of living in the given area at that time
- Fringe benefits offered or available
- Job requirements
- Hours
- The local labor market
- Financial condition of the organization
- Wage structure within the organization
- Wages for similar jobs offered in the same type of institution or industry
- Turnover history

The housekeeping department manager can only make limited recommendations within the above framework.

Many organizations, treating housekeeping as a necessary evil rather than as a vehicle to further its fundamental objectives, make it a practice to hire housekeeping personnel at the least figure that will bring enough applicants to fill the open positions. This can create problems of real significance.

The survey of wages being offered locally or in the same general region by similar organizations can be useful. Care should be exercised to interpret the responses on the basis of results achieved, the personnel turnover rate, etc.

Wage statistics prepared by local and federal organizations are also useful. Remember that the statistics indicate averages only for those organizations which contributed to the study. Their chief value is that they indicate the wage which would be paid by one of those organizations, and against which another organization may have to compete for a qualified worker.

DEVELOPING AN EMPLOYEE RATING SYSTEM

An employee-rating report is similar to the job description, as both are concerned with various job factors. The basic difference is that the job description is primarily used as a selection device, whereas the rating report is used to grade worker performance.

To be meaningful, any kind of grading must be numerical in order that comparisons may be drawn and rates of improvement noted.

The proper steps to set up a rating form are:

1. Devise a series of characteristics applicable to all workers in all positions.
2. Prepare a series of descriptions indicating the various degrees of excellence of each function.
3. Assign numerical values to each description.
4. Sum values to obtain a total grade.
5. Relate total grades and total grade ranges to overall worker performance.

Experience shows that the *whole* story cannot be shown in this way, and the numerical rating must be supplemented by notations and remarks of a general nature. Such a list of general comments might concern answers to such questions as these:

- Is the employee performing the task best suited to his or her ability? If not, what kind of work should the employee be doing?
- What are the employee's most desirable traits?
- What does the employee need to improve?
- To your knowledge, what is the employee doing to improve?
- What is the employee's attitude and record concerning safety?
- Any other comments?

A refinement of the above procedure is to weigh each factor depending on its relative total importance, rather than assuming that each is equally important. Thus, the top grade in one factor may be five points whereas the top grade in another factor may be twelve points, for example.

Through the rating report, each employee's ability and fitness in his or her present occupation, or consideration for promotion, may be appraised with a reasonable degree of accuracy and uniformity. The rating requires the appraisal of an employee in terms of *actual performance.* It is essential, therefore, that snap judgment be replaced by careful analysis. To make ratings carefully, follow these guidelines:

- Use independent judgment.
- Disregard general impressions of the employee and concentrate on one factor at a time.

- Study carefully the definitions given for each factor and the specifications for each degree.
- Remember instances that are typical of the employee's work and way of acting. Do not be influenced by unusual cases which are not typical.
- Make the rating with the utmost care and thought and be sure that it represents a fair and square opinion. Do not allow personal feelings to govern a rating.
- Avoid the subject of attitudes.

Generally speaking, a rating report is prepared by a direct supervisor and reviewed by his superior.

For the ratings to be of the greatest use, they should be as accurate as possible. A regular system of preparing notations concerning individual workers will improve accuracy, as these can be referred to as the rating report is made. Such information can come from supervisors' daily notes, commendations, complaints, suggestions, etc.

For an employee evaluation system to be workable, it must be simple. Figure 12-1 provides two different examples of grading systems: sample A is difficult and confusing, with eight different levels of performance; sample B is a meaningful and understandable grading system with only four levels.

TYPICAL RULES FOR CUSTODIANS TO FOLLOW

Before you can evaluate the performance of your custodial workers, you must ensure that they know the ground rules of your organization. The following list provides general rules used in many companies:

1. Custodians must be regular in attendance and must not remain away from work without permission from their supervisor unless illness or emergency makes it impossible for them to report to work.
2. Custodians must be ready to begin work promptly at starting time and must not leave their work area until quitting time.
3. Custodians should report any suspicious stranger to their supervisor.
4. Custodians are responsible for their tools and equipment and must either return them in clean and proper order to the storage area or turn them over to the next shift worker.
5. Custodians in regular contact with the public must wear the required clothing or uniform while on duty. Their appearance must always be neat.

Figure 12-1. Sample Grading Systems

System A: DIFFICULT AND CONFUSING

Exceptional
Excellent
Goes beyond requirements
Satisfactory
Needs improvement
Often does not meet requirements
Poor
Unsatisfactory

System B: MEANINGFUL AND UNDERSTANDABLE

A-Often goes beyond requirements
B-Generally fulfills requirements
C-Sometimes fulfills requirements
D-Generally does not fulfill requirements

	Proportion of work completed
A-Excellent	over 5
B-Good to very good	4 : 5
C-Fair (must improve)	half
D-Poor (unacceptable)	2 : 5

Absenteeism (such as Monday or Friday absence) must play a big part in the evaluation. **Don't over-evaluate!**

6. Custodians must be courteous to the public, to fellow workers, to callers, and to company officials with whom they come in contact.
7. Custodians must not disturb papers, drawings, or other documents which have been left on tables or desks, nor may they open drawers, cabinets, etc., unless specifically told to do so by the supervisor.
8. Custodians must keep their own lockers clean and share in keeping the locker area clean and neat.
9. Custodians should economize in the use of electricity by turning off lights not needed unless the supervisor instructs otherwise.
10. Smoking is prohibited in restricted areas.
11. If a custodian discovers a fire, he or she is to turn in the alarm immediately.
12. Custodians may be disciplined for repeated failure to pay just debts.
13. Gambling is not permitted in the buildings or on the grounds.
14. The use of liquor or intoxicants is not permitted and will be met with disciplinary action.
15. Custodians must report any accident to the supervisor.
16. Custodians should report to their supervisor any repairs needed for equipment, furniture, light fixtures, wiring, tools, and other items.
17. Each custodian will see that the windows in his or her area are closed when leaving.
18. The custodian must turn in to his or her supervisor any lost articles that are found.
19. Custodians are expected to follow the prescribed safety rules.

13

Guidelines for Developing a Successful Custodial Training Program

Only one in ten housekeeping and custodial workers has had any formal training in cleaning techniques; probably less than one in twenty has had such training on a regular or periodic basis. Yet the turnover rate is relatively high in housekeeping operations, and new methods, materials, and equipment are coming into use more and more rapidly.

Many housekeeping departments today are static organizations. There is considerable experience in outmoded methods; only minimal attention is given to supervisory ability and performance. Training classes may have been attempted, but were probably dropped after a few sessions because they were troublesome and results were not immediately observed. Under such conditions, the training of a new housekeeping worker may have consisted simply of handing him his cleaning tools and putting him on the job with a "buddy" for a few days, after which he is placed on his own. The new employee, in turn, less than a year later, similarly "teaches" another new worker.

Although your housekeeping program should indicate the training program procedure, it will eventually be the job of the supervisor to continue the training function. Even where the program has been fully implemented and is achieving the basic goals set for it, it is still desirable from time to time for training programs to be conducted by outside personnel. This brings a freshness to the approach, excites new interest on the part of the custodians and helps to keep the department up to date. No matter how mechanized the cleaning operation becomes in the years ahead, you will still depend on the individual custodian for the housekeeping work.

The investment in plant and equipment per worker is constantly rising. Entrusting expensive equipment to untrained machinists is an unsound and costly practice. An untrained custodian can also be costly. Finishes for paneling and furniture can be damaged by the use of the wrong cleaning materials. Wasted motion means that some housekeeping jobs are not going to get done properly or on time. Housekeeping tools and equipment are abused and become unnecessarily short lived. Accidents may occur, and unnecessary fire hazards will exist.

This chapter will show you how to achieve a sound training program—one that is carefully conceived and earnestly carried out.

HOW TO ORGANIZE YOUR TRAINING PROGRAM

The training program, to be effective, requires careful planning. These steps are involved:

1. Clearly define the needs and goals of the program. A training program should not be designed to entertain the custodians, but rather to give them the information they need to help them grow in their jobs.
2. Place responsibility for carrying out the program with an individual who, in turn, may delegate responsibility for various phases of the training. Full support must be given to the training leader.
3. Organize the program properly. Select meeting places, obtain training aids, determine training methods, and make arrangements with participants.
4. Set up the training schedule at least a month in advance and for a period of a number of weeks or months. Make sure the dates selected for various meetings don't conflict with other company activities, holidays, vacations, etc. Such a schedule should indicate dates and times, subject matter, participants, and the location.
5. Be sure that the personnel who are actually going to conduct the training classes are qualified to do so. Additional qualification may be gained through reading, or the teachers may undergo certain phases of the training themselves in such techniques as public speaking, psychology, training aids, etc. The trainer must know the subject to hold the attention and respect of the audience.
6. Reevaluate training methods at intervals to make certain the proper subjects are being covered, the custodians are actually learning from the classes, and the best methods are being used.

Training Program Objectives

The objectives of a custodial training program include:

- Safety to custodians and others.
- Protection of health.
- Conservation of capital investment through surface protection.

- Improved morale of all building occupants.
- Efficiency of housekeeping performance.
- Development of the relationship between supervisors and workers.

The best means of obtaining these objectives is with a personalized program developed for the individual organization. Have pictures of workers in your facility using equipment and supplies on the job. (The consultant should advise on additional or different types of equipment and material which would improve efficiency.) A "canned" training program typically shows other types of people doing work using the wrong equipment, and the trainer may spend a great deal of time apologizing for the information which is being given to them that does not relate to their situation.

Six Steps for Implementing the Program

These are the steps to follow in developing a training program:

1. Determine standard procedures to be used for the various cleaning activities (this involves a determination of equipment and chemicals as well).
2. Decide on what type of visual aids you intend to use. Slides provide great flexibility in making changes. They are available at a reasonable cost, and lend themselves very well to supervisory development, as well as to retention of information by the worker.
3. Prepare a script for shooting the pictures based on the standard procedure, and include such items as make-ready, safety precautions, clean-up, etc.
4. Select volunteer workers to act as models (remember to obtain a release form).
5. Take at least twice as many pictures as are needed, so that only the best shots will be retained. (If your organization has its own training department, you can do this in concert with that department.)
6. Key slides to the scripts.

Audiovisual Training

Most custodial workers receive audio-visual training on cleaning activities daily. Millions of dollars are spent annually by some of the nation's largest companies to provide this training, and it is given in a colorful, energetic, and enthusiastic way by some of our nation's best salespeople. I am referring to television commercials! Typically what you see on television, in terms of how to mop a kitchen floor or how to clean a bathroom sink, is quite different from the system of cleaning, as well as the chemicals and equipment involved, that you desire your personnel to use. (Television will show, for example, the push-pull system of mopping, which is very fatiguing and of limited productivity, and the use of scouring powder for restroom facilities and fixtures, which is abrasive and creates problems of its own.) A good example of the forcefulness of this training is the fact that a number of workers will bring products from their home to use on the job at their own expense!

Your training must be good enough and interesting enough not only to portray the proper system to the worker, and to show its benefit, but to overcome the exposure your workers have had to television and other forms of commercial advertising.

THE TRAINING TRIANGLE: A THREE-PART APPROACH TO GREATER PRODUCTIVITY

Three forms of training must be provided—using only one or two will limit your results. This is shown in "The Training Triangle," Figure 13-1.

Orienting the New Employee

First impressions are very meaningful and often never forgotten. This principle can be used to good advantage if the new custodian reporting to work is greeted and oriented properly.

The new custodian should be greeted by his or her direct supervisor with an informal discussion. The supervisor can help to make a new employee feel at ease by asking questions about the custodian's former work experiences, military duty, size of family, birthplace, or things that the two persons have in common.

The company's attitude toward housekeeping, the objectives of the housekeeping department, and the general nature of the housekeeping program are discussed. The importance of the individual custodian's work toward the attaining of overall company objectives is stressed.

A number of organizations provide a general orientation concerning the company, its products, and its place in the community before the worker reports to his or her department. Therefore, the extent of orientation within the housekeeping department will depend on whether or not a general orientation has been provided. In either event, the employee must understand the general operating rules of the company and the specific rules of the housekeeping department.

The new custodian should be properly introduced to the department manager and other supervisors, as well as to his fellow employees. In the large operation, initial introduction should be made only to those custodians

Figure 13-1. The Training Triangle

The triangle provides great stability.

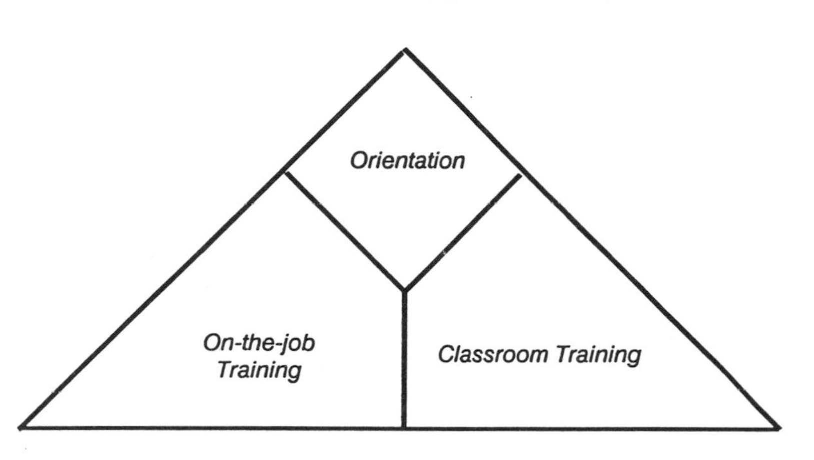

TYPE OF TRAINING	PERFORMED BY	BENEFITS
Orientation	Supervisor or Training Officer	• Positive first impression • Creates a lasting memory • Makes the worker feel important • Avoids some initial errors
On-the-job Training	Group Leader, Supervisor or Training Officer	• Develops the Group Leader • Opportunity for feedback • Creates worker confidence • Develops uniformity of methods
Classroom Training	Supervisor or Training Officer	• Develops the supervisor • Excellent morale builder • Certificates provide tangible recognition • Encourages a group spirit

in one supervisory section so as to avoid confusion. The introduction should include an identification of each employee's duties and how they affect the new custodian. The new employee should be taken on a familiarization tour of the plant. In the very large plant, show the employee a map of the entire facility and give an actual tour of the buildings in which the employee will work. The tour should include:

- Familiarization of the premises to prevent becoming lost and to put aside fears of the unknown.
- Observations of custodians performing their cleaning tasks.
- Pointing out examples of both good and bad housekeeping techniques and results.
- Housekeeping facilities, such as the central storage area and closets, cabinets, location of water taps and drains, etc.
- Personnel facilities such as rest rooms and locker rooms, water fountains, etc.
- First-aid facilities including first-aid cabinets, dispensary, eye-wash fountains, fire blankets, etc.
- Food service and vending areas.
- Fire warning and extinguishing equipment.

The new custodian should be shown his work station, or the route of his work tour. He should be given the program data and other literature affecting his work.

The new custodian is now ready for on-the-job training.

On-the-Job Training

Most workers want to receive specific work instructions. It is only when workers know exactly what is expected of them that they can do their best and feel secure.

The training of the worker in more than one job is often desirable, but the capacities of the individual must be considered. A person hired as a relief worker would naturally have to be trained for all the jobs that he or she might be called on to perform. In order to improve flexibility, employees can be rotated from one job to another at periodic intervals for on-the-job training under group leaders or other supervisors.

Where turnover is high, the supervisors will find themselves spending a great deal of time with on-the-job training if adequate instructions are to be given. This situation can be improved through the use of group leaders or a temporary specialist trainer, if properly qualified.

On-the-job training should not be considered complete after a given time interval but should be continued until the individual has demonstrated suitable understanding and proficiency.

Consider the four basic steps in the teaching process:

1. Preparation. Put the worker at ease. Explain the benefits of the methods to be shown him and explain that you don't expect him to get it exactly right without some practice. Discuss possible hazards of the work, safety practices, and the materials and equipment to be used. Be sure that the equipment is in good condition and that the materials are uncontaminated or unspoiled. Be sure the custodian's working facilities are properly arranged and outfitted. Make keys available, or set up a procedure for unlocking doors that are normally kept locked. Written instruction sheets should be reviewed.

2. Demonstration. Explain the operation to the worker. Proceed one step at a time and only as quickly as you feel that the information is being absorbed. Give an actual demonstration of the work, moving slowly and methodically. If the operation is difficult, first give a demonstration at slow speed, then follow it with a demonstration at normal speed. The "before-and-after" method can be used to demonstrate the results of good housekeeping technique. For example, clean one half of a lobby or rest room and compare it with the half not cleaned. At all stages in the training, give the custodian an opportunity to ask questions. Answer questions carefully and with patience.

3. Application. When he feels he is ready, let the custodian demonstrate the job back to you. He should also be able to tell you *why* the various steps are being taken. Encourage him to look at written instruction sheets if he feels this is necessary. Again, try and keep the custodian at ease by using patience and understanding. Do not stand too close. Compliment and encourage him. If he becomes confused, give him another complete demonstration, or take a break for refreshments. When you are both satisfied, congratulate him on learning this new skill and let him perform the work alone and unassisted.

4. Inspection. Check the work often, giving the custodian plenty of opportunity to ask questions. He will want to know from you whether satisfactory progress is being made. Do not hesitate to tell him what needs correction, but first compliment the part of the work that has been done properly. And remember, training should never involve criticisms. If the worker has not properly learned the job, it is best to blame yourself for incomplete motivation or instruction.

Arranging a Training Class

Training classes pay considerable dividends in improved productivity. The cost of training classes is primarily in the supervisory time necessary to organize and conduct the classes and the time of the workers actually spent away from their work because of the classes. Unless suitable arrangements are made for a training meeting, time will be lost, and retention will be reduced.

Use the following checklist to ensure a complete arrangement:

- Arrange the proper time and date with the individual responsible for coordinating meeting rooms or auditoriums. Provide the coordinator with a complete schedule of future meetings. Check back on each proposed meeting the day before to make certain that all arrangements have been taken care of.
- Check that proper announcements have been made and that all custodians who are to attend have been notified. Invite representatives from other departments who have an interest in cleaning, such as quality-control personnel, maintenance personnel, etc.
- If the meeting place is not a familiar location to all, give simple directions as to its location, or the group should go to that place as a body.
- Prepare physical conditions of the meeting place as well as possible, including comfortable heating or air conditioning, sufficient lighting, sound and noise control, etc.
- Be certain that there will be enough seats for everyone and that they are properly arranged.
- Obtain a podium or lectern for those who will address the audience. Of course, there should be a table, chairs, and a water carafe.
- Serve refreshments if the meeting is to be held to coincide with a break period. This would also be true if a meeting were lengthy, or held after usual working hours. Be sure that all materials and equipment to be used are on hand and in proper working condition.

HOW TO USE TRAINING AIDS EFFECTIVELY

Training aids promote learning because they supplement and relieve verbal instruction and information. Learning is definitely improved through increased information retention, interest, and appeal when the subject matter can be seen, felt, smelled, or heard.

Here are some useful training aids:

- Charts to emphasize key points in a presentation—The best flip charts contain only a few words in large print, are colorful, and contain illustrative cartoons which are sometimes humorous.
- Posters to convey key points or morals—These can remain on view throughout the session.

- Sample products and equipment to be examined.
- Chalk boards or sketch pads for drawings or diagrams.
- Literature, such as program instruction sheets, product labels, trade magazines, manufacturers' literature, etc.—Printed material may also be shown on a screen with a suitable projector. Figure 13-2 is a sample checklist that could be used as a follow-up to a training class on avoiding common mistakes.
- Laboratory-type demonstrations—These are very effective and can include demonstrations of surface tension, chemical hazards, emulsification, and spontaneous combustion.
- Slides and videos—These are the most useful training aids.

Test visual aids before the meeting. Here are some suggestions:

- Locate the projector and screen or video player where they will not obstruct vision and can be seen by the entire audience.
- Designate an individual to control the darkening of the room by closing blinds or pulling shades and turning out lights. Check these out in advance.
- Run a videotape for one or two minutes to make certain that everything is in proper order. This will require suitable electric current and extension cords.
- Let an experienced person handle the actual projection. Supply extra bulbs, fuses, or other items that may fail during the showing.

Slides require a live speaker. In general, this should be the first-line supervisor. Using a script, such as the excerpt shown in Figure 13-3, can give you self-confidence.

Published cleaning guides can also be an adjunct to training. Figure 13-4 is from Yale University's excellent "Pocket Guide for Cleaning Services."

CONDUCTING A SUCCESSFUL TRAINING CLASS

The first training class is extremely important. It should be the vehicle for orienting all cleaning personnel to the housekeeping program and will set the tone and pattern for all classes to follow. Do not let it be "the big one that got away!"

Although subsequent training classes may be broken up into groups of only fifteen to twenty persons, for greater effectiveness the first meeting should contain the entire housekeeping force, if at all possible. This will tend to emphasize the team effort viewpoint and will help to build personal pride in the housekeeping department.

Figure 13-2. Checklist of Common Cleaning Mistakes

Dust Mopping

- Swinging the mop, causing dust
- Tapping or raising the mop, causing dust
- Overtreating, making the floor slippery
- Undertreating; mop too dusty
- Letting mop get too "loaded" (vacuum, wash, or replace)
- Bumping furniture or walls
- Failing to police matting
- Bending over

Wet Mopping

- Overuse of detergent (suds, sticky film)
- Failure to change dirty water (redistributes soil)
- Marking baseboards; not "striping" the baseboard
- Mop not wrung completely (soil transfer, inefficient)
- Handle sticking out; hazardous
- Overuse of water
- Failing to wash out mop
- Uncomfortable work pattern

Spray-Buffing

- Application on a very dirty floor
- Not washing out the pad
- Not cleaning out the sprayer
- Letting the spray dry
- Overloaded pad
- Use as a cure-all
- Improper pad selection

Wax Stripping

- Using too concentrated a solution—"burning" the floor
- Leaving water down too long—loosening, curling tiles
- Letting the solution dry on the floor (redepositing)
- Failure to strip completely
- Failure to rinse after stripping
- Failure to clean baseboards

Figure 13-2. (continued)

Waxing

- Applying coat too thick (soft, slippery, soils)
- Putting more than one coat up to the wall
- Changing waxes without stripping or testing
- Waxing over a dirty floor
- Putting on a second coat before first has dried
- Getting wax on baseboards and furniture
- Rubbing, causing bubbles
- Pouring old wax back in drum (break emulsion)

Buffing

- Letting floor machine hit furniture, walls
- Not dust mopping first—scatters dirt
- Using wrong pad
- Fighting the floor machine

Rest Room Care

- Not cleaning under lip of toilet bowl
- Daily use of acid descaler
- Not dusting high up
- Letting waste receptacle stay dirty
- Reliance on deodorants

Room Care

- Not dusting under furniture
- Not dusting over lamps, doors, cabinets
- Door frame and hinge dusting
- Not cleaning wastebaskets
- No attention to handprints
- Ignoring Overhead

Figure 13-3. Sample Custodial Safety Narration for Slide Presentation

Slide No.	
1	Freshly treated dust mops can be prone to fire, so they should be stored in a sealed, fireproof metal container such as a metal trash can with a tight-fitting lid. Dust mops should not be stored out in the open where they could accidentally be set on fire.
2	Remember at all times while at work to bring your custodial cart with you wherever you go. This is important not only for safety, but for convenience and time savings as well.
3	Any time the cart is left unattended people will be tempted to tamper with the tools or chemicals it contains. This presents a safety hazard because these people often don't even know what these items are actually supposed to do or how they are properly used.
4	Keep the custodial closet clean, neat and well organized. Leaving things on the floor and out of place can result in a fall. Keep the area clear and prevent an accident from happening.
5	Keep the door to the custodial closet locked so that others will not enter without authorization. This may save them from an accident of their own. The equipment and chemicals in the custodial closet should be used only by those trained in the safe and proper methods by which they are to be used.
6	While cleaning offices and work areas, never disturb the papers sitting out on desks or tables. The danger aspect of this mistake may become obvious when the irate occupant of the office sees you reading personal materials.
7	Be very careful when opening sealed boxes. Use a tool designed for the job; never use a razor blade by itself. Cut in a direction leading away from yourself, and hold the box behind the area where you're cutting it, not in front, so if the knife slips it cannot cut you.
8	Use a broom or counter brush and a dust pan to pick up broken glass. Even the largest pieces should be picked up in this way. Never pick up a piece of broken glass by hand, no matter how safe it may seem. A large piece of broken glass may have smaller cracks in it, and it could break even further when you pick it up.
9	Damaged carpet that could present a tripping hazard should be reported to your supervisor immediately so that it can be replaced.

Figure 13-4. Sample Cleaning Guide

Washing Interior Walls

The purpose of washing walls inside buildings is to remove dirt, soil, oily film, tape and smudges. The majority of walls in this University are made of plaster. Please note:

a. We do not wash outside walls. Your Supervisor can arrange for the removal of graffiti by another department.

b. Wooden walls are not washed but are treated with lemon oil.

c. Walls that are scheduled in the near future for painting are not washed.

d. The work schedule of an employee will not include wall washing. This type of cleaning is done by special request from a customer.

e. We do not wash wall surfaces that are beyond reach from the upper step of a step ladder.

f. Washing walls includes interior doors, partitions and window frames and sills.

Safety Suggestions

a. Employees that wash walls must wear rubber gloves to protect their hands.

b. Anyone using a step ladder must be careful and not try to reach too far. The ladder must be moved as often as necessary.

c. Refer to the safety tips listed in the front of this manual for further information.

Operation

a. Assemble all equipment and supplies in the area in which the walls are to be washed. Put on rubber gloves.

From "Pocket Guide for Cleaning Services," by Yale University.

Figure 13-4. (continued)

b. Put a drop cloth along the wall to be washed to protect either the floor or carpet.

c. Fill the pail with warm water and add premeasured cleaner. Refer to your personal copy of the **Reference List of Authorized Products** for the name of the correct product to use.

d. Use a duster with an extension handle to first remove dust prior to washing the wall.

e. Submerge a cellulose sponge into the cleaning solution mixture in the pail and start at the bottom of the wall, working up to avoid running streaking. Wear protective rubber gloves at all times when working with a chemical and water combination.

f. Rinse the washed section of the wall with clear water from a second pail while using a different sponge.

g. Wipe the wall with a clean, dry cloth and proceed onto another section as the work progresses.

h. Change the water and solution mixture frequently as well as the contents of the pail of clean water to make sure that the walls are not streaking.

i. At the end of the wall washing task:

1. Empty the contents of the pails into the nearest utility closet sink while running the water at the same time.

2. Rinse out the sponges thoroughly and place in the storage room if they are still serviceable or throw them away if they cannot be used again.

3. Place the pails in an upside down position in the storage room to dry.

4. Return the ladder and all other equipment and supplies to the storage room and refer to the final chapter in this manual for further information on the storage of equipment and supplies.

Three Key Goals for the Orientation Meeting

The basic purposes of the orientation meeting are to *inform, reassure,* and *create enthusiasm.*

1. Inform. Begin the meeting with an honest description of its purposes. Make whatever introductions are necessary. Point out that housekeeping is much more important to the success of the company than most people realize and that management's realization of this fact has led to a program especially designed for the housekeeping department. Point out examples where planning and organization by qualified specialists, depending heavily on wide experience with persons just like those in the audience, bring many benefits to the housekeeping department, to their fellow workers in other departments, and to management.

2. Reassure. Point out that the training classes particularly, as well as the program in general, have been set up as a benefit to the custodians so that they may work *smarter* rather than *harder.* This can be done through the use of timesaving equipment and materials, by using ways of cleaning that are less tiring and more effective, by arranging the work so that some things are not done unnecessarily or too often, and by receiving cooperation from others. Point out that the program will be carefully paced and that improvements will be made carefully and without friction or confusion. If management has agreed to this point, guarantee that the program will not result in the loss of any jobs for anyone in the audience. Be sure that there is plenty of opportunity for questions on these points.

3. Create Enthusiasm. Show that it is through the custodians themselves and their spirit of interest and cooperation that the housekeeping department will achieve its goals and finally get the recognition and cooperation that it deserves. Describe how the program works, and the fact that the housekeeping department, in working with the help of the program, depends on the custodian for advice and suggestions, since the worker is "on the firing line" and most directly involved with the problems of cleaning.

Wind up the meeting with a short training subject such as "The Importance of Sanitation" or some sample demonstrations of how the most simple jobs might be done easier and more effectively. Finally, ask for questions or suggestions, or if there is any part of what has been said that requires repeating or clarifying.

What to Discuss at the Training Sessions

If custodians are to learn improved methods and later put them into use, you must conduct the training classes in a way that encourages learning. A friendly, informal atmosphere and approach will help to remove tension, suspicion, and anxiety. Before the meeting actually begins, talk to some of the individual custodians about their families or hobbies or compliment them about truly good work or attitude. Begin the meeting with a lighthearted joke that relates to cleaning, holding meetings, or the like. Do not belittle the importance of the meeting, but keep it informal.

By all means, introduce guests from outside the company or from other departments. This is also a good opportunity to introduce new members of the housekeeping department or to make announcements such as forthcoming retirements, holidays, etc.

Training classes should always be opened and coordinated by the housekeeping department manager. The actual presentations may be given by others, but no individual should speak continuously for more than fifteen or twenty minutes. In addition to the housekeeping department manager, talks and demonstrations may be given by:

- Management representatives
- Housekeeping department supervisors
- Custodians with special skills
- Supplier representatives
- Consultants
- Guests from other departments (methods, safety, medical, and industrial relations, etc.)
- Teachers from local universities

Limit training classes to an hour (forty-five minutes without training aids) for these reasons:

- The span of attention of the typical person is limited. This is true particularly when the course departs from tangibles such as equipment or materials and goes into abstractions such as cooperation and morale. Here the ability of the custodian to concentrate weakens.
- In any kind of training, information retention increases when training sessions are shorter and more frequent rather than being longer and more rarely held.
- Morale is improved, as well, with shorter classes held more often.
- It is possible for many of the workers, with a little extra effort, to "catch up" on their work after the loss of an hour. When more time than this is lost, it becomes very difficult to make it up.
- Training classes should be anticipated by the workers. If they are overly long they will tend to be boring and oppressive.

Hold meetings at regular intervals, such as every two weeks in the early stages, and later on each month. Give ample notice; select a time and location that will create the least possible hardship for the attendants.

The number of housekeeping training subjects is impressively large. The more important subjects, such as safety, floor maintenance, and the care of equipment, may require more than one session and may be repeated after a time.

Once a subject has been covered in a training class or demonstration, it cannot be put aside as "completed." The average person retains only a percentage—often as low as 25 percent or less—of what has been discussed on any given occasion. Training subjects will have to be repeated, the frequency and total number of repetitions depending upon the complexity of the subject, whether or not the subject is interesting or dull, the skill of the trainer, the ability of the audience to learn, etc. It has often been said that in order to teach people something you must:

- Tell them what you are going to tell them
- Tell them
- Tell them what you have told them
- Remind them of what you told them!

All classroom training should be followed up by on-the-job training and inspection.

The purchase of training certificates is well worth their small cost. These certificates may be awarded after twelve classes without absence, for example. See Figure 13-5.

All training sessions should provide a period of time at the end for discussion, questions, and final remarks.

Six Benefits of Using Written Tests as Part of the Training Process

The use of sanitation tests for custodians can provide several advantages:

- The tests are a support for the training program, as they measure the retention of information by the trainees.
- The tests can be used as guides in selecting group leaders or other supervisory personnel.
- Tests can be used to help implement bonus or incentive systems.
- They can be utilized to screen prospective applicants or personnel transfers from other departments.
- The tests help to boost custodial morale by creating interest, if they are not used punitively.
- They can be used as the basis for a worker qualification system, as previously described.

Figure 13-5. Sample Achievement Certificate

Certainly there is no question of a company's moral or legal right to use such tests on a basis that is optional for the employees. Where the tests are to be mandatory, such as a requirement to bid for a job in the housekeeping department, it also appears that this is perfectly correct. Figure 13-6 is a sample sanitation test.

SIX QUESTIONS TO ASK BEFORE HIRING A TRAINING CONSULTANT

In selecting a training consultant, ask the following questions:

- Does the consultant's field of specialty relate to your situation?
- Does the consultant have the depth of experience on which you can rely? Experience gives one a practical feeling for what can be done, how quickly, and how well it is received by the workers.
- What is the consultant's reputation? Is he or she known by a number of people who can provide references relating to the quality of the work?
- Does the consultant's personality fit yours? Do you agree in terms of objectives, limitations, and approach?
- Is the fee charged for the service reasonable?
- Is the specialist independent of influence by other organizations?

The above questions, while useful in considering the selection of a consultant, are not the only factors. For example, if the firm to do the work is only a year or so old, it may not last out the duration of your project. Preferably, you should look for an organization that would be at least ten years old, and therefore stable enough and with enough backup to assure you of continuity.

Your selection of a training consultant would be the same as if you were making a complete analysis and recommendation for the building sanitation operation. (You may wish to consider this, since training has an inter-relationship with other functions, such as the method of job assignment, procedures for cleaning, the equipment and materials involved in procedures, etc.) Most consultant work in housekeeping is performed by general management and accountant consulting firms, whose personnel know little about this subject. It is not enough for a training consultant to know how to evaluate an operation; the consultant should know the cleaning business. A training consultant should be able to answer test questions that you might ask, such as: "What benefits do the new high-speed floor machines offer in spray buffing with metal-interlock polymers?" Or, "What are the drawbacks of vertical scheduling in a building using area assignments augmented by project teams?"

Figure 13-6. Sample Sanitation Test

TRUE FALSE TEST

	True	False
1. A waxed floor is always more slippery than an unwaxed floor.	☐	☐
2. Lifting the handle of a floor machine while it is running causes it to move to your right.	☐	☐
3. A wooden floor that is warped and buckled has probably been cleaned with too much water.	☐	☐
4. The proper way to dust mop a corridor is to swing the dust mop from side to side in a figure "8" pattern.	☐	☐
5. Dirty baseboards are often caused by poor handling of mops.	☐	☐
6. The reason for a dust-mop treatment is to make the dust mop slide easier on the floor.	☐	☐
7. The proper way to use sweeping compound is to sprinkle it on the entire area to be swept, then sweep it up one section at a time.	☐	☐
8. The proper way to get dust off an uneven surface is to blow it off, using the exhaust from a vacuum, and then sweep it up.	☐	☐
9. Wax should always be applied from wall to wall.	☐	☐
10. Putting too much water on the floor can cause asphalt tile to curl up at the edges.	☐	☐
11. Quarry tile is probably the easiest type of floor to keep clean.	☐	☐
12. Solvent is a good material to use on asphalt tile floors.	☐	☐
13. The use of the wrong type of cleaner can cause asphalt tile floors to crack.	☐	☐
14. Thin coats of wax do a better job than thick coats.	☐	☐
15. A mirror is a good tool for checking rest-room sanitation.	☐	☐
16. Pouring wax back into its drum can cause the entire drum to spoil.	☐	☐
17. An acid-type cleaner should be used on rest-room fixtures daily.	☐	☐
18. If you wait a few minutes after applying a cleaning solution before beginning scrubbing, you can let the chemical cleaner do part of the work for you.	☐	☐

Figure 13-6. (continued)

	True	False
19. It is not necessary to wait until water-emulsion waxes are completely dry before beginning buffing.	☐	☐
20. A dusty or dirty light fixture gives out less light than a clean one.	☐	☐

In the above tests, only the following numbers are true: 3, 5, 10, 11, 13, 14, 15, 16, 18 and 20.

PRACTICAL PERFORMANCE TEST

Workers must demonstrate the proper care and use of the following:

a. Push broom (storage, reversing the block, combing, proper handling).

b. Dust mop (storage, treating the dust mop, proper method of use).

c. Wet mop and wringer (storage, selection of size, trimming the strands, method of use, and washing the mop and wringer).

EVALUATING HOW EFFECTIVE YOUR TRAINING PROGRAM IS

Do not expect results too quickly with a training program. Many workers have a "wait and see" attitude and may not respond right away until they are truly convinced that the purpose of the program is to help them find easier, safer ways to do the work (which it most certainly should!). Repetition and continuity are of utmost importance, as is professionalism.

It is often impressive to sanitation workers that management has obtained a consultant to work with them. It helps to convince them that they are doing important work. So does the printing of an attractive certificate that accompanies completion of the first phase of the training program. Custodial training should be a continual activity, and should not come to an end after a few months or even a year—this would simply convince some of the workers of what they might have suspected all along—that management was not serious about this.

The eventual evaluation of the results of the program would be made in terms of improvements in these areas:

- The absenteeism rate.
- The turnover rate.
- The quality of worker performance.
- The worker-supervisory relationship.
- Supply costs.

14

How to Motivate Service Workers

Motivating custodial workers is a problem, not so much because of the number of workers, but because of the special difficulties with this type of worker. Custodial work is sometimes perceived as a "necessary evil" not connected directly with the fundamental objective of the organization. The workers often feel forgotten. Custodial workers suffer from poor status and low image due to such factors as humor made at their expense, the soil and dirty equipment connected with their work, the use of people with little or no training, the failure to provide attractive uniforms and the like.

This chapter offers proven tips for motivating your service workers to provide the initiative and effort you need for a successful housekeeping and sanitation program.

Any serious attempt at motivation requires a number of changes from current practice. Be careful not to conflict with various legal requirements, such as those involving the equal pay law, equal opportunity regulations, union and civil service contracts, OSHA regulations, and the like.

THE "SHOTGUN EFFECT" TO MOTIVATION

There are all types of personnel performing service activities: employee characteristics such as age, gender, background, experience, and appearance vary. There are excellent service departments using all types of personnel. The superior service worker is one who has the advantage of an excellent supervisor and management.

Successful managers and supervisors have discovered that a complete motivation program is required—there are no magic one-time answers. A dozen courses must be pursued simultaneously—to attempt one or two activities at a time would require years for benefits to be apparent. This multiple motivation approach is called the "shotgun effect," and it can include the following motivation techniques:

- Establish a "Custodian of the Month" program.
- Provide attractive uniforms (selected by employees).
- Provide appropriate equipment.
- Provide effective classroom training, with certificates.
- Ensure fair and effective supervision.
- Distribute a departmental newsletter.
- Send out employment anniversary and birthday cards.

One problem with the shotgun approach is that it's all over the place (as the name implies): some employees may be awarded the "Custodian of the Month," others may receive classroom training, still others may benefit from effective supervision. But some employees may be overlooked with the shotgun approach.

The largest source of complaint from service workers, and the cause of most grievances, deals with equitable treatment. Thus, the first consideration in motivation would simply be "fairness." Each worker must be treated the same under the same circumstances. For example, an organization that has a published schedule of disciplinary actions is better off than one where the discipline can be meted out according to the supervisor's whim or state of mind. Another example involves sick leave policy: a policy is equitable only when all workers either receive, or are paid for, all sick leave each year. (Most policies benefit the poorer workers and penalize the better workers, since the latter do not use all their sick leave and a portion of it is taken away from them.)

Seven Tips for Motivating Workers

One of the biggest stumbling blocks in establishing an effective housekeeping and sanitation program is the low morale of housekeeping personnel. This generally is the result of a combination of these factors:

- Lack of status.
- Failure to obtain recognition from management, fellow employees, and the general public.
- Little or no chance for advancement within the housekeeping department.
- Lack of incentive and motivation.

Poor morale leads to low efficiency, abuse of materials and equipment, resistance to learning improved methods, and slovenliness. It is extremely difficult to effect meaningful improvements in housekeeping in the face of such attitudes.

Here are seven ways to motivate service workers:

1. Be sure that the first-line supervisor knows the name of each worker and uses it regularly. The names must be used in standard format—use either all first names or all last names—since to mix up first, last, and nicknames indicates a form of favoritism, even if not intended.

2. Titles such as "Porter," "Janitor," "Laborer," or "Mop Boy" are destructive, while titles such as "Service Operator" and "Maintenance Aide" are positive.

3. Administer a monthly recognition and award, which might be entitled "Service Operator of the Month." In a larger organization, each supervisor would nominate a worker, with the department head involving management in making the final decision. Weighted quality points can be used in making the selection, perhaps with perfect attendance being a requirement for nomination. The awards should be publicized (such as in the company newspaper) and a certificate awarded at a meeting.

4. Provide uniforms, which can be a positive motivator if the uniform is attractive and especially when the worker has been involved in selecting the color, style, and material. Colorful embroidered shoulder patches can show the name of the department, embroidered stars or other devices can show length of service and name badges can identify the worker individually.

5. Have the worker meet members of management or administration on being hired and periodically thereafter. This demonstrates to the employee that his or her job is meaningful to people in higher positions. Management should make periodic tours of inspection at meetings and award ceremonies.

6. Send a personal letter to the home of the employee when hired or transferred to the department. Letters can also be used to congratulate the employee after one year of service, as well as for other purposes. Letters stuffed in the employee's paycheck indicate that the subject—and perhaps the employee's recognition as well—is not worth another postage stamp.

7. Birthday cards or other cards can be sent by the first-line supervisor to the home of the employee (perhaps with a hand written note appended) involving such matters as length of service, the worker's family, his or her progress, and the like.

MOTIVATING WORKERS BY FORMALIZING CLEANING PROCEDURES

Another way to motivate workers is to develop a "user's guide" to tell workers how the housekeeping department functions and what is expected of the staff. The cover page might indicate the logo of the department, and the photograph of some workers, with their equipment. This equipment might include an automatic scrubbing machine, a wet vacuum, a double bucket mopping system, etc.

Include a different page for each of the following topics:

- An organization chart of the department, including the officer to whom it reports. Photos showing the key personnel are especially helpful, along with names and telephone numbers.
- A list of the activities performed by the department, at approximate frequencies.
- A list of tasks that cannot be performed by the department, such as cleaning privately-owned rugs, watering growing plants, etc.
- A list of those things that are against the rules for the occupants, such as moving their own furniture, or driving a nail in the wall to hang a picture.
- A list of problems and who to call for each, not only within the housekeeping department, but people in other departments, such as for a security problem.

Another helpful device is the "help wanted" card, which can be left on the desk of a person whose cooperation is sought to improve housekeeping in a given area. The card might indicate that help is needed in avoiding coffee being dumped into a waste basket, or waste not being properly separated for recycling, or marking furniture.

Large housekeeping departments sometimes publish their own newsletters. Such newsletters might be sent to certain key customers. An informal, home-made newsletter can be very effective, especially when the employees participate with cartoons, news items, etc.

How Quality Equipment Promotes Pride in Work

No one likes to do a job where he has not been given the proper tools for the task. In terms of equipment, ensure that the employee is provided with the proper quantity, size and type of equipment. Avoid having workers share equipment with other workers on the same shift, as this causes dissension and a considerable loss of time, and makes it difficult to hold a single employee responsible for the upkeep of that particular piece of equipment. The equipment should be attractive, and the investment of another 20% to obtain that is worthwhile. Don't cut workers short on supplies and chemicals!

Five Ways to Encourage Interest in Training Programs

Another type of "tool-for-the-job" is the training of the worker. Employees do not feel good about their work if they cannot perform it efficiently and properly. The basic rules for service worker training are simple:

1. Use pictures, including some of your own workers following the standard procedure, which show the work performed easily and safely, using efficient equipment and materials.
2. Involve the first-line supervisor in classroom teaching to augment on-the-job training (a training officer in a large organization might be used for general subjects and for coordinating the programs.)
3. Make sure training is continual.
4. Make sure workers receive recognition for attendance, including a certificate signed and presented by management.
5. For supervisory service personnel, use the role play technique. This is unbeatable as a means for supervisory practice.

How to Keep Workers Enthusiastic about Job Assignments

Job assignment is extremely important since it controls the way workers spend their time. Here are some guidelines:

- Make sure the assignment is based on easy, safe methods, using efficient equipment and materials.
- Make sure the assignment represents a reasonable day's work—not more and not less.
- Balance work loads so that no employee feels that he or she is asked to do more than another.
- Avoid the use of stopwatches in determining work loads—there are many other effective ways of determining time requirements.
- Do not use short-interval scheduling for service work—This can be extremely damaging to the morale of the workers. Trying to program each minute of the day suggests that the worker is a machine or a number. Instead, get the worker involved in what daily variations are necessary due to occupancy, weather, availability of equipment, trouble calls and the like. Instead of short-interval scheduling, for many types of service work, ask the

employee to perform all repetitive work within a geographical location, without requiring a detailed work time breakdown.

- Make sure area assignments are supported by personnel who perform project work (i.e., infrequent activities) and who therefore can relieve absent area-assigned workers.

- Rotate job assignments for those workers who become bored and who request such a change. However, do not rotate good workers who enjoy their fixed assignment and want to remain in it. Schedule regular rest periods; studies have shown that breaks actually help to increase productivity. Consider team assignments as a further means of offsetting boredom.

- Make sure your organization supports promotions, although not all workers desire promotion. The use of a working group leader is desirable because of the additional pay, title, and development as a possible future supervisor. Consider using group leaders only if appointment can be made on merit rather than on longevity.

The supervisor and manager can have a positive effect on worker motivation simply by setting an example—by walking briskly through a building, not reading newspapers or magazines in the office, picking up litter to demonstrate personal interest, fighting for the rights of the worker when he or she is verbally abused by others, and by pitching in in a real emergency (but not otherwise).

People support those activities which they help to design, or those things they help to create. You should encourage worker participation in as many situations as possible. Ask for suggestions for changing building design to promote more economical maintenance; in selecting chemicals, materials and equipment; in choosing uniforms, in determining methods, and so forth.

Pay and fringe benefits are not good motivators. These things simply make it possible for us to select from a better labor market and reduce the amount of turnover.

The principal function of the supervisor and manager is simply "to get the work done." Try to avoid the conflict of "Attitude versus Performance." It is difficult to act as a psychologist or psychiatrist in evaluating worker attitudes and psychological conditions, but you can evaluate his or her performance.

Motivation, just as in training, is an exercise in salesmanship—you are trying to convince someone to take the action that you feel is desirable. In motivation, remember the rules of selling:

- Stress the benefit to the "buyer." Tell workers what they are likely to gain from a change in performance, rather than what the benefit is to management.
- Know as much as you can about the subject being presented.
- Present the information in an interesting way.

- Choose a time and location that is comfortable to the worker to discuss this subject.
- Be enthusiastic.
- Most important of all: be persistent. Don't bring up the subject once, bring it up repeatedly.

HOW TO MEASURE YOUR SUCCESS: THE MOTIVATION CURVE

Don't become discouraged when you find that it is not possible to motivate all employees. The ability to motivate workers falls into the same distribution curve as shoe sizes and other conditions involving variables, as shown in Figure 14-1. Most people fall in the middle, can be motivated and will be motivated with a comprehensive program through the use of effective supervisors. People like to be motivated. At the thinner end of the curve there might be a few workers who are self-motivators and need no attention whatever except, of course, fair treatment and recognition where it is due. There are also people who cannot be motivated and they should simply be terminated as soon as this fact has become apparent, and as quickly as possible.

HOW TO DEVELOP GOOD PUBLIC RELATIONS FOR YOUR HOUSEKEEPING STAFF

Consciously or not, your housekeeping department is deeply involved in public relations. The principal purpose of any maintenance effort is to provide

Figure 14-1. The Motivation Curve

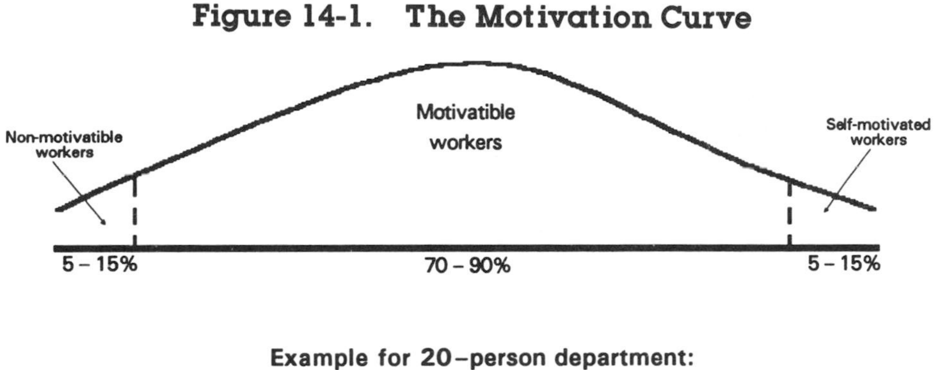

Example for 20–person department:

1 - 3	14 - 18	1 - 3

clean, comfortable and pleasant surroundings for several groups: building tenants and employees, visitors, and often, the community at large.

Your staff can increase its good image and earn respect from these groups if they:

• Direct extra attention to lobbies and entrances. Employees and visitors form an instant impression the moment they enter the building. Make sure it is a pleasant one. Give special attention to lobby floors and don't overlook walls and glass doors. Smudges and fingerprints at eye level are particularly offensive. Give them a quick spot check at least once daily with a spray bottle of detergent and a sponge.

• Concentrate on thorough cleaning and sanitation of all rest room facilities every day. Keep a spray bottle and sponge available and spot clean rest room walls frequently. Provide enough paper towel receptacles to prevent floor litter. Two receptacles side by side are better than one where it's needed and a second out of reach.

• Keep mirrors and sinks spotlessly clean.

• Watch carefully for litter and spillage in work areas around desks and vending machines.

• Clean walls, floors, doors and counters in reception rooms have tremendous effect on your building's prestige and should rate at the top of the cleanliness standard. Do not overlook parking lots, entries and other outdoor areas.

• Use "Caution" signs to warn visitors of areas being cleaned during the day. This tells people you are concerned about their safety. It also puts cleaning operations on a level with other company activities and boosts custodial morale.

• Investigate every complaint received by your department, and take appropriate action where called for. Be sure and get back to the complainer to state what action was taken.

Guidelines for Improving Personal Relations with Your Custodial Staff

You must form good relationships with the people who work for you and with people further up the chain of command. Here are suggestions to generate good will in both directions:

• Soften criticism with praise. Give criticism in a friendly and helpful way. Show the worker how to do a job and don't be impatient if he doesn't remember the first time or two. Keep showing him until he gets it right. Be patient and helpful; you'll earn his good will and get as much cooperation as he's capable of giving.

• Give orders in a courteous, friendly manner. Sharp speech, careless criticism or abusive remarks destroy subordinate morale and may well limit supervisory usefulness.

• Praise an employee as often as you can. Most supervisors fail to commend workers often enough. Don't take the attitude that "Joe doesn't do anything worth praising." Inspect his work area and FIND things that merit praise. An employee who doesn't do anything worth praising may need to be replaced, not just ignored.

• Keep subordinates informed. A department bulletin board is an excellent way to do this. A good place for it might be near the time clock. Don't assume there isn't enough news. MAKE news out of personal events. Sample notes you can put on your bulletin board regularly are:

"Compliments to the third floor crew for their excellent work cleaning the corridors."

"Best wishes to Mary Smith who has just become engaged to Bill Brown. The wedding date is October 2nd."

"Mrs. Jones, our executive vice president, complimented us the other day on the high level of cleanliness in the executive suite. Good work!"

Keeping Upper Management Informed

You must also keep building management informed. This communication should include:

- Projects under way
- Projects planned for the future
- Projects recently completed
- Special problems to be solved
- Ways the manager can help

A busy manager has no way of knowing what is going on in the housekeeping department unless he is told. Do this in a standard way. Use a monthly report, concise but complete. Put into writing even matters which have been discussed personally. This gives both management and supervision a running record of maintenance staff activities. It lets the building manager know which personnel deserve promotion or praise. It keeps him aware of problems and, best of all, it lets him know how much work the maintenance staff is doing.

How to Instill Pride in Workers

Don't overlook the relationship between your housekeeping department and the community. Not only does the community judge the company by the appearance of its buildings, but also by the conduct of its employees. Set standards of personal appearance for custodians equal to those for other employees. Too often custodians are given no feeling of pride in their work. Consequently, they may travel to their jobs in dirty, ragged or unkempt clothes. They will not wear dress-up clothes on the job, of course, but a dressing room or shower room at work makes it possible for them to appear on the street in clean dress. Uniforms for custodial help can do wonders for morale and improve the attitude of other employees toward them. Cleaning workers encouraged to be neatly dressed both on and off the job soon raise their standard for the building itself.

Efforts to improve worker spirit do more than improve work output. Remember, highly critical employees can damage the company image. Some people may judge the entire operation by one disgruntled employee who circulates criticism among his friends. He may also make it difficult for you to get good employees to work for you.

15

How to Handle Employee Absenteeism

Absenteeism is a serious problem, not only among building service workers, but in general. Millions of workers are absent each Monday and Friday. The annual cost in the United States alone is tens of billions of dollars—the costs to management include wages, social security benefits, overtime premiums and substitutes, unemployment and workmen's compensation, loss in production, replacement and training costs—while the cost to the public is in terms of higher taxes and higher prices.

A small decrease in the absentee rate for your workcrs would mean thousands of dollars in annual saving.

NINE REASONS WHY WORKERS ARE ABSENT

Many workers, and many union and civil service representatives, see absenteeism as an "earned" benefit. Where management has looked the other way during good times, this viewpoint may appear to have been sanctioned.

There are many causes of absenteeism, and even the following list is not complete:

1. Illness. Not all absenteeism recorded as illness is for an actual illness; this category also includes a feigned illness, one that is imagined, one that is perhaps coming on, or an illness in the worker's family.
2. Accidents, both on the job and off.
3. Problems with transportation, public or private.
4. Personal activities.
5. Personal problems, such as financial or marital problems.

6. Problems in morale—workers who do not feel good about their job and the way management treats them do not feel like coming to work.
7. Permissiveness—the worker feels that management doesn't care.
8. Desire for attention.
9. Unfair policy, which tends to drive workers away from the job.

Some unions ask for a contract which provides a certain number of personal days. That is, if the employee does not feel like coming to work, he can simply say "to hell with it" and be paid for that day!

HOW TO IMPROVE EMPLOYEE ATTENDANCE

If you were to study an organization or a department with a history of low absenteeism, you would probably find these factors in effect:

- The workers have the opportunity to discuss job problems with their immediate supervisors.
- The immediate supervisor takes time to discuss personnel matters.
- The supervisors have developed a feeling of unity and each is handling problems in a basically similar way.
- Attitudes of confidence and team spirit have been developed.
- The workers are satisfied with their opportunities for promotion and upgrading.
- The supervisor recognizes the importance of the workers and recognizes their work efforts.
- A training program is in effect.
- Workers feel that they have good working conditions, adequate supplies, and appropriate equipment.

To improve your absenteeism rate and thereby overall performance, find out precisely what the situation is:

- Find out what your absentee rate is. Figure 15-1 shows a sample absenteeism record.
- Clarify your organization's official policy concerning absenteeism and disciplinary action.
- Determine to what extent management is willing to enforce its rules and disciplinary procedures.

Figure 15-1. Sample Absentee Record to Help Spot Patterns of Absenteeism.

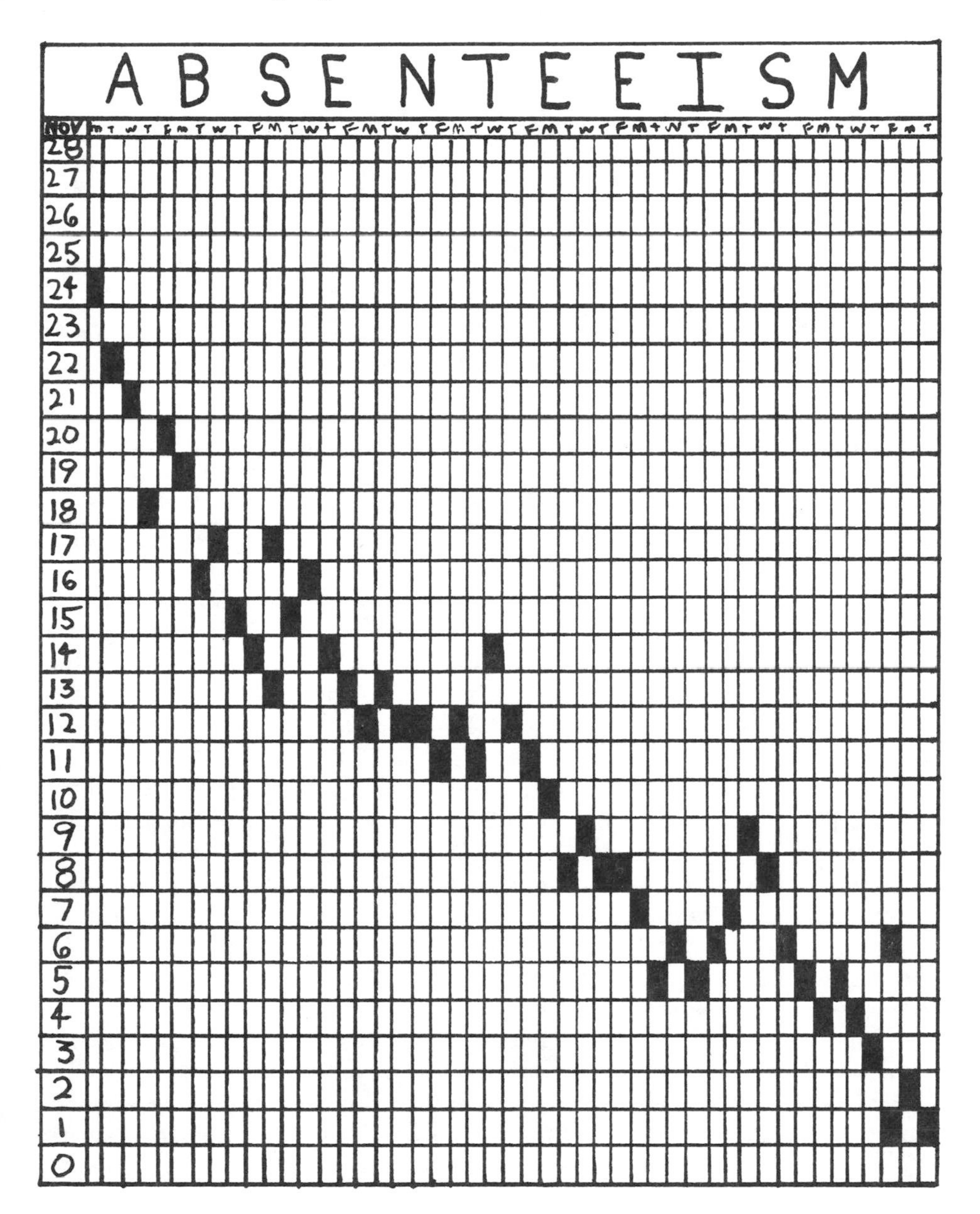

- Establish a good system of maintaining attendance records if the existing system is not adequate.
- Inform employees of the campaign to improve absenteeism, stressing the benefits to them in terms of job security, improved earnings, protection of sick leave benefits in case of illness, and equitable treatment.
- Examine working conditions and institute any necessary improvements.
- Display an active interest, involving as high a level in management as possible—this display of interest should be continual and not just a "one-shot" thing.
- Discuss absences with individuals, and make it clear that you expect good performance.
- Take disciplinary action without fail where called for in company rules.

Improving an absenteeism rate is not a question of taking one or two steps to solve the problem. Certain details do improve the situation somewhat—for example, when a supervisor takes phone calls concerning a worker who is absent, rather than a clerk taking the phone call, this tends to decrease absenteeism. Another idea is to provide a reward for a month of perfect attendance and send the reward to the spouse of the worker.

As part of the program for improving attendance, keep in mind that the first-line supervisor is one of the keys to better attendance, and that management should work with supervisors in improving their ability to handle this situation. Figure 15-2 provides a form for comparing absenteeism rates for supervisory groups.

The first step a supervisor can take is to set a personal example in terms of his or her own absenteeism. Probably the one word that is most meaningful in supervisor-employee relations is the word "fair." When workers are treated unfairly, or when they feel that supervisors are playing favorites, absenteeism is a natural result. Workers want to know why and need to be given adequate information. They want to be involved in goal-setting and problem solving, and this should be encouraged when possible. Throughout, the supervisor should express a positive expectation with respect to absenteeism.

The absentee problem must begin with management. Management must express its intent with respect to enforcing company rules—or changing the rules if necessary—and management must be involved by providing adequate publicity and information concerning employee absence. Any changes in working conditions, supplies or equipment must be authorized by management.

It is best if management publishes the disciplinary guidelines for infractions of absenteeism rules as well as other rules and also a clearly worded sick-leave program.

Figure 15-2. Sample Absenteeism Report, by Department

ABSENTEEISM AND VACATION REPORT

Supervisory Group Absenteeism Rates

Supervisory Group	Month	YTD
1. ______	______	______
2. ______	______	______
3. ______	______	______
4. ______	______	______
5. ______	______	______
6. ______	______	______
7. ______	______	______
8. ______	______	______
9. ______	______	______
AVERAGE	______	______
VACATION DAYS	______	______

DEVELOPING A FAIR SICK-LEAVE PROGRAM

Most sick-leave programs encourage absenteeism. The programs are administered in a way that penalizes the loyal employee and rewards the poor employee.

The typical sick-leave program provides a certain number of days available for sickness, but in practice awards the paid time off for almost any reason—or for no reason. Those organizations requiring a doctor's certificate find that this does not solve the problem, because many people are able to obtain such certificates. At the end of the year, unused benefits are either taken away (from the loyal employee), or accrued to some long-term benefit (which most workers never expect to use or do not take into consideration). At retirement, they may lose all or part of their accrued leave—but the worker who took off Mondays and Fridays got it all!

Consider a sick-leave program with the following arrangements, which would treat all employees perfectly equitably:

- No sick-leave is paid for the first day of absence (some organizations do not pay sick-leave for the first two days of absence). This does not lessen the number of sick-leave days paid—it simply requires the worker to fund his or her own absence for a very short period of time.

- Two weeks before Christmas (when everyone needs money) unused sick leave should be split, half being paid to the employee in cash, the other half being accrued to a long-term benefit. The long-term accrual should not exceed a specific number of days (such as 90), and once that figure has been reached, then all unused sick-leave is paid in cash.

The above arrangement treats everyone equally and fairly, while discouraging workers from taking "sick-leave" days before or after a weekend or before or after a pay day. The worker must spend his or her own money, so to speak, to take off a day.

Sometimes sick-leave is a crutch for the worker, indicating that he does not know how to manage his own money and therefore it must be saved for when he needs it. A tremendous amount of paper work would be saved if the employee obtained the value of the sick-leave in terms of wages or salary, and then funded his own absence (using insurance, perhaps, for long-term purposes). If this were done, the absenteeism rate would drop greatly. But probably management would not be interested in such an approach, thinking that once the value of the sick-leave were given to the worker in terms of additional compensation, then later on a demand would be made to reinstate the sick-leave without affecting the new compensation rate.

HOW TO CONTROL ABSENTEEISM: DISCIPLINE AND TERMINATION CHECKLISTS

In managing absenteeism programs, it is helpful to use two checklists. First, before administering discipline, review the employee's record of absenteeism. Make sure the following steps have been taken:

- Have the rules been explained to all employees?
- Have prior verbal and written warnings been given?
- Are the records available concerning the employee's past record of absenteeism?
- Has management enforcement of its rules been consistent?
- Are there unusual or mitigating circumstances in the specific case?
- Is the disciplinary action compatible with past action for similar cases?

Use a second checklist in preparing for termination based on absenteeism. This checklist should answer the following questions:

- Is the termination policy reasonable?
- If so, has it been properly communicated to the employee?
- Are you being objective in terms of treating the employee as any other employee would be treated?
- Has the employee been properly reprimanded verbally and in writing in the past and thus prepared for this action?

Additional items may be on the checklist, such as providing an opportunity for the employee to resign, consideration of a suspension rather than a termination, etc.

Long-term improvements in absenteeism require action on a number of fronts beginning with management and the development of a realistic and equitable sick-leave policy, the involvement and management of many aspects of the problem, through the first line supervisor in terms of the fundamental principles of fair and equitable treatment to employees. When you improve your absenteeism rate, you have improved a number of other things as well, basic to the effective operation of the organization.

16

Promoting Job Safety and Fire Prevention

Accidents take an enormous annual toll in human misery, business expense, and social losses. Safety engineers point out that most accidents are preventable. A large percentage of accidents in facilities are related to *housekeeping* practices.

This chapter will help you target specific areas in your housekeeping program to prevent costly accidents and promote safety.

CALCULATING THE HIGH COST OF WORK-RELATED ACCIDENTS

The total extent of the cost of any given accident, to the individual, industry, and society, is surprisingly high. Money is spent, as the result of an accident, many months or even years after the occurrence. Such costs and expenses represent a complete and total waste in terms of:

- Payment of compensation claims.
- Payment of noncompensated medical expenses.
- Lost income of the injured employee.
- Lost time of fellow workers who stop to help the injured worker or merely to watch.
- Loss of time of supervisors, staff personnel, and management in investigating the accident, helping the injured worker, training new workers, etc.
- Damaged equipment or materials directly involved in the accident.

- Impaired quality on the part of other workers because of emotional involvement.
- Related accidents by other workers because of emotional concern.
- Decreased production rate due to diverted interest or upset of workers.
- Lost production because of machine stoppage.
- Decreased effectiveness of the injured employee after his return to work.
- Lost business or good will resulting from a decreased production rate, poor quality, etc.
- Legal expenses and costs of litigation.
- The general loss to society and industry because of reduced skill, potential, and capacity and loss in earning power.

Accident prevention can be a far more profitable investment than is usually realized.

Some supervisors, who are not familiar with the principles of insurance, consider compensation insurance costs as a fixed expense. Generally speaking, however, most liability contracts include a feature called the *experience rating.* This is an incentive feature which lowers premiums as the result of a low-accident rate, but also increases the premiums following a high-accident rate.

Any serious safety or fire prevention campaign must directly involve housekeeping. The Bureau of Labor Statistics of the United States Labor Department analyzed fifty-seven different accidents in boiler-shop and foundry-type operations. Of these, approximately 15 percent were *directly* traced to poor housekeeping practices.

Either directly or indirectly, poor housekeeping practices probably contribute to well over half of all industrial accidents. Some safety engineers feel the figure is as high as 75%.

Not only must a housekeeping program document contain specific instruction sheets summarizing safe housekeeping practices, but also each section of the program should have safety "built in" in methods, materials, etc. Training and promotion relating to housekeeping safety must be continuous.

HOW TO PROMOTE SAFETY AWARENESS

Most firms have, or should have, a safety director, safety engineer, or individual who has been assigned overall safety responsibility and authority. Such personnel generally jump at the opportunity to work with the housekeeping department in improving general safety practices and conditions. The entire housekeeping program document, for example, should be reviewed by the safety officer.

The duties of the safety engineer, or the safety coordinator who handles this work as a part-time function in the smaller facility, should include:

- Conducting monthly safety meetings.
- Keeping accident records and making accident reports.
- Calculating safety records and making safety awards.
- Interviewing workers who have had accidents.
- Coordinating safety publicity.

Union officials are usually very cooperative where improvements are considered in the interrelated fields of safety and housekeeping. Encourage their assistance and suggestions.

Employee relations personnel will also provide the same positive response. They will assist in promotion and publicity campaigns, and in other activities.

Personnel outside either the safety department or housekeeping department may be given specific safety duties to perform. For example, security personnel may be provided a check list of potential unsafe conditions which are to be reported in writing.

Safety and fire prevention should be promoted, similar to the promotion of good housekeeping. For example, The National Safety Council points out that "the worth of the safety poster has been demonstrated so often that there can be no question of its value in any industry program. Invariably, it proves an effective medium for attracting the employees' attention in getting across a safety message that will stick."

Good advice and direct assistance in the safety and fire prevention program are gladly furnished merely by asking workmen's compensation insurance companies and fire insurance companies. They are prepared to furnish posters, show films or videos, conduct training classes, provide promotional literature, and make inspections and reports. Professional groups, such as The National Safety Council, National Board of Fire Underwriters, and the National Fire Protection Association, are anxious to be of help. The Government Printing Office also has valuable material on safety and fire prevention.

HOW BODY MECHANICS CAN PROMOTE JOB SAFETY

Proper use of the body protects the worker from injury, strain, and undue fatigue.

The basic use of the bones, ligaments, tendons, and muscles is often called "body mechanics," or, more technically, kinesiology.

Work Avoidance: Eight Questions to Ask

As in so many human activities, the questioning attitude may save a great part or all of the physical effort. Consider these questions:

1. Can the load be shared? Example: Two workers, rather than one, lift a box.
2. Can the load be reduced? Example: Buy supplies in one-gallon jugs rather than fifty five-gallon drums.
3. Can the work be performed mechanically? Example: Rather than manually handling buckets of water and swinging a mop, utilizing an automatic scrubbing machine.
4. Can the load be moved in another way? Example: A load can be moved on rollers, rather than lifted.
5. Should someone else do the work? Example: A person more physically suited might move a load, rather than someone who is slight or frail (although this may appear to conflict with the equality of work, it is a very practical choice in some cases).
6. Can the load be handled in parts? Example: Rather than lifting a case of six one-gallon containers, remove two, one in each hand.
7. Can the force of gravity be used to advantage? Example: Slide a box down a plank rather than carrying it.
8. Is a specific device available to perform the work? Example: Use an automatic stair-climbing machine rather than manual labor to move furniture.

How to Use the Muscles Efficiently

The basic thing to remember in body mechanics is to use the largest muscle possible to do the work. The smaller muscles, such as those of the back or arm, are much more susceptible to strain or injury than the large muscles, such as those in the leg. A simple example is with the use of the dust mop: in cleaning a corridor, swinging the mop from side to side not only creates a great deal of dust, and fatigues the worker, but it is possible to cause a strain of the muscles. By placing the handle of the mop stick on the hip, or locking the elbows against the body so that the hands don't move, one can walk straight ahead, using primarily the large muscles of the leg.

The same is true in lifting: you need to lift primarily with the large leg muscles by squatting down and bringing the load near the body, rather than attempting to lift with the small back and arm muscles by bending over and picking up a weight.

In considering the wonderful machine that is the human body in its general terms, we observe two things: First, the body is bilaterally symmetrical (that is, the right side is generally a mirror-image of the left side), and secondly, the appendages (hands, arms, etc.) work in radials, so that the hand pivots around the wrist, the forearm around the elbow, etc. Work is simplified when the hands, for example, work in opposition and in circular motions rather than straight or reciprocal motions. A simple example of this is the erasing of a chalk board: the most difficult way is for the worker to use one hand and eraser in straight-line patterns—it is much easier (and quicker) to use two erasers working in circular patterns, the right hand moving, say, counterclockwise and the left hand clockwise.

TIPS FOR IMPROVING WORKERS' SAFETY AND REDUCING FATIGUE

Here are some guidelines for improving safety and reducing fatigue in body activities:

1. Wear protective devices, such as gloves and safety shoes when handling loads.
2. Make sure the path is clear, without obstructions or projections, before attempting to move something heavy.
3. Be sure that when moving the object, the workers' vision is not obstructed.
4. When two people share a load, decide in advance who will be group leader, so that only one person is giving signals.
5. Where two people are involved, try to assign workers who are of similar size.
6. Study the load; when lifting it, be sure something is not knocked over, or when putting it down, be sure that something else does not fall.
7. Give the load adequate physical support: attempting to use too small a moving device, or a rack that is not strong enough, could be serious.
8. Wherever possible, push rather than pull.
9. Try to move a weight on wheels or rollers rather than by sliding (wouldn't this save a lot of damaged floors!).
10. Use a device where available, or where it can be purchased, such as a desk-moving device, or file cabinet mover.

11. Avoid lifting and moving on slippery floors, which may be caused by excessive dust, water, paper clips, etc.
12. When carrying a can or box, keep the load close to the body: the further away the load is from the backbone, the less mechanical advantage the worker has.
13. When lifting above the shoulders, use a sturdy platform so as not to unduly extend the arms and strain the back.
14. Do not use the top step of a step ladder, since the legs are not braced.
15. Be sure the feet of a straight ladder are secured.
16. Take a secure grip on the load before the full weight is carried, and try not to change the grip while the load is being carried or moved.
17. For those containers that are moved regularly, consider attaching handles or other devices that simplify lifting.
18. Avoid twisting the body when carrying or moving a heavy load.
19. In lifting loads from the floor or a low position, be sure to bend the knees, not the back.
20. When placing a load on a horizontal surface, put it down on the edge of the surface, and slide it back, to avoid holding the load out from the body.
21. Move slowly and gradually—avoid rapid or jerky motions (remember inertia: starting something into motion is much more difficult than keeping it in motion).
22. If a low load to be lifted is small enough, straddle it with the legs and then follow the correct lifting procedure (again, this brings the load closer to the backbone).
23. If two people are moving an object on a stairway, if going up, make sure the lower person is facing the load and the upper person facing the load or facing away from it—but going down, both persons should face the load.
24. In carrying long objects, such as pipes or ladders, use two people even if the objects are not heavy, to prevent striking someone in the face, or damaging surfaces.
25. Remember that lifting for a heavy person is more dangerous than for a person of normal weight, for two reasons: the load is further from the backbone, and there is a greater strain on the heart.
26. Encourage workers to use good posture—walking or sitting with the backbone nearly straight. This improves health in general, while better conditioning the body for physical exercise.
27. To avoid accidents, remove any grease, oil or other slippery material from the object to be carried, as well as from the hands or gloves.

28. Do not attempt to move or lift a load that appears beyond your safe capacity, or that is in any way hazardous.

Effective and safe cleaning activities not only appear to be easy, they *are* easy.

The sanitation worker who is struggling with the floor machine, who is trying to mop the corridor in a single pass, who is flinging the dust mop around the floor is straining, using a lot of energy, and getting little work done—if not actually creating more work to be done later. It is the worker who controls the machine, swings the mop in a comfortable pattern, pushes the dust mop in a straight line that gets the job done and with no discomfort or injury to himself. Consider: Three out of ten accidents which cause a temporary injury are the result of moving objects in the wrong way.

SAFETY STANDARD REGULATIONS FOR HOUSEKEEPING PERSONNEL

Safety should be promoted positively through its universal benefits. It is necessary to supplement the informal activity, however, with a formal statement of the company's safety rules and regulations.

Accident prevention consists of performing every operation, whether large or small, the correct way. It puts into practice the intelligent and common-sense methods of getting the job done. No activity is so important that an unsafe condition or act should be permitted in its performance.

Management's basic responsibilities with respect to safety are:

- To provide equipment which has been properly investigated for safety, and equipped with necessary guards, shields, and other safety devices.
- To arrange all processes and operations in a safe fashion.
- To train and motivate supervisors and employees in safe operating practices.
- To provide regular safety inspections and condition reports.
- To investigate all accidents in order to prevent recurrence.

Worker responsibilities include:

- To use the safety equipment provided.
- To be aware of and follow safety rules and regulations.
- To be on the lookout for hazardous conditions, and report them.
- To caution fellow workers about unsafe conditions or practices.

The following is a typical list of safety regulations for housekeeping personnel. Many of the items are applicable to other personnel as well. Special regulations concerning floor, materials, and equipment safety are treated later in the chapter.

1. In wet weather, avoid dangerous falls by using rubber mats or runners near entrance areas. This will also prevent the tracking of mud and water onto clean floors. Put out and maintain mats at doors assigned to you.
2. Report necessary light bulb or fluorescent tube replacements in critical areas such as stair landings, near doors, at aisle intersections, etc. Poor vision can cause nasty accidents.
3. Watch out for vehicles; assume that the drivers do not see you and get out of the way.
4. Plug in a cleaning machine only immediately before using, and unplug immediately after using.
5. Be sure to report any accident to your supervisor.
6. When lifting heavy objects, keep your knees bent and the back straight. Be sure you know the proper method of lifting. Get help when necessary.
7. Avoid horseplay on company property. Horseplay can result in serious injury.
8. Be familiar with and use emergency treatment measures where necessary. For serious injury, get first aid or medical help. The injured person should be moved only if unavoidable.
9. Apply artificial respiration if breathing has stopped. In case of chemical splashes, wash off immediately with plenty of water.
10. Familiarize yourself with the potential safety hazards of your work area. Report safety hazards and recommended improvements to your supervisor.
11. When entering a tank, pit, or other enclosed space, obtain inspection and approval by your supervisor, who must make certain that proper ventilation is provided, respirators and other protective equipment are used, and assistance is available.
12. Be sure to pay attention to all safety signs and notices.
13. Smoke in designated areas only.
14. Stay out from under hanging loads.
15. Run through the facility only in an emergency.
16. On stairs, watch your step and use the handrail.
17. Carefully step down from platforms, stacks of material, or other high places, rather than jumping.

18. Use a brush to clean your clothes, rather than compressed air. Flying chips can cause eye injuries.
19. In emptying waste receptacles, dump the material out rather than reaching in. This will avoid cuts.
20. Disconnect fans and other electrical equipment before cleaning them. Pull on the plug rather than the wire.
21. Drain opening should be done only by a licensed plumber.
22. Don't take chances with safety. In any questionable situation, always check with your supervisor.

TIPS ON HANDLING CLEANING CHEMICALS

Here are some suggestions for improving the handling of cleaning chemicals:

- Pour and use solvents in well-ventilated places. Avoid using gasoline to clean anything. Replace the cap on all solvent containers after each use.
- If any liquid gets in your eyes, even just dirty water, flood the eyes with plenty of tap water immediately.
- Avoid using carbon tetrachloride for cleaning purposes. This chemical has been directly responsible for a number of deaths through inhalation of its fumes.
- Keep the protective equipment which you have been provided in good condition and wear it whenever it is needed. This includes face shields, goggles, hoods, gloves, aprons, respirators, etc.
- Use safety shoes when moving drums, working with furniture or heavy equipment, or in areas where objects might fall.
- Store inflammable materials in accordance with fire regulations.
- Where strong chemicals are used, keep the skin covered for protection against chemical irritation.
- Dry the hands and face carefully after washing to further prevent skin irritation.
- If it is necessary to dilute an acid, be sure to pour the acid into the water, rather than vice-versa.
- Carefully mark poisonous materials such as arsenical weed killers, and various types of rodenticides and insecticides. Also, make sure such chemicals are carefully controlled and used by designated personnel only.
- Avoid the hazards of broken glass by buying materials in plastic containers wherever possible.

- Use acids for descaling bowls only on a periodic basis, and then only by specially designated, trained, and equipped personnel.
- Provide conditions to permit safe storage. It should not be necessary to lift heavy loads, stack materials too high, etc.
- Replace harsh caustic or acidic products by neutral or milder products wherever possible.

EQUIPMENT SAFETY CHECKLIST

The following guidelines pertain to the use of housekeeping equipment and the general use and cleaning of other types of equipment:

- Watch how you carry your equipment, mops, brooms, etc., so that no one is injured by handles sticking out, etc.
- In areas having low ceilings, take special care to avoid hitting sprinkler heads with mop and broom handles in order to avoid the severe water damage that a sprinkler leak would cause.
- When cleaning stairs, be careful where buckets and equipment are placed so as not to cause someone to fall.
- Keep your equipment and maintenance area neat and in good order.
- Keep equipment tightened up. If a particular bolt or nut keeps working loose, obtain a special lock washer or fastener.
- Be sure wiring is in good repair and not frayed, knotted, or cut.
- Make sure electrical equipment is grounded when in use, particularly when there is water on the floor.
- Keep ladders in good repair and use them properly.
- Inspect tools regularly to make certain they are in safe operating condition. Personnel washing outside windows must have their safety devices inspected daily by their supervisor.
- Disconnect fans and other electrical equipment before cleaning.
- Be sure that switches on equipment are in the "off" position before plugging in the wire.
- Keep greasy or oily cloths or dust-mop heads in closed metal containers.
- Use floor-machine handles in the lowered position before starting the motor.
- Manually secure floor-machine brushes and attachment plates to the machine before the motor is turned on.
- Keep equipment out of aisles and traffic lanes to avoid tripping hazards.

- "Pay out" only the length of electrical wiring needed, rather than an excessive amount.
- Never wear loose clothing, neckties, rings, or other items around moving machinery.
- In driving a vehicle, obey all the plant traffic rules.
- Use machinery only if authorized.
- Use the right tools and equipment for a job. Do not use a knife blade for a screwdriver or a brush block for a hammer, for example.
- Don't throw tools and equipment from one worker to another.
- In using scrapers or other sharp tools, apply the force in a direction away from the body.
- Wear a helmet when your cleaning job must take you under construction or maintenance work.
- When riding on a vehicle, stay seated.

HOW TO ENCOURAGE FLOOR SAFETY

Most rules concerning floor safety are directed to preventing falls, which is normally considered the most serious type of accident exposure. Falls cause most lost-time accidents and occur more often than any other type of accident. The following guidelines can help workers avoid falls and other painful accidents related to floors:

- Use "Wet Floor" caution signs freely whenever floors are being wet cleaned, waxed, sealed, or stripped. Signs should be placed so that they will be visible from all avenues of approach. Do not rely on "wet floor" lettering on mop buckets.
- Keep floors stripped according to schedule to prevent build-up of brittle wax, which causes slipperiness.
- Mop or vacuum rain, snow, or other liquids immediately to dry the floor.
- Remove oil drippings or grease spots by vacuuming, scraping, or mopping, or by using a nonflammable oil absorbent.
- Dry clean floors regularly to remove slipping hazards.
- Monitor heavily trafficked areas in addition to performing the basic maintenance, to remove slipping hazards such as paper clips, toothpicks, etc.
- Do not use solvents in the maintenance of composition floors. Their chemical action will cause softening and roughening of the tiles.

- Avoid natural soap in floor cleaning, as the film remaining may become slippery. If natural soaps are used, rinse them carefully.
- When working on a wet floor, walk carefully and take shorter steps than usual.
- When it is necessary to apply liquids to a floor during usual business hours, clean the floor in sections, and dry each section carefully before proceeding to the next.
- Be sure that mats and runners lay flat. Wrinkles or turned up corners cause tripping accidents.
- Report defective flooring, loose handrailings, bad stair treads, and dangerous projections from walls to your supervisor so he or she can arrange for their repair.
- If any furniture must be moved, be sure this is done before the floors become wet. Attempting to move heavy objects on a slippery floor can cause back strain or injuries due to falls.
- Be sure that all drawers or doors are closed before moving furniture.
- Provide furniture moving equipment.

There are also a number of things that *management* can do to improve floor safety:

- Design floors and stairways to have an antislip surface when possible.
- Correct existing slippery surfaces. Particularly bad slip problems at critical concrete areas can be lessened by etching the concrete with a mild muriatic acid solution. This should be done very carefully as overetching will cause the floor to become rough and more difficult to clean. Etching should always be done under the direction of a supervisor. Use carborundum strips on stair treads or ramps.
- Load floors within their capacity limits. Sagging will cause unevenness, which may lead to tripping.
- Repair worn floors quickly. This includes loose or curled tiles, cupped wood, holes and cracks, etc.
- Keep carpets in good repair. Use a nonslip pad under small rugs.
- Put a general rule into effect that the person causing a spillage should be responsible for either removing it himself or, for a serious problem, obtaining help.
- Provide adequate lighting to prevent persons walking into obstructions, slipping on foreign articles, or being struck. A good time to check for lighting problems is at night.
- Carefully set up and check on arrangements for ice removal. Ice removal materials and equipment should always be available for the worst possible condition.

- Stipulate that all workers must wear low-heel shoes with suitable matting. Provide nonslip shoe coverings if needed.
- Purchase floor finishes, waxes, and coatings which have adequate slip-resistant qualities.

Using Appropriate Floor Finishes to Reduce Slipperiness and Accidents

Most persons associate the shine on a floor with slipperiness. This is such a strong association that people actually slip more often on a floor that is thought to be slippery than on a floor that is thought to be unslippery, although the two surfaces may be absolutely identical in their coefficients of friction! Actually, shiny floors are not necessarily slippery and slippery floors are not necessarily shiny.

Contrary to some belief, high-grade floor maintenance may be achieved by using waxes and other floor-surfacing materials without increasing the slipping hazard. In some cases, the slipping hazard is actually reduced.

Properly compounded and approved floor-coating materials may be used with assurance that they will not increase the slipping hazard, particularly if proper application techniques are used. Investigations by independent organizations have shown that the slipping hazard of floor finishes has been highly overestimated. Studies of the National Safety Council and Underwriters' Laboratory have both indicated that many manufacturers are marketing floor waxes and floor-surfacing materials that provide a higher coefficient of friction than the bare floor itself.

The basic cause of floor slipperiness, then, is not the application of wax where a quality product is used. The Underwriters' Label (UL) is a worthwhile guide when selecting waxes, as this approval indicates a wax that falls within their minimum slip-resistance standards. Many of the newer polymeric floor coatings, which contain little or no natural wax, have particularly good slip-resistant qualities. Where a good floor coating material is used, most falls will be caused by wet floors, foreign objects, bad physical repair, poor lighting, and other such hazards.

FIRE PREVENTION TECHNIQUES

Fires can be considered as a special class of accidents. Although an accident which involves a breach of safety normally only directly involves a few persons, a major fire can—and half the time does—put a company out of business. Fires, like safety accidents, can be prevented.

Rubbish and waste materials of various kinds contribute to a large number of fires. Storage of any useless material, from rags, rubbish, waste paper, and wood to obsolete flammable liquids should be discouraged. Storage in out-of-way places such as attics, basement areas, under stairs, and in other hard-to-use locations should be avoided if possible.

Proper and regular disposal of waste products is a basic and elementary rule of good fire prevention and housekeeping. Metal receptacles, preferably covered, should be utilized in critical areas, for many of the plastic receptacles on the market today actually contribute fuel to fires, or create noxious fumes.

The maintenance of a high standard of cleanliness and order is one of the first and basic elements of good fire prevention. Here are some ways to encourage fire prevention:

1. Provide sprinkler protection in all necessary areas.

2. Hire a watch service consisting of well-trained, alert individuals.

3. Post fire extinguishers of suitable type and size, in well-marked locations throughout the plant.

4. Enforce smoking rules completely.

5. Train all employees in fire-fighting practices and in the use of the fire protection first-aid extinguishers throughout the plant.

6. Inform employees about job hazards in handling flammable materials and other hazardous operations.

7. Create a fire brigade, properly trained, organized, and equipped, utilizing periodic drills. The local fire department should inspect the facility periodically. The fire brigade should be trained with help from the fire department and the insurers.

8. Have weekly self-inspections of the facility by trained personnel, covering fire-protection equipment, electrical equipment, and housekeeping. Reports should go to management for review and correction, and copies should be kept for the insurer's representative.

9. In conjunction with the insurers, set up proper special protection from industrial hazards, such as welding and flammable liquids handling.

10. Gain cooperation from the fire protection engineers who visit the plant from the insurance company.

11. Carefully handle cleaning compounds, polishes, floor oils, insecticides, etc., that are hazardous because of a tendency to heat spontaneously or because they contain flammable solvents. Before any products are used, identify their properties so that proper precautions may be taken in handling.

The number of fires caused by poor housekeeping in well-protected industrial plants is both surprising and unnecessary. Fire protection engineers say that "a clean plant is a fire-safe plant."

The Williams-Steiger Occupational Safety and Health Act, otherwise known as OSHA, has gone through numerous interpretations and changes, and will probably continue to do so for some time. It was passed to protect employees of organizations involved in interstate commerce from unsafe or unhealthful working conditions. The definition of who is involved in interstate commerce is so broad as to include most organizations.

The Act covers many types of operations and maintenance, including custodial. Building service management and supervision are well advised to be aware of the provisions of this Act, which contains requirements ranging from the avoidance of slippery floors to the visibility of electric cords on floor machines. Somewhat overdone, as many government activities are, OSHA regulations nevertheless do represent a benefit to workers, unfortunately with an additional possible cost to management. At least the Act points another finger at the importance of proper cleaning activities.

HOW TO REGULATE WASTE COLLECTION, RECYCLING, AND DISPOSAL

General waste handling is a subject involving local and federal regulations, public and employee relations, fire and safety, etc. It has seen the development of specialized equipment, contracted services, consultants, professional associations, seminars, and publications.

Waste collection requires an intensive, detailed investigation. First, name a coordinator for waste collection activities. This coordinator may have other functions, depending on the size of the organization; these functions might include hazardous waste handling, and general safety and fire prevention.

General waste collection and disposal usually involves a custodian dumping waste baskets into a hamper or container (either on the custodial cart or as a separate wheeled item), and then later dumping into a bulk container or incinerator. Larger organizations may use custodians whose only job is to collect the larger waste containers. When an organization is in an austerity situation, office workers and others may dump waste baskets into a larger container.

Even where not regulated, recycling should provide for the separation of #1 paper—typewriter paper, photocopy paper, etc. A rectangular container is placed either on or alongside a desk and the employee places the paper in it without crumpling. The custodian can collect #1 paper and other waste on alternate days (unless the container is nearly full), thus not using more time. This paper can be sold to recycling companies.

Where confidential papers must be protected, they can be placed in large locked containers, that are dumped into recycling vats under your overview, thus saving the expense and considerable fire hazard of shredding.

It is inevitable that more and more items will require recycling through legislation for smaller organizations. There is great public pressure to do this. If you have not already done so, you may have to provide the space, equipment, and the custodial time to make all or most of these separations:

- #1 paper
- Computer paper
- Newspapers and magazines
- Cardboard
- Clear glass
- Colored glass
- Solid plastic
- Foam plastic
- Wood
- Aluminum
- Ferrous
- Garbage

Promotional logos, campaign posters, and other activities can help the recycling effort.

HOW TO DISPOSE OF HAZARDOUS WASTE

The disposal of hazardous materials is an important subject for every organization. The subject is covered in many magazine articles and books; it is regulated by local and federal laws (involving agencies such as OSHA and EPA); it has become an issue involving ethical, moral, insurance, ecological, safety, public relations, and other factors.

The extent of the problem may be realized from the fact that over half a million products are classed as hazardous chemicals.

Hazardous waste disposal has also generated new societies, seminars, periodicals, service companies, consultants, and equipment manufacturers.

The activity may involve not only the Housekeeping Department, but also Safety, Employee Relations, Training, Legal, Public Relations, and in larger facilities perhaps a separate department or coordinator.

It requires—demands—your time to pursue this subject as a serious and separate activity.

Minimum requirements include:

- Identifying and labeling offending containers
- Inventorying all such materials

- Keeping MSDS (Material Safety Data Sheets) that are readily available in an emergency
- Developing an employee and supervisors' awareness and training program
- Preparing and maintaining written material concerning publicity and training
- Keeping records of all related activities
- Providing written regulations and procedures.

17

Money-Saving Tips for Purchasing Cleaning Supplies and Equipment

When you give workers a job to do, it is wise to provide them with reasonable facilities, equipment, and materials for the work. Your job is to make sure that this universal principle is applied to the operations of the housekeeping department. For example, you could not expect a painting crew to work effectively or safely unless they were properly supplied with a suitable preparation area and paint locker. They need equipment such as ladders and scaffolding, spray guns and accessories, brushes and rollers of all types, tarpaulins and shields. They require materials such as pigments, vehicles, thinners, varnishes, stains, etc. The custodian must similarly be equipped with adequate working and storage facilities, equipment, and materials.

This chapter examines the process of purchasing cleaning supplies and equipment to help you save equipment selection time, keep a more efficient inventory supply, and reduce spending costs.

HOW TO EVALUATE SUPPLIERS

The sanitary supply industry consists of literally thousands of firms, most of which are quite small but many are multi-million dollar operations. The yellow pages of the telephone directory in any large city will turn up scores if not hundreds of manufacturers, distributors, and jobbers (those who sell products manufactured by others, but often under their own name). The competition is fierce, one reason for this being the comparatively large percentage of sales commission.

As in all types of businesses, there are both reputable firms and salespeople prepared to serve their customers, as well as disreputable firms and salespeople out to make a "quick buck" by any means. The typical supply firm has hundreds of products, perhaps several brand names of each type. For example, a given company may market a half-dozen detergents, all generally for the same purpose, but at different concentrations, viscosity, and ingredients. The claims made for chemical products are often exaggerated, and it sometimes appears to the buyer that every salesperson has "magic products" to offer.

No wonder you are dazzled and confused!

The problem is further compounded by television commercials, which tend to perpetuate obsolete products and refute statements of reputable supply salesmen. For example, television commercials still extol the virtues of pine oil as a disinfectant; yet for a number of years this product has been eclipsed by other types of disinfectants which perform much better, and without the drawbacks of pine oil, as will be discussed below.

Many managers buy chemical supplies on an emotional basis, making decisions on factors that have little or nothing to do with performance. For example, some buyers are only interested in soaps that are colored green, possibly because their grandmothers used green soap when they were children; others like only the odor of lemon; some buyers are impressed with viscosity or thickness, thinking that the more syrupy a liquid the more concentrated it is; and many are impressed with how much foam a detergent will create. Yet, it is possible to formulate a chemical solution that is green, has a lemon smell, is so viscous that it can hardly be poured, and fills the room with foam, yet is such a poor detergent that it cleans practically nothing!

Obviously the color and odor contribute nothing to cleaning—the other factors are a little more elusive and many buyers will rely on them even after agreeing that they may be of no consequence whatever. For example, it is possible to thicken water through the addition of cellulose derivatives, which have no active ingredient at all from a detergency standpoint. And there are many cases where soap suds would be a detriment rather than an advantage, such as in an automatic scrubbing machine, where the foam would choke out the vacuum system (some of the best detergents have no foam at all).

Ideally, you should purchase tested products from reputable suppliers, on an annual commitment basis in order to obtain a quantity price, with the products delivered in convenient and safe containers.

Nine Factors for Evaluating Suppliers

Suppliers can be evaluated on these factors:

1. Reputation in the industry. Consider such items as how the firm is regarded by its competitors, and by its customers, and whether any adverse publicity has arisen concerning shady or illegal dealings.

2. Whether or not the firm is engaged in interstate commerce—this is especially important when considering germicides, insecticides, and other federally controlled products.

3. Delivery—does the supplier have a local outlet, and can he provide emergency deliveries when the need arises? (Remember that emulsions, such as waxes, can be ruined when a long delivery causes the material to be exposed to repeated freezings.) The manufacturer typically has better delivery than the jobber, since it doesn't have to go through a second pair of hands.

4. The performance of the supplier, in terms of fulfilling commitments.

5. The qualifications of the salespeople that represent the company: their experience, training, and understanding of their own product line (many of the sales reps in the sanitary supply industry concentrate on selling a very small percentage of their own company's products, and know very little, if anything, about other than their own product line).

6. Price (this must not become the overriding factor to the exclusion of all the other items).

7. Laboratory control, to assure uniformity of product and its performance (again, the manufacturer has the edge over the jobber).

8. Whether or not the manufacturer or seller is willing to provide demonstrations.

9. The type of packaging offered.

As with the purchase of any type of material, the purchase of sanitary chemicals should follow three basic criteria: cost, performance, and delivery—but the other factors mentioned above can be important.

As with equipment purchasing, standardization has an important part to play. If products are continually changed in order to effect a small cost reduction, the savings can be lost several times over in terms of retraining personnel, errors that may be committed, deterioration of quality, and the like. You should change products only if a significant cost reduction is indicated.

Five Methods for Selecting Cleaning Products

1. Reliance on a Reputable Supplier. A reputable firm represented by a knowledgeable salesperson can save a great deal of time in product selection. For the smaller housekeeping operation, this may well be the best approach, where time is not available for the more lengthy and complex procedures described below. For other than the smallest operation, it might be desirable to purchase from two or three suppliers, so as to retain the advantages of competition.

2. Product Testing. For the larger operation, where the purchase of chemicals represents a sizable investment, it is economically feasible to devote enough time to this subject to be more selective, and to test products under various conditions supplied by manufacturers. Purchase samples of one-gallon size or larger for testing, to eliminate any obligation to the supplier. Practically speaking, you want to purchase products that will do a given job—your interest lies little in the field of chemical specifications except in terms of any hazard that specifications might avoid. Most supervisors have neither the chemical background nor the means of testing samples to be sure that specifications are fulfilled. To them, the product test tells all.

In chemicals testing, as with equipment evaluation, involve the actual workers who will use these materials—if they are part of the decision-making process, they will be much more apt to use the chemicals selected. A number of suppliers will offer to apply the materials themselves, such as a floor finish, to give the customer a test—this should be very carefully controlled. Instead, ask the supplier to provide instructions in the use of the material, and then have your own housekeeping personnel actually perform the test application since, after all, they must perform it in the future. Figure 17-1 is a testing report for new products and equipment.

3. Purchase by Specification. Where chemicals are purchased on the basis of chemicals specification, the custodian typically complains of the regular deterioration in quality from year to year. The General Services Administration, which is responsible for maintaining public buildings for the U.S. government, has a continual struggle with this problem. The composition of the product, such as its solids content, pH, content of certain chemicals, etc. may have a limited effect on a given product's performance—and only performance specifications are really meaningful. Yet, performance tests are difficult to establish—such tests for cleaners, waxes, and floor sealers have not been fully standardized or accepted. Laboratory tests to simulate such performance are sometimes useful but often erratic.

Where bid purchasing is required by law, such as for governmental bodies, it is understandable that specifications are necessary, but the specifications should include both chemical and performance specifications, and surely a system of rejection should be possible through performance failure, even where the chemical specifications are in full accord. Further, where specification buying is necessary, at least two varieties of certain materials should be available for the user to select from, especially for such key chemicals as floor finishes, hand cleaners, and disinfectants.

If your organization is not required to purchase on a bid basis, avoid specification buying of chemicals (although it is perfectly applicable for other types of materials in other departments).

Figure 17-2 shows a sample equipment specification, in this case for a two thousand-RPM buffer. Figure 17-3 shows a sample chemical specification, in this case for a concentrated neutral detergent.

4. Purchase of Certified or Approved Products. Various organizations will develop approved purchasing lists for use by their member organizations. There are some serious problems with such an arrangement—for one, the standards are often minimal, and the products approved may not

Figure 17-1. Sample Equipment/Product Testing Report

TESTING REPORT FOR NEW PRODUCTS AND EQUIPMENT

Type of Equipment/Products(s): ______________________________

Trade Name: ______________________________

Approximate Cost: $______________

Manufacturer or Distributor: ______________________________

Address: ______________________________

____

Phone: ______________________

Recommended Use: (Attach brochure if available) ______________

____

Outstanding Features: ______________________________

____

____

____

Weak Points: ______________________________

____

____

____

Evaluation: ______________________________

____

____

____

Tested by: ______________________________

Title: ______________________________

Branch: ______________________________

Phone: ______________________ Date: ________________

Figure 17-2. Sample Equipment Purchasing Specification

<u>2000 RPM HIGH SPEED BATTERY POWERED BUFFER</u>

<u>USES</u>:

Spray-buffing and polishing resilient tile and terrazzo floors.

<u>PURCHASE GUIDELINES</u>:

- [ ] Pad pressure adjustable, load meter.
- [ ] Pad speed 2,000 rpm.
- [ ] Pad size 20" (51cm) diameter.
- [ ] Motor 2 1/2 hp permanent magnet DC.
- [ ] Drive 12 groove poly-V belt.
- [ ] Three 180 amp-hour, 12 volt batteries.
- [ ] Welded structural steel frame. All metal framework and housings protected with epoxy coating.
- [ ] Sheet metal with molded A.B.S. hood housing. Plastic hood completely opens to expose the entire interior for easy battery access, charging, and servicing.
- [ ] Two 20 cm (8") dia. ball-bearing cast aluminum rear wheels with neoprene tread. Front swivel caster, 9 cm (31/2") ball-bearing.
- [ ] Bumper rollers.
- [ ] Two circuit breakers.
- [ ] Master on/off control.
- [ ] Mounted compressed air sprayer for spray-buff solution.
- [ ] 110 volt battery charging unit

Figure 17-3. Sample Chemical Specification Guide

Specification for Concentrated Neutral Detergent

SCOPE:

1. This specification covers one grade of a liquid concentrated compound suitable for wet cleaning of both painted and unpainted surfaces where hard or soft water prevails.
2. An effective cleaning agent for use on wood, rubber, asphalt tile, terrazzo, marble, concrete floors and other surfaces.

REQUIREMENTS:

1. Compound shall be composed of synthetic organic detergents, sequestering, suspending and other cleaning agents.
2. The ingredients shall be assembled to form a homogeneous liquid with no more than a trace of suspended matter. It shall be biodegradable and mildly perfumed.
3. The compound shall be non-caustic and contain no soap.
4. The compound shall be completely soluble in distilled water at room temperature.
5. The compound shall contain no free alkali or ammonia.
6. The compound shall contain no free oil, abrasives or other harmful ingredients and shall not be irritating to the skin.
7. The compound shall contain no more than 87% by weight, of matter volatile at 105°C.
8. The compound shall be stable and not lose its original effectiveness or otherwise deteriorate when stored for nine months in a closed shipping container at room temperature.
9. The synthetic organic detergent shall be non-ionic or anionic type, and formulated in such a manner as to assure its being packaged in mild steel-type containers.
10. The pH of the compound shall be no higher than 9.9 (Beckman pH meter) at a 1% concentration in distilled water.
11. Emulsification of grease, oil and dirt—very good.
12. Free rinsing—excellent.
13. Foaming—moderate foam with excellent stability, in the presence of grease and oil.
14. The undiluted concentrate shall have a viscosity of not less that 450 cps. or more than 850 cps. at room temperature (20°C.).

Figure 17-3. (continued)

15. **Cleaning Efficiency: A one to one hundred and twenty-eight (1-128) dilution of the product shall exhibit a cleaning efficiency of not less than 80% when tested as described in paragraph 4.4.6 of Federal Specification PC-431a. In solution, the product shall provide adequate but not excessive suds.**
16. **The product shall be safe for use on all surfaces when used as directed.**
17. **Hard Water Tolerance: The product shall meet all other requirements of the specifications in water up to and including 750 p.p.m. of hardness.**

---

represent the best investment in terms of utilization of labor although they may represent the more economical purchase from a "dollars per gallon" standpoint. Then, such approvals are only as good as the criteria for their establishment, and the laboratories utilized for their policing (non-existent in a number of cases). No system may be in effect to cross-check and upgrade such lists, yet most sanitary supply manufacturers will change their formulas every couple of years or so and no provision may be made for the introduction of new and improved products from the same or other suppliers. From the viewpoint of the chemical supplier, it is a costly and unnecessary expense, and many reputable manufacturers refuse to submit products for such listings, while other manufacturers find this a means of having marginal products listed on par with top-quality performers. A product that performs well in one locale, may not perform at all well in another. The certified products list has too many drawbacks to make it a valuable purchasing tool unless a great deal of time is spent on a continual basis in its establishment and updating. A better service would be to list suppliers who generally produce inferior merchandise, do not operate in a reputable fashion, or who have no laboratory control over the quality of their products.

Figure 17-4 is a form used for equipment purchases, to control cost and to standardize on type of equipment.

5. Independent Testing Agencies. For the larger purchaser, the results of an objective organization can save a good deal of time in product evaluation, but in the final analysis, it should be supplemented with your own product evaluation.

Figure 17-4. Sample Record of Equipment Purchases

HOUSEKEEPING EQUIPMENT

QUANTITY	EQUIPMENT	MANUFACTURER	PURCHASE PRICE	COST

VENDOR VERIFICATION

Quotation Provided by: ______________________ Date: ______________________

Telephone: ______________________ Signed: ______________________

Company Name & Address:

______________________ ______________________

______________________ ______________________

______________________ ______________________

Special Considerations in Selecting Products

When selecting products, consider these facts to help you make your decision.

• Location and size of the central storage area (and satellite central storage areas) may control the quantity, type, and packaging of materials under consideration.

• The size of custodial closets and cabinets (often designed below the barest minimum requirements) may impose some purchasing limitations. For example, the type of shelving may eliminate the possibility of purchasing a certain product.

• The size or weight of a container may represent a safety hazard which must be avoided.

• The material from which the container is constructed may affect your decision. For example, there may be a scrap or re-use value in steel containers, glass bottles represent a safety hazard that should be avoided, and plastic containers may not be digestible in the waste grinding system etc.

• Shelf-life can be a strict limitation on some products, especially emulsions which may tend to break after a period of time (don't forget to rotate products on a "first in, first out" basis).

• The weather may dictate that certain products be purchased only in a given season due to freezing which may occur in harsh climates.

• If the product is flammable—especially if it is a "red label" product—only certain quantities in certain containers may be stored at any one time, under controlled conditions.

• Products representing safety hazards, such as strong acids and caustics, must be carefully controlled in terms of quantity and packaging, and storage conditions.

• Many organizations find a considerable benefit in arranging an annual purchase commitment for products, at least those which represent a sizable percentage of the total chemicals cost. Under this arrangement, investigation of new products may only occur once a year for a period of a couple of months, before the next annual purchase commitment is established. Ideally, this arrangement obtains a considerable price reduction on the basis of the quantity purchased, but it is an estimated quantity, and delivery is made according to the customer's needs and storage capacity.

• Services available from suppliers may be the deciding factor where other things are merely equal—and they very often are in the sanitary supply industry.

• To increase discounts due to quantity purchase, some organizations arrange a joint purchase where two or more buyers pool their orders (this can be done just as successfully for equipment as it can for chemicals).

• The supplier who is willing to provide demonstrations of the use of his product would naturally receive preference over that supplier who avoids this.

• Purchase from suppliers who have sample and smaller demonstration quantities available for purchase, as well as pertinent literature available.

• Some buyers seek chemicals marketed with "tie-ins," such as where a free electric sprayer is given with each drum of deodorant. Some buyers appreciate this, since they are limited in purchase of capital equipment but not in the purchase of expendable supplies! Obviously, nothing of this sort is ever free, and the product is priced to cover the cost of the "tie-in." When certain suppliers devote much of their sales effort to these tie-ins, one is led to suspect that the emphasis of such a company is not in the production and marketing of quality chemicals. Where tie-ins are of a personal nature, such as a "free" radio, they may be unethical or illegal.

• Sometimes purchases are made to beat a price deadline, due to increased freight rates, higher raw material costs, or the like.

• The system of stock distribution and control, tied in with packaging mentioned above, may also be a factor in the product selection.

HOW TO DISTRIBUTE SUPPLIES MORE EFFICIENTLY

Distribution of supplies can be handled either on a pick-up or delivery basis. On the pick-up system, the custodian presents the list of requirements at the central storage area and returns with these materials to his or her work station. In the smaller operations, the custodian may actually remove the materials from the central storage area, or, even more desirable, one of the custodians may be assigned this work as a part-time duty. If group leaders are used in the housekeeping operation, one of them should handle this function.

The delivery system is a much better means of supply distribution. Thus, each custodial closet or cabinet is properly stocked periodically against a supply list posted on the door. The custodian using the closet always has sufficient materials and equipment with which to perform his job and never needs to spend time obtaining them. In an extensive operation, a special vehicle may be required for this purpose. In some cases, the driver of the vehicle can also perform the function of stock clerk, but in the largest departments both personnel may be required.

In larger facilities, a full-time supply room attendant or stock clerk would perform these functions:

- Maintain perpetual inventory and product-consumption information.
- Make minor repairs and lubricate mechanical equipment.
- Launder and treat dust mops and dust cloths.

- Repackage bulk materials into smaller containers.
- Fill orders for the housekeeping department, as well as other departments using cleaning materials.
- Dilute concentrated detergents, liquid hand soaps, etc.
- Prepare requisitions against the purchasing department for the housekeeping department manager or a designated assistant.
- Receive materials into the housekeeping department.
- Keep the central storage area clean and properly organized.
- Perform other functions as assigned, such as special pick-ups or deliveries of housekeeping materials.

Figure 17-5 shows a typical requisition for supplies.

However chemicals are purchased, they should be distributed only in properly labeled containers. Each container should be labeled as follows:

- Product name
- Chemical content
- Product use
- Instructions for use
- Dilution instructions
- Hazard warnings

Such labeling is a federal regulation under the Occupational Safety and Health Administration, and is commonly known as the Right to Know Act.

14 WAYS TO IMPROVE YOUR INVENTORY CONTROL

A well-organized inventory control procedure can provide these important benefits:

- Assure that housekeepers will have materials and equipment on hand with which to perform their work properly.
- Prevent an overly large inventory that ties up excessive money and space.
- Minimize time and effort in the requisitioning and ordering process.
- Uncover wastage, failure to use the proper product, or pilferage, by pinpointing consumption in various areas.
- Provide a basis for comparing efficiency.

Figure 17-5. Sample Supply Requisition

JANITORIAL SUPPLY ORDER FORM

DELIVER TO:

Station No. ____________________

Requested by: ____________________

Building: ____________________

Date: ____________________

Authorized by: ____________________

Room: ____________________

Date: ____________________

Building No. ____________________

Order Date: ____________________

INSTRUCTIONS TO ALL JANITORS: Use this form to order supplies. Fill in amount of supplies needed in "Quantity Ordered" column and turn in to your supervisor. When supplies are received, note any shortages, and return form to your supervisor.

DESCRIPTION	STOCK NO.	PKG. UNIT	QUANTITY	
			ORDERED	RECEIVED
Chalk				
Erasers				
Ammonia				
Bowl Cleaner				
Floor Soap				
Glass Cleaner				
Rug Shampoo				
Disinfectant Detergent				
Cleanser				
Hand Soap				
Spray Bottle (16 oz.)				
Wax Bag				
Poly Bag for Wastebasket				
Trash Receptacle				
Insect Spray				
Mop Dressing Oil				
Head Mop (12 oz.)				
Head Mop (24 oz.)				
Pads Spray Buff				
Pads Strip				
Tissue (case)				
Towels (case)				
Dust Cloths				
Wiping Rags				
Sponge				
Scrub Pads				
Floor Finish				
Rubber Gloves				

Here are 14 ways to control inventory:

1. Set up a simple record-keeping system and follow it. Use a card index file or simple looseleaf notebook. Make a card or separate page for each product purchased. For example, use a separate stock page for floor cleaner or all-purpose detergent, disinfectant, floor wax, sweeping compound, dust-mop treatment, floor seal, liquid hand soap, bowl cleaner, etc. As for equipment, typical items that would require separate pages might be dust mops, wet mops, push brooms, wax applicators, mop buckets, utility trucks, floor machines, scrubbing pads, etc.

To keep track of all of these different items, the record sheet should indicate:

Name of the product.

Supplier.

Minimum stock.

Quantity to order each time.

Average use per month.

There should be a space to record the date that new orders for the items were placed as well as the date shipment was received. The stock sheet should also show what quantities of the item have been taken out of the central storage area and distributed to various custodial storerooms on different floors or different areas of the building.

2. Know the correct methods of using each product and correct rate of usage. This prevents waste and incorrect use. Larger organizations should develop consumption standards for repetitive jobs so that consumption for new work as well as continuing work can be estimated. You have a right to expect guidance on product use and work methods without charge from suppliers. You should purchase the material from manufacturers who can provide suitable data on the correct frequencies and probable consumption rates of products.

3. Train workers not to overuse products. An untrained worker will often use a full cup of detergent per pail of water where the instructions call for one-half cup. He will use the most concentrated recommended dilution for the lightest duty jobs, or will add another cup "for good measure." The only reason for the custodian overusing the products is his belief that it will make the work easier. Only when he is convinced that it is *not* to his advantage to overuse materials will such wastage stop. Consumption control records alone will not stop it because the custodian can utilize less water in order to increase the concentration. Proper training in use-dilutions will explain and demonstrate the *extra* work caused by overuse of detergents, for example, because of the formation of too much suds, the

need for excessive rinsing, and the formation of a soap film which collects dirt.

4. Train custodians to read labels or instruction sheets and to follow the directions carefully. When the correct methods are used, as well as the proper use-dilutions, the materials will be serving the user at highest efficiency.

5. Utilize use-control aids to provide accurate concentration. A graduated plastic measuring cup is safe and easy to use and eliminates guessing in measuring. Make the cups with a fill-to-line where they are to be used for a single product. Proportioning or simple metering devices predetermine the concentration and mix automatically. The use of such equipment contributes to the custodians' sense of status. The expense, however, may be considerable—but some suppliers offer these devices on memo charge.

6. Minimize spillages and leakages by having small containers filled from drums as a function of central storage and by the adequate distribution of funnels to all custodial closets. When drums are stored on racks, they must be supplied with faucets which do not leak. Place a bucket under each faucet to catch any minor dripping. Use drum pumps with the drums standing vertically; this avoids all possibility of leakage and dispenses measured amounts of the material.

7. Dilute on arrival certain concentrated materials (such as liquid hand soaps) at the central storage area and mark them "ready for use." This provides an accurate dilution and eliminates the task of making a dilution at the use point, thus conserving time as well as materials which may be lost through spillage or failure to dilute. If a container has been purchased which requires dilution with an equal amount of water, a simple procedure is to take an empty container of the same size and fill it half-full of water. The same container is then filled full from the container of concentrated material. The half-full container of concentrated material is then filled full with water (the diluent may be mineral spirits or some other solvent in the case of insecticides, etc.). Water under pressure should never be run into a concentrated detergent as this causes a great amount of foaming and will waste a good deal of time. The water should be poured through a large funnel, to which a hose is attached that reaches to the bottom of the container.

8. Purchase products in controlled packets, and issue a given number of packages to perform a certain job, thus avoiding wastage. The economics of using such controlled packages should be studied comparing packaging costs to housekeeping wages. Pilferage is more difficult to control with such packages, and the packages contribute to litter.

9. Avoid misuse through the careful marking on arrival of all containers with the nomenclature used in the housekeeping department. Each storage area should have a given position for every type of material. Color-coding containers helps when dealing with illiterate workers or with a high turnover situation. The misuse of a product not only wastes the material

that is used, but also often requires an additional product of another type for its removal. More important is the damage to surfaces or injury to workers which might occur.

10. Establish a simple but complete inventory control system to provide a record of materials distributed and pinpoint consumption at the use point. This will help you track pilferage, appropriation by other departments, and wastage and then correct the cause. The formal system on paper should be supplemented with a sight check periodically, such as twice monthly. A form for this purpose is shown in Figure 17-6.

11. Keep all custodial closets, cabinets, and other supply areas locked when not in use. Housekeeping materials are particularly subject to pilferage and personal use because of their universal applicability in business as well as in the home. The key should be retained only by the custodian using the closet, the supervisor, and the housekeeping department manager. Any individual assigned to distribute supplies to all closets must also have a key, of course. If materials are issued on a "pick-up" basis, empty containers or worn-out supplies should be turned in to justify replacement. On a "delivery" basis, the absence of containers or minor equipment should be reported.

12. Avoid breakage by purchasing materials in plastic containers rather than in glass containers.

13. Prevent loss of materials because of aging or spoiling. You can do this by purchasing so that supplies do not remain on hand for too long a period of time. The housekeeping department should have a one- to three-months' supply of housekeeping materials on hand, depending on the type of material and consumption rate. Orders for general replacement of housekeeping materials should be made monthly. Older materials should be used first.

14. Arrange adequate storage facilities properly to prevent damage and loss due to wetting, overheating, freezing, soiling, or contamination.

CHOOSING CONTAINERS AND PACKAGING

Housekeeping products are now packaged in a wide variety of containers. The choice of container sometimes has a direct effect in aiding or hindering the custodian in his or her work.

There are 7 basic types of containers.

1. Closed-Head Steel Drums. This is the most common bulk container for shipping liquid products. The typical drum has a varnished inner coating to prevent rust by the action of water vapor. Special baked-on phenolic resin liners are available for mildly corrosive chemicals. Mild acids will require a plastic lining and plastic plugs. Strong acids should not be shipped in this

Figure 17-6. Sample Supplies Inventory Record

APPROVED MATERIALS, INVENTORY & ORDER POINT

DESCRIPTION	INVENTORY	ORDER POINT

Inventory Conducted by: ____________________ Date: __________

Ordering Instructions:

type of container as denting may cause lining failure which will result in leakage.

For many years the standard bulk container has been the 55-gallon drum. A drum of wax or detergent will normally weigh about 515 pounds. Sixty-five gallon drums are also available from certain suppliers. Even though these may in some cases offer a slightly lower unit cost than the 55-gallon drum, they should not be used. Generally weighing some 610 pounds, they are awkward and dangerous to handle.

Smaller drum sizes are available. Thirty-five or 30-gallon drums are known as "half drums," while 20- or 15-gallon drums are known as "quarter drums."

In all these cases, even though a lower unit price is obtained because of the quantity purchased, the hazard of using heavy containers far outweighs the economy. Many organizations that have suffered regular lost-time accidents due to handling drums have not only eliminated these accidents when purchasing in smaller containers, but also have lowered their total chemicals budget by avoiding waste and over-consumption.

2. Cans and Pails. The middle range of containers for liquids are cans and pails, which normally do not require the use of a faucet. The most common sizes are 12, 10, 7, 6, and 5 gallons. Most of these containers are fitted with a metal or plastic pouring spout. Tilting racks are available for these containers, but they have not gained wide acceptance. Probably the best combination of unit price and ease of handling is available in the 5-gallon container. These containers are also available with the special corrosion-resistant linings used in the drums.

3. Cans, Jugs, and Bottles. These are the smallest containers for liquids. They are available in a wide variety of styles, types, and materials, ranging in size from 2-gallon cans to tiny glass ampoules. Because of the greater safety in the handling of plastic containers, you should not accept glass containers as a container for housekeeping chemicals except for those materials not compatible with plastic, such as certain types of solvents.

4. Aerosol Containers. A number of chemical products are available in aerosol containers, but these are extremely expensive per unit quantity consumed, and also there are ecological problems concerning some propellents, and disposal of the cans. Typically, a pistol-grip small plastic spray bottle gives good performance at low price. Aerosol cans, however, may be considered for infrequent insecticide use and for nitrogen propellent in order to freeze and remove chewing gum.

5. Fiber Drums. These are primarily used for packaging powdered chemicals. These containers are measured in gallons of volume and range from 70 or more gallons to just a few gallons. In the lower-size range they are known as fiber pails or tubs. Fiber containers must be stored so that they will not become wet, as this will cause most products to cake and thereby become useless.

Some fiber containers, with a plastic inner bag, are used for the shipment of liquids. This has met a good deal of resistance because when such containers are wet by accident they may collapse. They may also be punctured relatively easily by drum trucks having sharp forks.

6. Open Head Steel Drums. This type, as well as fiber drums with plastic liners, are used for the shipment of paste or jelly-type soaps. Some of these drums contain built-in solutionizing attachments, where water is run into the drum and then drawn off as it dissolves a certain amount of the soap. This type of container is not often seen now, as paste soaps are falling into disuse, and also because of the variability of the concentration and the fact that complete solution may not take place.

7. Special Packages. Special containers are also available for specific uses. Multiwall bags are used for inexpensive bulk materials such as sweeping compounds or oil absorbents. Chipboard cartons are used for packaging small amounts of powdered materials. Fiber or plastic shaker-top containers or metal or plastic cans and cups are used for powdered and paste-type cleansers and polishers.

Carboys of plastic or glass are used for shipping acids and other highly corrosive materials. These are generally shipped in an overpackage of steel, fiber, or wood. Some of these combinations require a returnable container arrangement which is costly in freight and handling costs. The combination of plastic carboy and fiber or wirebound wood overpackage, in the size range of 10 to 15 gallons, provides a suitable container that is economical enough for a single-trip use.

Figure 17-7 shows the principal forms of sanitation chemicals packaging, while Figure 17-8 shows recommended packaging for a basic chemical inventory.

Figure 17-7. Types of Chemicals Packaging

	PRO	CON
Aerosol can	*Sprays with force (example: insecticides) *Compact-easy to carry	*Very expensive *Propellant may be air pollutant *Pilferable *Hazardous to incinerate
Premeasured packets	*Convenient *Eliminates measurement	*Expensive *Empty package creates litter *Difficult to adjust quantity *Pilferable
1-gal jug with dispensing pump	*Pre-labeled *Convenient measurement	*May be more expensive than larger containers
5-gal bucket	*Reusable bucket *Excellent for waxes and sealers	*Weighs about 45 pounds *No measurement system

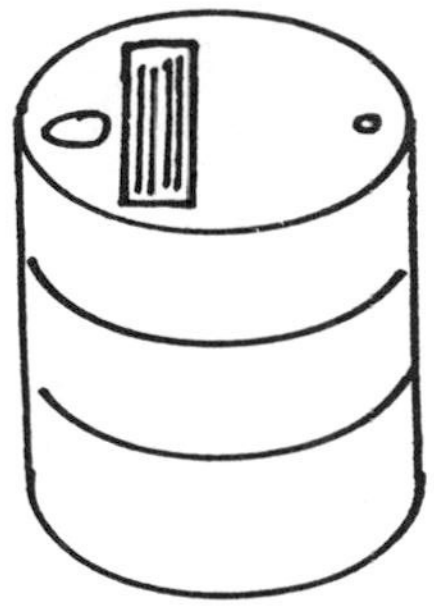 55-gal drum	*Cheapest unit price	*Hazardous to handle (weighs about 515 pounds) *Last gallon or two may be unusable *Takes time to put in other containers *Labeling problems
 Dispensing station	*Looks professional *Dilution pre-set	*Large investment *Maintenance cost *Not portable (travel time losses)

Figure 17-8. Recommended Packaging for Various Chemicals

Type of chemical	Recommended Packaging
Acid-type bowl cleaner	1 qt. squeezable container with flip-top cap
Auto scrub cleaner	5 or 6 gallon container of concentrate
Carpet shampoo	1 gallon container of concentrate
Carpet stain remover	1 gallon container of concentrate or pressurized, aerosol containers
Concrete & terrazzo seal	5 or 6 gallon containers
Degreaser	1 gallon container of concentrate
Dust mop dressing	5 or 6 gallon containers
Floor finish	5 or 6 gallon containers
Floor finish remover	5 or 6 gallon containers of concentrate
Germicidal detergent	1 gallon container of concentrate
Graffiti remover	Pressurized, aerosol container
Gum remover	Pressurized, aerosol container
Glass cleaner	1 gallon container of concentrate
Lotion-type cleanser	1 qt. squeezable container with flip-top cap
Neutral detergent	1 gallon container of concentrate
Soap and scum remover	1 gallon container of concentrate
Spray-buff solution	1 gallon container
Stainless steel cleaner & polish	1 gallon container of concentrate

These types of chemicals are adequate for almost all cleaning tasks.

Part THREE

OPERATIONS

18

The Three-Stage Approach to Preventing Soil and Trash Accumulation

In all forms of maintenance activities an ounce of prevention is worth a pound of cure, as the old saying goes. In custodial work we are concerned with time, which typically represents over 90% of a building cleaning budget. We might alter our saying to claim that an hour spent in soil prevention might save ten hours later in removing that soil.

This is not always readily apparent. For example, one company questioned the importance of using entrance matting. However, when the maintenance manager put down a single length of matting 4 feet wide by 12 feet long, by the next day, it had collected 12 pounds of dirt—or one pound for each linear foot of matting! Without the matting, that soil would have been distributed throughout the building, not only on the floors but on all surfaces, as floor soil tends to become airborne after a time.

This serious and important subject can have humorous overtones. At another company, a custodial worker objected to entrance gratings because the catch-pan underneath kept filling up with dirt which he had to remove! (Apparently it did not occur to him that it is easier to shovel dirt out of a pan by the kilogram than to clean it up by using a mop or other manual tool to remove it by the microgram.)

Another story concerned a custodian responsible for the lobby of a building—he was assigned to keep carpet matting in place. One day his supervisor, not seeing the matting at the entrance, asked the custodian where it was. The worker replied that he had rolled it up and put it in storage because, since it was raining, the matting would get dirty! Without the matting, of course, it was the lobby floor which was getting dirty and slippery.

STAGE 1: PREPARING THE BUILDING ENTRY TO PREVENT SOIL ACCUMULATION

Each building entrance should serve as a barrier for reducing the volume of soil brought in by foot traffic. The same would be true inside a building, where areas change from a heavily soiled situation to that such as from a manufacturing area to an administrative area, or a shop area to a cafeteria.

You should do as much as possible to prevent soil accumulation in your facility. A good place to begin is the walkway to a building. For example:

• If you have input to the design and construction of the walkway, suggest that the walkway be canted or tilted to one side. Or, suggest crowning the walkway so that it is higher in the middle than on the sides (although this method is more expensive to pour and finish). Either way, water and mud are less likely to collect on the walkway and be tracked into the building.

• Finish the walkway with a rough surface to trap soil before it is tracked into the building.

• Install a canopy over the walkway to prevent rain or snow from falling directly onto the walk.

• Use snow-melting devices so that snow is not tracked into the building; these can also prevent accidents.

• If there is an exterior stair landing, this might be an ideal place to install a grating, especially if it is protected by a canopy. Since one of the problems with a grating is removing accumulated water, it is easy to provide a drain from the grating to the outside through the landing. Note: gratings should not be connected to plumbing systems, as they tend to foul them up.

In many cases, the space between the two sets of entry doors is the best location for the grating, and this preferably should be drained to the outside as well. The grating should be as large as possible, preferably covering the full area of the entryway. The grating bars should run perpendicular to the line of traffic—the bars should not be spaced more than 1/10″ apart to prevent any possibility of narrow shoe heels being caught in the spaces. Some types of grating are provided with plastic ribs, which act as squeegees, to remove as much soil as possible from the soles of shoes. Naturally, the grating must be recessed to present a level flush with the floor so that there is no tripping hazard.

• Immediately behind the grating, put down some form of matting to trap that small quantity of soil and moisture which the grating did not receive. Ideally, this would be a type of carpet runner or removable carpet squares. This provides a 2-stage soil prevention system. The matting should preferably be in a recess, again to avoid tripping accidents. Keep this matting clean and replace it when necessary. The matting should have good absorption characteristics to retain moisture in large quantities. It should be large enough and properly located so that each person passing over it will take at least two steps (preferably three) across the mat with each foot. This would require a mat to be at least 12 feet and preferably 16 feet long. Be sure that the mat is wide enough so that a pedestrian will not step off it when entering the building.

STAGE 2: INSTALLING FLOOR MATTING AT ALL ENTRANCE AREAS

From a housekeeping standpoint, the purpose of floor matting is to entrap soil at entrance areas, or where the soil is produced, so that it is not tracked throughout the facility. Matting is used because it is easier to keep dirt out than to remove it once it is inside.

Door Mats

The most common type of mat is the door mat. Door mats are available in the following types:

1. Rubber Link. These mats are very often made from old tire carcasses, but they may also be manufactured from special molded rubber links. This type of matting, because of its many openings, will collect a good deal of soil. It has the disadvantages that it is difficult to clean and may cause accidents by catching women's heels.

2. Tire Strips. Similar to above, but without openings. They don't trap much soil, however, and they are difficult to clean.

3. Molded Rubber with Perforations. The openings in this type of matting perform the same function and present the same hazards as the link mat, but to a lesser degree.

4. Molded Rubber with Squeegee-Type Ribbing. This is the safest type of matting because it presents no tripping hazard. The design tends to wipe water and soil from the bottom of the feet and collect it in the grooves.

5. Carpet Surface. This type of matting is suitable only for use at inside doors, such as the entrance to a private office.

6. Coco-Mats. Coco-mats are safe if recessed and particularly efficient in collecting water. Their chief drawback is they are very difficult to keep clean.

7. Woven Grass. This type of mat is chiefly decorative. It quickly becomes unsightly and is difficult to keep clean.

8. Gelatin Mat. This is a very specialized mat for extremely efficient removal of fine soil particles, such as in entrances to dust-free areas.

9. Molded Plastic. This is durable and not difficult to clean.

10. Plastic Fiber. This type of mat is attractive and very effective. It is available in open or solid back types.

Runners

Runners are matting material manufactured in length, for use in corridors, aisles, behind counters, etc. They are generally made of the same material as door mats. In addition to these, there are special types of runners:

- Runners made of grease-resistant materials.
- A decorative type of vinyl runner, in a variety of colors and patterns.
- Steel link runners, which are useful for the removal of metal chips from the soles of the shoes.

Special Types of Matting

Many types of special matting are made. Here is a sampling of the wide variety available:

1. *A "tacky mat"* consists of a stack of plastic film with a sticky side up, somewhat similar to a large stack of jumbo cellophane tape! This type of mat has the advantage of being able to remove great quantities of very fine dry soil and is useful for critical areas such as industrial super-clean rooms. When one surface becomes "loaded" it is simply peeled off, exposing the next layer.

2. *A "sponge mat"* is a device sometimes used at the entrance to hospital operating rooms or research facilities involving bacteria, where the pedestrian steps into a stainless steel pan containing a sponge which is impregnated with a disinfectant (such as a synthetic phenol). The reason for using this device is that wet disinfectant is most effective in bacteria control. For less severe conditions, a simple soil remover is sufficient since most bacteria are attached to small soil particles.

3. *The "mechanical mat"* is a device consisting of parallel metal bars between which are nylon bristles. When someone steps on the mat (or crosses through an electric eye), the mat is actuated and the bristles begin to vibrate, thus literally scrubbing the bottom of the shoes, with the soil falling

between the grating into a catch-pan beneath. This is a rather expensive device requiring considerable installation time—ideally, it should be installed with new construction or renovation. It is very efficient when quite a bit of soil must be removed in a small space (such as the entry from a factory producing products using carbon black, chemical dust, or the like) before entering an administrative area.

4. *A "removable grating"* might be considered a special form of matting because this mat consists of slats with a carpet or plastic surface spaced to form a grating. The "mat" is constructed in such a way that it can be rolled up and removed for cleaning. The floor underneath must be protected so that the soil falling between the spaces does not grind the floor with foot action on the slats.

5. *Carpeting selected areas* is another way to prevent the transfer of soil from one floor to another. The best examples of this include the carpeting of elevator floors and stair landings. Such mattings will be easier to maintain if both the interior of the elevator and the stair landing contain an electrical outlet.

6. *The laminar airflow entrance* can also be considered a special type of matting. In this case, a flow of air from above passes over the pedestrian who walks across a grating and the air is pulled into the grating beneath. This provides a cleaning action not only for the shoes but also for removing soil from the clothing as well. This is a device used at entrances to prestige buildings, industrial clean rooms, some department stores, aerospace activities, etc.

Special Mats

To complete the matting picture, we have these special types:

- Decorated entrance mats, containing a logo or trade-mark, various kinds of wording, or colorful designs.
- Matting may be cut to special sizes and shapes or fitted to special situations, such as around columns or fountains.
- Molded rubber stair-tread coverings may be considered as a type of matting.
- Shelf mats prevent the damage of fragile materials such as laboratory wear. Their design also allows the draining of washed parts.
- Fatigue mats are placed around machinery and other places where personnel must stand for greater comfort.

NOTE: All matting should be removed and cleaned regularly, since if placed on a finished floor it can permanently damage the floor by the action of moisture and bacteria beneath it. This is especially true of matting with rubber or plastic backing. You should also develop a system for scheduling when to clean and replace mats in every location throughout the facility. Make a list of all mats in the facility, what type they are, what size they are, and how often they should be cleaned and replaced. Your custodial staff can then use this as a cleaning checklist.

STAGE 3: PREVENTING TRASH AND WASTE ACCUMULATION

Other approaches to preventing soil accumulation pertain to ordinary trash and waste. Here are 7 ways to prevent trash and waste from accumulating in your facility.

1. Place or mount waste and ash receptacles wherever people are likely to discard waste or cigarettes. Don't wait for cigarette burns to appear on the floor to decide to install ashtrays or urns! Normally, both a cigarette urn and a waste receptacle (which indicates that a combination fixture is most desirable) are needed wherever there is a change of activity, such as at a doorway, in an elevator lobby, in the elevator itself, on a stair landing, at a water fountain, by each telephone, between every two urinals, etc. Wall-mounted urns and receptacles have the advantage that they do not impede floor-cleaning activities and are less apt to become damaged or to stain the floors.

2. Use plastic waste liners in all receptacles where wet waste, food, or drink is likely to be deposited. The liners avoid the costly washing of the receptacles, they eliminate odors and vermin problems, and they maintain the cleanliness of the receptacles with a minimum of cost. Liners need to be replaced daily in health care and research facilities, but in administrative-type areas, they should be replaced only when necessary (typically about once per week, often less frequently).

3. Encourage the use of large drinking cups with lids for drinks that are carried from one area to another. The frequency of spills should decrease.

4. Mount soap dispensers so that any dripping is into the bowl of the sink, not on the floor, the faucet or on the handle.

5. Make sure waste receptacles are large enough so that they do not become over-full, with a resulting waste on the floor, especially in restrooms. Further, make sure the push-plates to waste receptacles are kept clean; otherwise, people will refuse to use them and will throw their waste on the floor.

6. Use roll-type paper toweling, which tends to create less litter than folded-type toweling. In the latter case, toweling may fall to the floor (particularly where too small a size is being used)—and people tend to pull out two

or three towels, often dropping one. The roll-type toweling dispenses the paper in a single sheet (which is also less expensive).

7. Use high-quality floor sealers and finishes, which can significantly improve the maintainability of an area, while reducing airborne dust. *Specific advantages:* spills mop up more easily; the floor is more resistant to heel marks, scuffs, and grit; and the floor can be maintained at a high level of appearance with less frequent scrubbing and buffing than if inferior materials, or none at all, are used. The use of sealer on a fairly rough floor such as concrete also provides a surface which can be maintained with simpler equipment.

Case Study: a client in the garment industry had an entire carload of clothing rejected by a military service because the clothing was covered with a fine film of dust, not apparent to the naked eye. It was discovered that the plant factory had a concrete floor surface and the wheels of the carts were grinding the surface into dust, which then became airborne and sifted onto the fabric. The sealing of this floor with the proper solvent-type sealer not only reduced the dusting (which also comes from the pores of the concrete) but created a surface which could be cleaned with a treated dustmop rather than having to use more sophisticated and expensive equipment.

19

How to Control Graffiti

There seems to be such an instinct for wall writing that we must address how to control graffiti, rather than how to eliminate it. Several books have been written on graffiti as an art form, and some psychiatrists have even studied the psychology of the graffiti writer, in order to determine the reasons for creating graffiti. But custodial maintenance workers are more concerned with how to reduce the amount of graffiti, and they need to know how to remove the disastrous effect it can have on walls, doors, furniture, and other surfaces.

14 WAYS TO CONTROL THE AMOUNT OF GRAFFITI CREATED

Graffiti cannot be completely prevented, but it can be reduced considerably by providing surfaces that are more difficult to mark:

1. Carpet an elevator wall or an entrance door. This yields a surface that is extremely difficult to mark but is easily maintained and can be very attractive.

2. Wax restroom stalls of painted metal, using the same metal interlock polymer that can be used on the floor. This helps because it is difficult to write on a waxed surface.

3. Install a tough vinyl wall covering, not completely white and without deep or sharp fissures. This is especially helpful if it is impregnated with a chemical soil retardant.

4. Use laminated plastics such as "Formica" and "Micarta." These are also much easier to maintain than stainless steel or any other surface.

5. Cover concrete block walls and the joints between them with epoxy resin, which presents a glass-like surface difficult to mark and easy to clean. This is even better than glazed ceramic tile, since the grout between the tiles is so easily damaged. Whenever ceramic materials are used, however, the grouting used should not be white but of a colored material.

6. When installing new ceilings that are reachable, make sure they are not made of mineral materials, but of perforated metal pan, which can be both washed and painted.

7. Paint walls with enamel rather than flat paint.

8. Remove graffiti rapidly. This is by far the most effective means of graffiti control. Graffiti writers desire broad exposure and permanence to their writings—if you demonstrate to them that what they have written will not remain longer than 24 hours, then they may well give up. Rapid removal also eliminates the "growth" of drawings or poetry, since it does not provide time for people to embellish on what has already been written.

9. Ensure that all facility areas are well lighted. Good illumination is the next best approach to graffiti reduction. The human tendency is to do bad things in dark places, and a higher lighting level tends to dampen the graffiti spirit. Some managers may balk at this approach because they are concerned with energy conservation, but perhaps there are better areas to save lighting. Further, the illumination level can be intensified by using more light-reflective paints, keeping cleaner light fixtures, etc., without actually increasing the wattage.

10. Use graffiti wallpaper, where the pattern on the paper or vinyl surface is practically solid graffiti, the entire area being covered with words and pictures in various colors, so that any additional graffiti simply becomes lost to view!

11. If you can't beat them, join them. Provide writing surfaces in rest room stalls and on walls, by hanging note pads or chalk boards. An especially successful approach to this is a "graffiti board" in a public corridor, such as in a public school. People are invited to write or mark on the board, which is cleaned off and repainted periodically.

12. Try a more creative approach, such as the use of plywood walls surrounding construction sites, where artists are literally invited to paint the surface. An interesting case was seen in the fine arts building of a university campus, where the students were invited to paint and decorate the interior of the building—following this, the students themselves made quite certain that neither those in the building or visitors to the building ever marked one of "their" surfaces.

13. Punish offenders. It is possible that arrest and prosecution can have a beneficial effect, and perhaps the least that can be achieved from this is to require the person who commits the act and causes the damage to pay for it, or remove it personally.

14. Do not rely on the use of signs that say "Please do not write on the wall." This is the least effective way of controlling graffiti—in fact, it may have no effect at all. This is taken by some graffiti writers as an invitation to write on the sign!

REMOVING GRAFFITI MARKINGS

The removal and correction of graffiti markings can be an expensive and time-consuming activity. This might involve cleaning—perhaps more than one cleaning for effective removal. It may also require the use of poultices to draw foreign matter from porous surfaces, repainting, and sometimes even completely replacing the surface.

Various approaches are taken toward this problem. In one public school system, the arts classes are called on to paint over graffiti with contemporary designs. In some industries, rest room doors and even partitions have been removed where graffiti has become a problem very expensive to cope with. This removal might conflict with various legal requirements, but sometimes these laws are "bent": management claims that the doors or partitions have simply been removed temporarily for repair! Some college housing authorities, as well as public apartment buildings, assess a penalty for damage, including graffiti, which is deducted from an advance deposit.

There are many similarities between graffiti and vandalism, not only in the psychological causes but in the prevention and control. Usually, if you take action to limit graffiti, such action should have a beneficial effect on vandalism, and vice versa.

How to Care for Rest Rooms

Of all the areas possible to be constructed in a building, the only required type, by law, is the rest room—not only must there be rest rooms, but in sufficient quantity and with enough fixtures to serve the needs of the number of people who might use the building.

And the legal requirements go further: the rest room must be maintained in such a way as to be safe and healthful under OSHA and other regulations.

The care of rest rooms provides a microcosm of the entire custodial field, as it involves specific equipment, chemicals, training, procedures, as well as a number of special requirements such as those involving health and hygiene.

CHOOSING ECONOMIC AND PRODUCTIVE REST ROOM SUPPLIES

Here are some guidelines to choosing rest room supplies that are economical and facilitate easy maintenance:

- Hand soap—A wide variety of hand cleaners are available, but for a single item the most acceptable as well as economical is a liquid soap dispensed in a foam or lather form.
- Hand towels—For toweling, the roll type paper towel where the length is controlled through the turn of a crank or the push of a button provides very good economy, and gives much less litter than folded towels. Electric hand dryers consume too much energy and are not hygienically desirable, as germs are blown from the dried skin into the air.

- Toilet paper—Roll tissue is much more desirable than folded tissue; and a double dispenser is required to eliminate complaint calls.
- Waste receptacles—The waste receptacle should be of large enough size to accommodate the waste, and more than one installed if necessary. If a push plate is used, it must be kept clean so that the receptacle will be properly used.

Here is a brief review of some rest room design features:

- *Urinals* should be of the flooded open throat type, so as to avoid stoppages and odor problems connected with strainers.
- *Floor drains* should be in the proper location to permit the use of spray cleaning and to handle overflows.
- *Fixtures and partitions* should be wall or ceiling hung to keep floors clear for cleaning.
- *Floors* should be of ceramic material with dark sealed grout; resilient materials should never be used.
- *Proper ventilation* should be supplied.

Many custodial managers and supervisors—not to mention the cleaning personnel themselves—are frustrated with the problem of graffiti in rest rooms—more of this wall-writing appears in rest room facilities than in all other areas of a building combined. See Chapter 19 for detailed guidelines; some of the salient aspects of controlling graffiti are:

- The principal factor in control is rapid removal; the rule should be to remove graffiti within 24 hours.
- Better lighting means less graffiti, and the lighting should be arranged to illuminate the inside of the stalls as well as the balance of the rest room (merely keeping the light fixtures clean would help a lot).
- Surfaces should be used which are more difficult to write on while being easier to clean, such as epoxy resin on concrete block, or waxing metal partitions.

CONTROLLING ODOR

The control of odors in a rest room represents a problem that has several possible approaches.

- Do not use masking chemicals. The time and money spent on these should instead be devoted to cleaning and disinfection. Avoid the use of drip fluid, deodorant discs, and similar items.

- If a deodorant is to be used, use the chlorophyll type, perhaps injected into the room with a timer or through a wicking device.
- Control odors by removing bacteria, which is the basic cause of all odors other than transient odors. Custodians can remove bacteria by cleaning thoroughly, using effective disinfectants, and sealing surfaces to permit their proper cleaning.
- Ensure adequate ventilation to eliminate transient odors. Ventilation can be improved in many cases by installing more open grilles, cutting off the bottom of doors, putting a larger grill in a door, or removing the door altogether if a proper blind is provided.

REST ROOM CHEMICALS: DECIDING WHAT WORKS BEST

The complex subject of disinfectants need not be fully discussed here; let's briefly consider only the four disinfectants that are most commonly used and their application to rest room care:

1. Quaternary ammonium disinfectant—This is the most useful type. It has the advantage of superior odor control and is easy on the skin. It cannot be used successfully, however, where the soil is too oily or greasy.
2. Phenolic disinfectant—This is the next best choice. It can be used on oily soils and may be a somewhat better germ killer, but is harder on the skin.
3. Iodine-type disinfectant—This is too hard on the skin for this purpose, since it contains acid.
4. Pine oil disinfectant—Although highly touted on television, this is a very poor germ killer, and is irritating to the skin.

Other chemicals and materials should be selected with care for rest room service:

- *For scouring or removing stains,* the proper material is the lotion-type cleanser, which resembles milk of magnesia. The powdered cleansers contain particles that are too large and sharp for most uses, and cause a good deal of damage by scratching. No scouring powder should be used if it can be avoided, but rather use a liquid germicidal detergent for daily care based on one of the above disinfectants.
- *For removing mineral build-up periodically,* use acid bowl descalers. These materials should never be used on a daily basis. It may be desirable to handle acid descaling on a project basis by specific personnel.

- *For removing scum and scale* from shower stalls and other surfaces, use a special chemical. Like bowl cleaner, this should be available for this periodic use. Some cleaning procedures substitute high-pressure cleaning for this, or combine the two methods.

Cleaning Toilets

Although toilet seat covers are taken by many to be a benefit, if not a sign of courtesy to the public, by others they are taken to be an indication that the restroom is not properly being cared for—this is in much the same way that people view the use of deodorant blocks and drip fluid. A properly-cared for rest room facility does not require paper seat covers, where the facility is both cleaned and policed at the proper frequency.

Silicone-based stain retardants are injected into flushing water either from a device suspended in the flush tank, or through a mechanism installed on a pipe and fed through a hole drilled in the pipe. It is true that the coating on the fixture surface is smoother, and does not tend to collect foreign material as easily as the more irregular surface of the fixture. However, if a fixture is properly cleaned, the soil is properly removed. Further, the use of the soil retardant may lead to the failure to clean and disinfect the fixture properly, because of excessive reliance on the retardant. Where fixtures are cleaned at the right frequency and with the proper technique, investment in this system is not required.

FIVE ESSENTIAL REST ROOM CLEANING TOOLS

Workers must have the proper tools with which to do their jobs. The right tools for rest room cleaning will depend on the method to be used, the size of the rest room, and many other factors. Equipment to be considered would include:

1. *The custodial cart* makes it possible to transport minor equipment items, supplies, chemicals, paper goods and the like from a work area to the rest room. In a large rest room this may be brought inside, otherwise stationed at the doorway.
2. *Mopping equipment* is important, including the type of mop, bucket and wringer. The mop should be of the type with an easily removable head for laundering, the bucket should be of plastic or stainless steel, and the wringer should be the downward-pressure variety.
3. *Floor machines* are required for periodic scrubbing, and use either wet vacuums, or a squeegee to remove the water to a floor drain.

4. *Pressure-sprayers* are useful for removing scale build-up in showers and for general cleaning where the rest room is designed to permit this efficient operation.
5. *Manual tools* are also required, such as a plastic pail, bowl mop, cloths, etc.

The question often arises concerning the cleaning of a rest room by a member of the opposite sex. This is being done successfully by many thousands of cleaners in many types of facilities. The first requirement is to be sure that the room is not in use, and this can be done by calling into the room, and even flicking off the lights momentarily to see if there is any response—after that, a sign should be placed to show that the room is being cleaned, and the door blocked by the custodial cart.

34 TIPS FOR CLEANING REST ROOMS MORE EFFICIENTLY

Although a given procedure for a specific type of rest room would have to be devised for that individual case, the following are general procedures for custodians. These instructions apply to rest room facilities, locker and athletic areas which are equipped with toilet facilities, and medical and laboratory areas where basins are provided for wash-up.

1. Clean the fixtures in these areas thoroughly and properly. Make sure these facilities are cleaned daily.

2. Mix the cleaner-disinfectant solution in a plastic spray bottle. During the normal rest room cleaning operation, a bucket of cleaner-disinfectant solution will be used to mop the floor. The plastic spray bottle (and small plastic pail) can be filled from this bucket.

3. Prop open or lock the door to the rest room and post a sign or some other indication that cleaning is in progress and that the rest room is not to be used.

4. Clean the basins in the rest room last, not first (as is usually done). Clean the basins *after* the mirrors, lights over the mirror, towel cabinets, soap dispensers, and other items in the area of the basins.

5. Turn on the faucets to rinse the interior of the basin.

6. Close the drain to make sure that the cloth will not catch on the drain plug. Spray the basins liberally with cleaner-disinfectant from the plastic spray bottle and wipe with a damp cloth. Wipe the hardware also.

7. Open the drain and spray cleaner-disinfectant into it. Wipe the drain plug and the rim of the drain with the cloth.

8. Spray some cleaner-disinfectant into the overflow outlet. Turn on the water and rinse the bowl.

9. Wipe the hardware dry with a clean cloth or paper towel to prevent spotting.

10. Wipe the skirts or sides of the basin with a damp cloth or, if they are heavily soiled, spray them with cleaner-disinfectant and wipe them.

11. Do not neglect the bottoms of the basins and the pipes and valves under them. Normally, these do not need to be cleaned every time the basin is cleaned. Wipe them with a damp cloth approximately once for every five or six times that the basins are cleaned.

12. Rinse cloths thoroughly in one of the basins whenever they begin to show a build-up of soil.

13. After the basins are clean, begin cleaning the commodes and the partitions. Begin the cleaning by flushing the fixture.

14. Spray the top of the toilet seat with a liberal amount of cleaner-disinfectant solution and wipe it thoroughly with a damp cloth (or dip the cloth in the pail of cleaner-disinfectant and wipe the seat).

15. Wipe the seat thoroughly dry with a clean cloth.

16. Raise the seat and spray the underside with cleaner-disinfectant and wipe it with the damp cloth. Also dry it with a clean cloth. Spray and wipe the entire exterior of the fixture.

17. Spray a liberal amount of cleaner-disinfectant into the bowl. Using the bowl mop, thoroughly scrub all surfaces above the water level, paying particular attention to the underside of the flushing ring.

18. Flush the fixture and follow the water line as it decreases with the bowl mop, scrubbing in a circular motion. After the inside of the bowl has been thoroughly cleaned, flush the fixture one more time to insure that all chemicals have been removed from the surface.

19. Use an acid type bowl cleaner to clean the interior of the commode to remove buildup from minerals in the water. This needs to be done only occasionally, not every time the rest rooms are cleaned. Be extremely careful when using this product. Avoid any contact with skin or eyes. To use the acid type bowl cleaner, first wet the bowl mop and then pour the bowl cleaner onto the mop. Always hold the mop over the bowl while applying bowl cleaner to it.

20. Scrub vigorously under the flushing ring and at the water level to remove all build-up, rust and scale. Flush the fixture and follow the water level down with the bowl mop, vigorously scrubbing in a circular motion. Flush the fixture one more time to remove acid from all surfaces.

21. After cleaning each commode, damp wipe all pipes, valves and connections.

22. Before leaving the stall, damp wipe all partitions, including the tops. They may be either sprayed with cleaner-disinfectant and wiped with a damp cloth, or dip a cloth in a pail of cleaner-disinfectant solution and then wipe all surfaces of the partitions. Remove any writing or drawing on the partitions as thoroughly as possible.

23. After cleaning the commodes and partitions, clean urinals with a method similar to that used for commodes. First flush the fixture; remove any waste which does not flush through.

24. Spray the wall areas above and beside the urinal with cleaner-disinfectant and wipe them with a damp cloth.

25. Spray and damp wipe all exterior surfaces of the urinal, including the base and areas underneath the urinal.

26 Spray the interior of the urinal with a liberal amount of cleaner-disinfectant solution. Beginning at the top of the urinal, scrub thoroughly with the bowl mop, particularly the underside of the upper flushing ring.

27. Flush the fixture and follow the decreasing water level down with the bowl mop, scrubbing vigorously. Flush the fixture one final time to remove any remaining cleaning solution or residue.

28. As with the commodes, periodically clean the urinals with an acid type bowl cleaner. Begin by flushing the fixture and wetting the bowl mop. While holding the bowl mop over the urinal, pour the bowl cleaner onto the mop.

29. Scrub the interior of the urinal thoroughly with the mop, especially around the edges and under the upper flushing rim. Scrub until all build-up, rust and scale is removed.

30. Flush the fixture and scrub the lower part with the bowl mop as the water level decreases. Flush the fixture one final time to remove any remaining cleaning solution or residue.

31. If there are partitions between the urinals, spray them with cleaner-disinfectant and wipe with a damp cloth (or, wipe with a cloth which has been soaked in the pail of cleaner-disinfectant).

32. Damp wipe pipes, valves, and other hardware connected to the urinal with cleaner-disinfectant. Then, wipe dry with a clean cloth or paper towel to prevent spotting.

33. Do not allow any traffic in the rest room until all rest room cleaning has been completed.

34. For clean-up, spray the bowl mop liberally with cleaner-disinfectant solution and rinse it. Spray and rinse all cloths. Wring the cloths as dry as possible and store them so that they will be easily recognized by their color as equipment which has been used to clean rest room fixtures. If the cleaner-disinfectant solution is to be left in the plastic spray bottle for later use, be sure that the bottle is properly labeled before it is stored.

21

Organizing Custodial Facilities for Maximum Productivity

To maximize the productivity of custodial workers, make sure your organization's custodial facilities are well organized. There are two primary areas to organize: the central storage area and custodial closets and cabinets. This chapter provides guidelines for locating and arranging these areas.

HOW TO ARRANGE THE CENTRAL STORAGE AREA FOR MAXIMUM EFFICIENCY

The central housekeeping storage area is the heart of the inventory control system, and its operation should be under the direct supervision of the housekeeping department manager or a designated supervisor.

The size and arrangement of the central storage area must be determined functionally. Although some useful estimates have been published which indicate the size of the sanitary-storage facilities as a percentage of the total area to be maintained, these are useful only as generalizations and can lead to real problems if directly utilized. The housekeeping department does not want to suffer the inefficiencies of operating with inadequate facilities. On the other hand, neither does it want to tie up more of the organization's money in these facilities than is warranted. Therefore, you should determine the size and layout of the storage area on the basis of the needs it must serve. For long-term economy, the layout must be flexible enough to permit rearrangement or expansion based on future needs.

Choosing the Location for Central Storage

The central storage area, preferably, should be located as nearly at the center of the housekeeping activity as possible, and adjacent to the housekeeping department offices. Where a separate central storage area cannot be set up adjacent to the department offices, designate a section of the general maintenance supply area for this purpose. If this is done, the subarea should be separately fenced off and locked.

Some companies provide separate buildings for maintenance stores and operations. This is done in new construction where the cost inside the main facility is too great or where the location might interfere with operational flexibility. It may also occur in a renovation or relocation project where more area is required for activities within the main building. Such a building might include supplies and work areas for painters, welders, riggers, plumbers, electricians, and other trades. A subdivision of such a building would be very useful for the housekeeping department stores and offices as well.

Organizing Equipment and Materials

The location of equipment and materials in the central housekeeping storage area should be carefully organized. To facilitate this, use the following guidelines:

- Mark all equipment and materials properly.
- Make sure each drum pallet or large piece of equipment has a fixed location with an appropriate identification sign.
- Mark each shelf for storage of smaller items similarly.
- Post a complete list of the minimum materials and equipment to be kept on hand in this area.
- Keep paper goods (e.g., tissue, towels) enclosed, either in their original containers or in closed cabinets provided for this purpose. This helps prevent soiling.
- Remove obsolete or otherwise unusable materials from the storeroom. This provides more shelf space, improves the appearance of the storeroom, and helps prevent injury to custodial workers as well as cleaning errors (e.g., using the wrong materials).

The ideal central storage area requires:

- *A location or stall* for each drum or barrel. If the containers are stored vertically, a drum pump should be supplied. If they are stored horizontally, drum racks will be required as well as a method of placing the drums on the racks.
- *Shelving* for small containers, case goods, and hand tools. Adjustable steel shelving is most desirable for this purpose.
- *Tool holders* for hanging mops, brooms, and other items with handles.

- *A pegboard* for equipment attachments such as hoses, wands, etc.
- *A work table* for the inspection, lubrication, and minor repair of equipment. This table may also be used to spread out and treat dust mops and dust cloths.
- *Fireproof metal containers* for treated cloths and dust mop heads. One container should be designated for soiled items, and the other for clean and freshly treated items.
- *Adequate lighting.* Electrical outlets must accommodate all the various types of equipment for testing purposes.
- *A utility sink* for the cleaning of equipment, washing hands, and other uses.
- *Adequate ventilation* and heating.
- *A floor drain* so that large equipment may be sponged or hosed off while resting on the floor.
- *An industrial type washer-extractor laundry machine.* This is useful for larger operations.

Figure 21-1 is a layout for a central supply and work area, which might be considered a minimum in a smaller operation.

Figure 21-1. Sample Layout of a Central Supply and Work Area, for a Small Custodial Department

HOW TO STRATEGICALLY LOCATE CUSTODIAL CLOSETS AND CABINETS

Locating and stocking custodial closets and cabinets properly provide great opportunities for conservation of actual work time. A properly equipped utility station provides custodians with the materials and equipment they will need to perform cleaning work, avoiding the need to walk a great distance to obtain these supplies or spend time checking them out.

Not only do insufficient facilities require the custodian to travel back and forth between the work area and the central storage area, but also, equally important, provide an excuse to be absent from the work station.

In the past, the selection of utility areas for custodians has been a negative procedure. That is, the housekeeping department was given areas which could not be used by others. Even where specific custodial closets have been designed in new construction, they have at times been appropriated by "more important" departments—that is, any department other than the housekeeping department. Therefore, many "custodial facilities" are often areas under stairways, in corners of rest rooms, locker rooms, or boiler rooms, or piping access spaces.

It is significant what certain architectural guidebooks have to say on this subject. In one case, there is no indication of minimum standards either for central storage areas or custodial closets (the latter are referred to on plans by architects as "j.c.," for "janitors' closet"). Less than a third of one page indicates the dimensions of "cleaning equipment" such as a washboard, radiator brush, wash tub, and a household vacuum, none of which are used in the typical industrial or institutional cleaning operation. The cause of good housekeeping would advance materially by the very simple inclusion in standards books of a custodial cabinet, closet and central storage area.

In general, no custodian should have to walk more than 250 feet nor have to go to another floor level to obtain supplies and equipment. It is desirable to supplement the role of the closet or cabinet in bringing the tools to the work by the use of mobile carts specifically designed for this purpose. These are discussed in Chapter 23, on manual equipment. The effectiveness of the utility areas can be further supplemented by the strategic location of water taps and drains throughout the facility.

Where you cannot find room for a walk-in closet, or where the area is too small to warrant the use of such a closet, a cabinet may be used. The simplest approach to this is to use "combination cabinets" sold by most office equipment supply firms, although a special cabinet made for the purpose would be more desirable.

It is interesting that hospitals have found that the custodial closets can be a very serious source of cross-infection if not properly set up and maintained. Where these areas are not given proper attention, the damp corners and dirty

Figure 21-2. Custodial Closet General Specifications

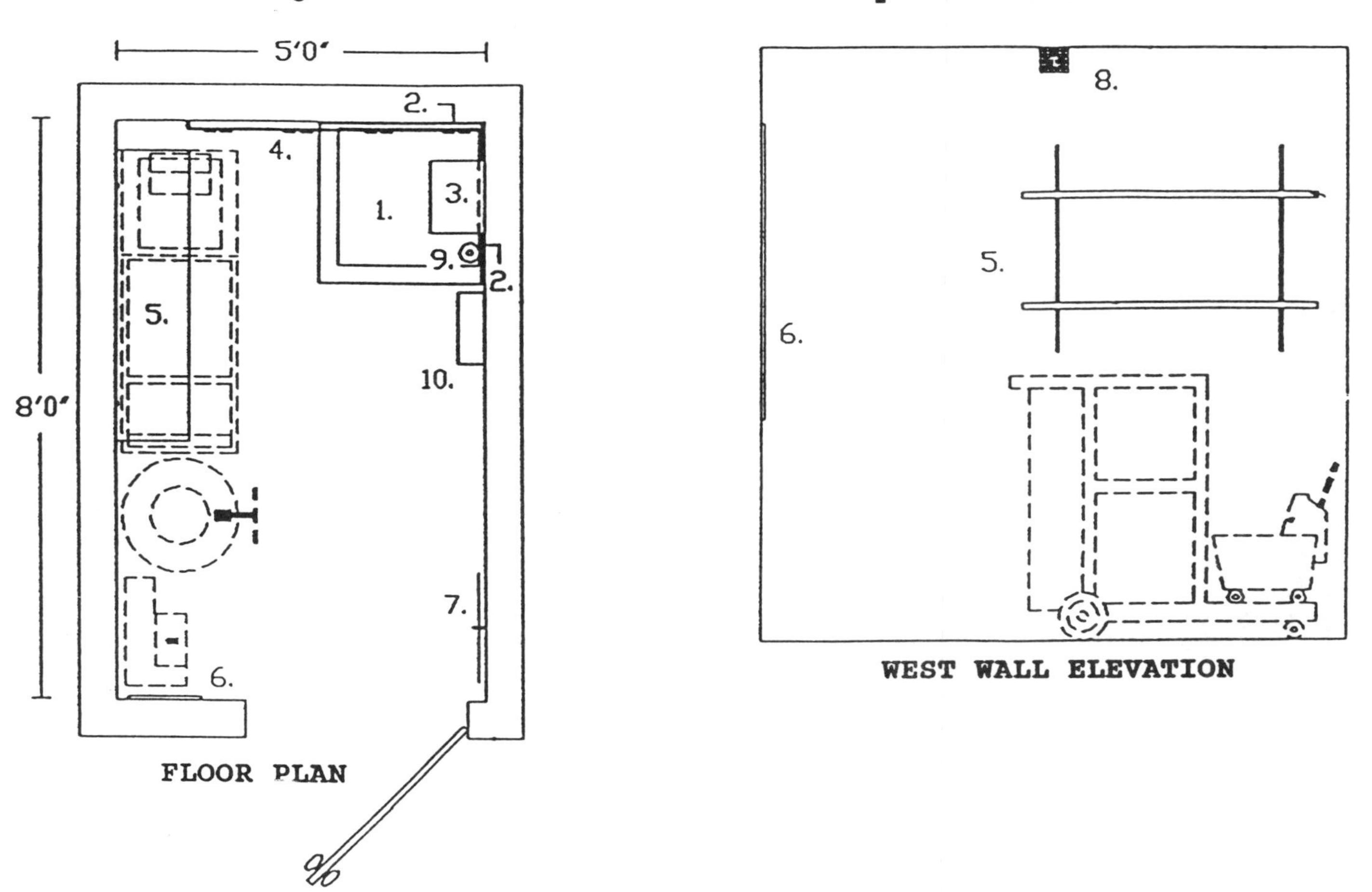

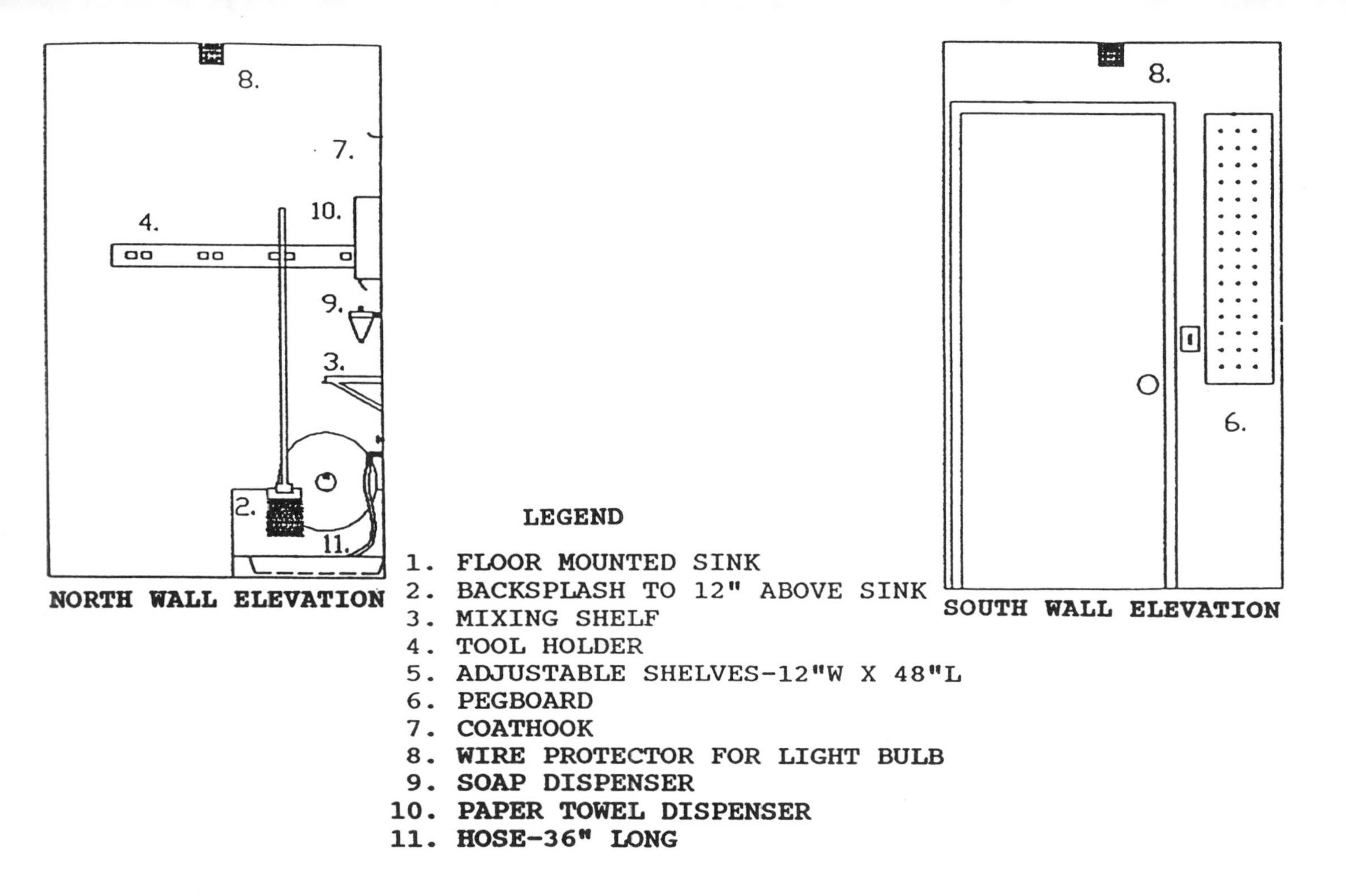
8.
7.
10.
4.
9.
3.
2.
11.
NORTH WALL ELEVATION
8.
6.
SOUTH WALL ELEVATION
LEGEND
1. FLOOR MOUNTED SINK
2. BACKSPLASH TO 12" ABOVE SINK
3. MIXING SHELF
4. TOOL HOLDER
5. ADJUSTABLE SHELVES-12"W X 48"L
6. PEGBOARD
7. COATHOOK
8. WIRE PROTECTOR FOR LIGHT BULB
9. SOAP DISPENSER
10. PAPER TOWEL DISPENSER
11. HOSE-36" LONG

equipment become a breeding place for bacteria. The materials and equipment stored in the supply closet are taken to every area of the facility, including such sensitive areas as rest rooms, dietary areas, and medical services. Thus, the proper organization and maintenance of the supply closets becomes essential in a sound hygiene program. Furthermore, the condition of custodial closets is usually an indication of morale and attitude toward the job. It is important to inspect these facilities regularly.

To meet the requirements for good layout and operation of an ideal custodial closet, follow these six guidelines:

1. Make sure your custodial closets can accommodate storage of mechanical housekeeping equipment. The "janitor closet" of yesteryear, which was suitable only for the storage of a mop bucket and certain hand tools, is of little use today. Figure 21-2 depicts a custodial closet of reasonable size, well organized to handle the requirements of one or two custodians. Figure 21-3 represents a minimum custodial closet, where space is not available for an efficient layout.

2. Organize the custodial closet carefully and mark items properly, so that there is a given location for each item, making the absence of any item immediately apparent.

3. Assign a number to each custodial closet and cabinet; the number should be clearly marked on the door, for simple reference.

4. Keep the door locked. Make sure the people using the closet or cabinet know that they are responsible for its contents.

5. Use color coding to identify equipment and supplies with specific custodial closets or work areas. The closets may be referred to by color, rather than by number. For example, all equipment from a "green" closet should be marked with a spot or a line of green, using paint or colored tape.

6. Prepare a supply list for each closet to insure its proper supplying and equipping. All equipment which cannot be stored in custodial closets *must* be returned to the central storage area. The minimum tool requirements for a specific custodial closet are:

- Measuring cup
- Funnel
- Plastic pail
- Plastic liners for custodial carts
- Plastic liners for trash receptacles
- Synthetic fiber cloths
- Gallon jug of neutral detergent concentrate with dispensing pump
- Gallon jug of germicidal detergent concentrate with dispensing pump
- Gallon jug of glass cleaner concentrate

Figure 21-3. A Minimum Custodial Closet, Where Space Is Not Available for an Efficient Layout

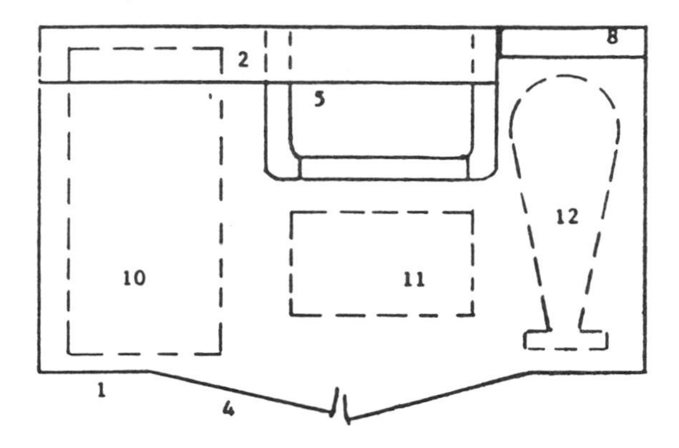

Plan View

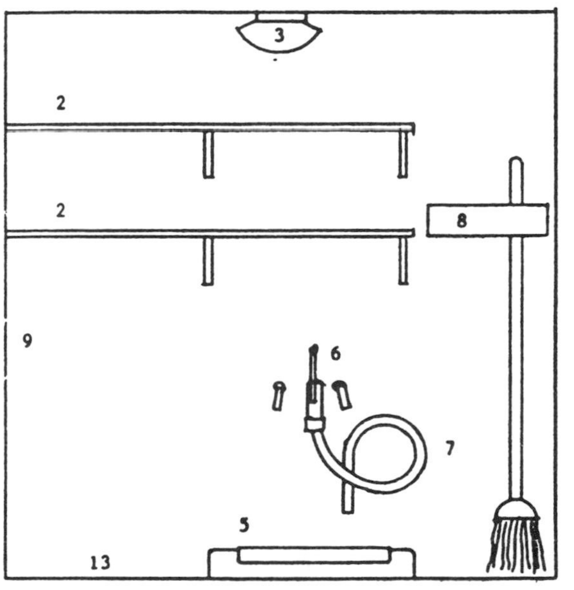

Front View

LEGEND

1 Dimensions: 8' long, 4½' deep (36 square feet)

2 Shelving 10" deep, with bracket supports

3 100-watt lamp, with door-hinge switch

4 Two 30" doors, pierced for ventilation, and lockable

5 Utility floor sink, with stainless steel lip cover. Note off-center

6 Bibb faucet with support hanger

7 4-foot length of hose

8 Tool holder

9 Walls ceramic to 4', painted enamel (including ceiling) above 4'

10 Location for custodial cart or waste hamper

11 Location for 2-bucket (or 3-bucket) mopping outfit

12 Location for floor machine or vacuum

13 Floor -- concrete, ceramic, or terrazzo (not resilient)

- Gallon jug of carpet shampoo concentrate with dispensing pump (only required in closets in areas with carpet)
- Gallon jug of ready-to-use spray-buff solution (only required in closets in areas which are spray-buffed)
- Gallon jug of degreaser concentrate with dispensing pump (only required for areas with showers).

Each custodial cart should be equipped with the following:

- Spray bottle of glass cleaner
- Spray bottle of neutral detergent
- Spray bottle of germicidal detergent
- Bottle of lotion-type cleanser
- Can of carpet spot remover
- Container of metal polish and cleaner
- Synthetic fiber cloths
- Rubber gloves
- Lambswool dusting tool
- Utility brush
- Percolator brush
- Radiator brush
- Hand-size abrasive pads
- Putty scraper
- Safety glasses.

22

Tips on Selecting and Maintaining Powered Custodial Equipment

Relatively few organizations have properly mechanized the housekeeping department. This is true even of those organizations which have shown remarkable advances in the mechanization and automation of other activities. The dollars spent annually on housekeeping equipment is certainly less than half of what should be spent. Although improvements in housekeeping will normally only be gained over a period of a number of months, the acquisition of correct equipment can bring startling results in a very short period of time.

As a word of caution, the provision of housekeeping equipment will not provide overall housekeeping efficiency in itself. Mechanical equipment thrown into a disorganized housekeeping operation soon becomes ineffective or completely inoperative. Without training and supervision, for example, it is not unusual to see work being done manually while a brand-new piece of equipment rests in the storeroom.

4 BENEFITS GAINED FROM ACQUIRING LABOR-SAVING CUSTODIAL EQUIPMENT

Custodial personnel must be equipped with sufficient labor-saving equipment to enable them to carry out their respective tasks efficiently. Management has the responsibility to properly equip each worker for the job for which he or she is held responsible.

Housekeeping equipment, if properly utilized, often pays for itself in less than a year's time. Although this justification for purchasing new equipment is adequate, there are other advantages in its acquisition:

1. *Improves Efficiency.* Properly selected equipment generally does a job better and more economically than can be done manually. Both a mop and a

vacuum will dry a floor, but the vacuum will remove the water faster and leave the floor cleaner, for example.

2. *Reduces Worker Fatigue.* Fatigue is reduced through the use of power equipment, raising the general productivity of the custodian even when not using the equipment in question.

3. *Enhances Employee Morale.* Equipment, particularly of the mechanical type, is a status symbol. The purchase of modern housekeeping equipment becomes a tremendous morale builder, not only improving the attitude of the custodian, but also enhancing the prestige and status of the housekeeping department as a whole. This aspect can be further developed by selecting attractive equipment (streamlined, chrome plated, etc.) and by marking individually assigned equipment with the custodian's name.

4. *Promotes Safety.* The use of an extension wand and pipe tool on a dry vacuum, as an example, avoids the need for ladders and scaffolding. Floors which are dried more quickly and thoroughly with a wet vacuum are less likely to cause an accident from slipping.

11 STEPS TO ASSESSING MECHANICAL EQUIPMENT NEEDS

A complete program of housekeeping mechanization will include all these factors:

1. *A survey of needs* based on an evaluation of the work to be done in the various maintenance areas. This survey must not be on the basis of the number of machines alone, but should also include all other factors. A large number of small floor machines, as a case in point, does not necessarily indicate a high level of mechanization. For areas where larger floor machines may be utilized, the smaller ones may only be operating at 50% efficiency. Or, the purchase of an autoscrubber may have tied up funds that could better have been utilized in another way, because of inability to utilize the autoscrubber in congested areas or winding corridors.

2. *Development of equipment specifications.* Remember: equipment and materials *combined* amount to only a small percentage of the housekeeping dollar. Thus, it is poor economy to provide other than the best equipment available. Of course, *best* is not synonymous with *most expensive.* If the use of a 20-inch floor machine is called for, it would be unsound to settle for a 15-inch machine on the basis of "economy." Some of the features you should consider when determining specifications are:

- Appearance
- Size

- Weight
- Horsepower
- Electrical characteristics
- Attachments
- Maneuverability
- Flexibility
- Finish
- Safety features
- Noise level
- General ruggedness and quality
- Warranty

3. *Selection of the supplier.* Check the reputation of the supplier and its ability to furnish service and parts. The equipment furnished should permit ease and rapidity of maintenance and should be manufactured over a long enough period of time to permit general standardization.

4. *Purchasing the equipment.* Equipment should be obtained by the purchasing department under the arrangement that best suits the organization. Equipment may be purchased outright, leased, rented, or perhaps obtained on a product tie-in.

5. *Training custodial workers.* The manufacturer or its representative should demonstrate the equipment and train selected custodians to use it. Encourage your supplier to provide additional retraining from time to time (this should unquestionably be provided where standardization on one brand of equipment has taken place). Where equipment is purchased on the basis of price alone, the supplier likely will be unable to provide service of any kind. Supplier training should only be a supplement to regular internal training, which will involve:

- Selecting the proper equipment for the job
- Selecting and using attachments
- Storing equipment when not in use
- Caring for equipment
- Following safety procedures

6. *Equipment repair.* Preventive maintenance will keep the equipment in condition to provide a long life of efficient service. Although basic repairs are performed by maintenance personnel, certain duties should be performed within the housekeeping department. You should assign these duties to someone with mechanical aptitude. This might be the central stock clerk, or the group leader of a special projects team. Within the housekeeping

department, the equipment should be properly lubricated, painted, cleaned, tightened, and adjusted (unless other departments perform these functions).

7. *Equipment inspection.* Make sure equipment is inspected regularly to be certain that it is in safe and correct operating condition. Equipment should be checked at least weekly for cleanliness, general condition, and lubrication.

8. *Keeping parts in stock.* Once efficient equipment has been purchased, downtime must be held to a minimum. In addition to selecting suppliers who can provide parts on short notice, keep spare parts on hand in the housekeeping stock area. Your department should purchase new parts which are subject to breaking or malfunction, such as attachment plates, switches, v-belts, pinions, and electric plugs. Also, worn-out equipment should be dismantled, and parts in good repair should be cleaned, properly marked, and put into storage.

9. *Providing storage facilities.* Provide proper storage and service facilities for all equipment to prevent damage, soiling, and loss, as well as to permit ease of inspection.

10. *Equipment attachments.* Provide suitable attachments to give as much flexibility as possible. Attachments should be carefully controlled by permanently issuing them to specific custodial closets. Color-coding is also helpful.

11. *Resurveying equipment needs.* A periodic reanalysis of equipment needs and conditions is very worthwhile. Do this every one to three years, depending on the size of your operation. The survey should cover these points:

- Extent of use—Investigate idle equipment on hand and correct any problems.
- Equipment condition—Replace inefficient or unsafe equipment.
- New developments—Investigate and evaluate new equipment coming on the market.
- Additional requirements—Find out if the housekeeping department is maintaining additional areas or performing additional or changed functions that could benefit from additional mechanization.

An inventory of equipment, as shown in Figure 22-1, provides the information for capital assets, replacement of equipment, and other functions. Of course, this can be computerized.

HOW TO JUSTIFY THE COST OF PURCHASING CAPITAL EQUIPMENT

Assuming that an organization has sufficient working capital and intends to remain in business, it will purchase capital equipment which will return its

Figure 22-1. Equipment Inventory Form

Building ______________________ Storage Room(s) ______________________

QUANTITY, DESCRIPTION OF EQUIPMENT ASSIGNED	SERIAL NUMBER	DATE ACQUIRED AND TAGGED	DISPOSITION & DATE

own cost within a specified period of time. Probably no responsible organization will permit a pay-out of longer than five years; most companies restrict the pay-out to two or three years, but vary the time depending on the type of equipment.

Capital purchases are normally a management decision because of their fixed nature. They do not vary with the production rate as wages or materials can be varied and thus a decision to purchase equipment must be "lived with" until the equipment is completely amortized. In order to properly control fixed expenses, many organizations require the approval of a special committee or board for the funds purchase of capital equipment. These boards normally have formulas as a guide to decisions on capital purchases, involving such factors as wage rates, interest charges, amortization periods, pay-out period, salvage value, etc. The group or person responsible for passing on capital equipment purchases deserves, and should be provided, a factual presentation of the benefits of the equipment. As many companies operate with a limited capital expenditure budget for any given year, the probable return from the purchase of housekeeping equipment will be compared with requests or recommendations from other departments.

5 Ways to Convince Management That You Need Custodial Equipment

Special consideration may be given to the purchase of capital housekeeping equipment if you make these points:

1. The housekeeping effort does not suffer obsolescence—it must go on despite any changes in the nature of the organization or the physical facility.

2. New housekeeping equipment is a great morale builder.

3. Housekeeping wage rates keep going up.

4. Although the economic analysis of the equipment will be based on the speed of the job, housekeeping equipment often also provides a better or more thorough job, sometimes difficult to analyze.

5. Housekeeping equipment is normally quite long-lived if it is of quality manufacture and properly cared for. Many floor machines and vacuums, for example, have been in use well over ten years.

Consider leasing housekeeping equipment when it cannot be purchased because of restrictions on capital expenditures. Although this does raise the total cost of the equipment, it has the advantage that moderate payments can be made on a periodic basis.

It is not necessary to save half a person's time in order to justify the purchase of a piece of equipment. The National Association of Bank Auditors and Controllers points out that, on the basis of a given annual salary a daily time saving of only ten minutes justifies the purchase of equipment costing only one fifth that salary!

Once the equipment has been obtained, it must be kept in proper repair. Figure 22-2 provides a form for an equipment repair log.

FLOOR MACHINES

Single-disc floor machines. These account for roughly half the dollar value of all powered housekeeping equipment. The basic purpose of the floor machine is the cleaning of many types of floor surfaces through a scrubbing action, but the same machines, with different brushes or attachments, are also widely used for buffing, scarifying, rug shampooing, and sanding. Their wide range of uses, sizes, and attachments makes them the most versatile piece of equipment available to the custodian.

Double-disc machines. In this type of machine, the self-propelling feature of the single-disc machine is lost, although it requires less skill to maneuver. Because it has more working parts, the dual-brush machine is more likely to give trouble. Finally, the size of the double-brush machine is misleading. Where the horsepower and brush pressure of a machine are equal, the efficiency will be proportional to the area of the brush in contact with the floor (a function of the square of the diameter). Thus, a double-brush 16-inch machine has two 8-inch brushes covering a total area of 75.38 inches which takes into account the 4-inch hole in the center of the brush where the attachment plate is located. On the other hand, a single 16-inch brush, also taking into account the 4-inch center hole, covers 188.49 square inches. This is a ratio of almost exactly $2^1/_2$ to 1, yet both pieces of equipment are sold as 16-inch floor machines!

Cylindrical brush floor machines. These have the advantage of concentrating the pressure over a small floor area. This machine does not have the flexibility and maneuverability of the single-disc machine and is comparatively rarely used in the smaller sizes; but is useful on irregular-surface floors. In the larger sizes, it becomes an autoscrubber, a variety of which is the common street-cleaning machine.

Uses. Floor machines can be used for other than buffing, scrubbing, or wax stripping. For example, a floor machine can be fitted with a scarifying attachment. Although most floor machines are used with synthetic pads these days, brushes may also be fitted for some applications, such as an uneven floor.

Selecting Floor Machines

When selecting a floor machine, the most important factor is its size, in brush diameter, and this leads to the most common and costly mistake which can be

Figure 22-2. Custodial Equipment Repair Log

EQUIPMENT ITEM ______________________________

SERIAL NUMBER ______________ DATE PURCHASED ______________

ASSIGNED TO ______________ ASSIGNMENT NUMBER ______________

DATE	ACTION TAKEN

made in selecting a floor machine—the choice of a machine which is too small. A 17-inch machine will cost more than a 15-inch machine, but will be approximately 30 percent more efficient. The small annual cost difference can be repaid in the performance of a single floor-cleaning project. Over the life of the machine, the additional cost becomes ridiculously low.

Too many floor machines are purchased on the basis of the smallest machine which it is possible to get by with. A much better economy is to purchase the largest possible machine which can be used within the limitations of the size of the area and its congestion. The typical floor machine in industry is 16 to 18 inches in diameter—it should be 20 to 22 inches in diameter.

When determining machine size, consider the following issues:

- The total area to be maintained.
- Congestion within the area.
- The weight and portability of the machine—This is particularly important when the machine has to be transported up and down stairs or between buildings.
- Larger floor machines are less likely to wobble and "buck" than the smaller machines.
- For uneven floor areas, it will not be possible to use the larger machines, as they will span over depressions.
- The weight per square inch generally decreases above the 16-inch size—Therefore, for heavy-duty operations such as sanding, grinding, or scarifying, a 15- or 16-inch machine should be used.
- For carpet shampooing, the machine should also be in the middle-size range.

The ideal floor machine would be completely flexible, having variable speed, pressure, and brush diameter, as well as built-in conversion systems to change from scrubbing to polishing. Some machine designs have some or all of these refinements.

Other factors to consider, in addition to the size and weight of the machine:

- *Horsepower:* No floor machine should be purchased with less than one-half horsepower, and this would be suitable only below 16 inches. Three-quarters horsepower is adequate through the 18-inch size; one horsepower through the 21-inch size; and one and one-half horsepower through the 24-inch size. Proper horsepower is required for an effective weight-rotational speed ratio.
- *Motor type:* Most floor machines use either the *capacitor*-type or *induction-repulsion*-type motor. Either is acceptable although the capacitor type has the advantage of not requiring a separate starter, which may give trouble.

• *Electrical characteristics:* Most floor machines are manufactured for use with "house current" (110 volts, 60 cycle a-c). Special motors are available for higher a-c voltages, or direct-current use.

• *Design:* In order to scrub or polish under furniture, the type of machine known as the "divided-weight" type is used, where the motor is placed back from the brush over the wheels, the wheels carrying a portion of the motor weight. Thus, the brush housing projects and is able to get under low-hanging furniture. This is opposed to the "concentrated-weight" machine, the much more usual variety, where the weight of the motor is entirely supported over the center of the brush.

• *Wheel retraction:* In the "concentrated-weight" type machine, the wheels must be off the floor to permit the full weight of the machine to rest on the brushes. Some wheels are spring-loaded and can be retracted by the operator while in a standing position by the application of foot pressure; other wheels must be retracted manually. Some designs keep the wheels in a fixed position.

• *Wheel size:* Larger wheels provide easier maneuverability when the machine is being transported, and are less likely to become damaged.

• *Switch:* Most machines are now equipped with a heavy-duty deadman type switch. Other types of switches are dangerous.

• *Adjustable handle:* Those machines which have an adjustable handle provide the greatest comfort to the operator, regardless of his height.

• *Ruggedness:* This includes such things as the quality of bearings, gears, and belting, as well as the type and design of the housing and other parts. Also important is the quality of the motor and its permissible temperature rise, as well as its electric cable.

• *Noise level:* This is particularly important when the machine is to be used in administrative areas during normal business hours.

• *Speed:* Standard speed (175 Rpm) is best for scrubbing and stripping; high-speed (250-300 Rpm) is useful for spray-buffing; and ultra-high speed (1,000 Rpm and up) is helpful in spray-buffing open areas. There is little advantage in going beyond 2,000 Rpm, and also more danger of burning out the motor.

• *Power type:* This includes electric, battery-operated, and propane-powered equipment.

13 Tips for Maintaining and Using Floor Machines

Proper care and use of floor machines will extend their life and efficiency. It will also provide a safer operation. Here are some pointers:

1. Floor machines should be plugged into a suitable wall outlet and properly grounded. Do not plug into a light socket as this may overload the

circuit or lower the efficiency of the motor. If an extension cord is used, it should be sized at least as large as the cord on the motor, and preferably a size larger.

2. Water or other liquids should not be splashed onto the machine. If acids or caustic solutions come into contact with the machine, they should be washed off immediately.

3. The machine should be kept clean. A dirty piece of equipment is not likely to receive good care. The electric cord should be wiped clean each time it is wet to keep from marking floors or furniture.

4. If a solution tank is used, both the tank and the feed lines should be flushed at the end of each day of use.

5. The brushes or driving blocks should be attached to the machine manually. If a brush or block is attached by placing the machine on the brush and then starting the machine, injury or damage will be likely to occur.

6. When the machine is not in use, it should be rested on the wheels, rather than on the brush. If the machine is permitted to rest on the brush, the bristles of the brush will flatten out at the point of greatest pressure and cause the machine to wobble in use. This reduces the machine efficiency and puts a greater strain on the operator.

7. Where a floor machine is used in a large medical area, it should be properly color coded and marked for that area and not used elsewhere. The machine should also be washed at least once per week in a disinfectant solution. These procedures will greatly lessen the possibility of cross-infection from this source.

8. The cord should trail behind the operator when the machine is in use. It should never be wound around the handle, thus making the deadman switch inoperative.

9. A machine should be transported and operated in such a way so as not to strike walls, doors, or furniture.

10. In transporting the floor machine, it should be handled carefully. It should never be forced over a rough surface, allowed to bang about in a truck, or bounced down stairs.

11. Care should be taken when operating the machine to keep it off the cord. If the cord winds on the brush, it may damage both itself and the machine, as well as shock or otherwise injure the operator.

12. Machines should be stored in a clean, dry room that is properly lighted. Cords should be wound on the hooks provided on the handle, and brushes properly cleaned and hung up to dry. Machines should be lubricated as required. Overlubrication should be avoided.

13. To prevent shock, the use of rubber gloves or rubber shoes is recommended when working with solutions.

Floor Machine Accessories

Attachments. One useful floor-machine attachment is the solution tank. Usually mounted on the machine handle, this tank dispenses a cleaning solution onto the floor. The tank, control valve, and feeder tubing should all be corrosion-resistant.

Carpet scrubbing is the perfect application for the solution tank, but opinion is divided on its use for general floor scrubbing. Although theoretically a very sound idea, it is practical only under certain conditions. The solution tank does have the advantages of avoiding the necessity of changing from mop and bucket to floor machine, and the placing of the solution at the point of use requires less stepping in the detergent solution.

On the other hand, the mop in the hands of an experienced worker permits more accurate placing of the solution so that it does not run and allows the solution to remain on the floor before scrubbing begins, so as to gain the advantage of the chemical cleaning action of the detergent.

Finally, where large areas are cleaned, it is quicker to prepare a mopping solution in a pail than to prepare and pour one into a solution tank. The latter operation often involves spillages or foaming over. When not in use, the machine with solution tank cannot be rested on its wheels but must remain with the weight on the brush.

Some floor machines are fitted with a small vacuum to remove dust created by the buffing operation. This is sometimes assembled as a homemade job. Up to this point, it has seen rather limited use.

A ring or "skirt" helps to prevent splashing on baseboards, doors, and furniture.

Grinding and sanding attachments. These are available in the middle-size range. They are not applicable to the smaller machines, which do not have sufficient weight and power, or for the larger machines, which do not have sufficient pressure. These attachments utilize specially cut carborundum stones or sandpaper. Floor machines used for light-duty grinding and sanding will be damaged if so used except for a short period of time. Such attachments should never be used on vinyl-asbestos floors.

Brushes. Floor-machine brushes are available in a considerable variety of types for each of the various machine sizes. Brushes are measured by the bristle diameter, as spread under the weight of the machine. It is this diameter which determines the size of the machine, rather than the diameter of the wooden or metal backing of the brush. Floor machines can utilize smaller brushes than the size of the machine, but not larger brushes. Smaller brushes are actually desirable when the floor machine is utilized for scarifying, sanding, grinding, or heavy-duty scrubbing of problem areas over a considerable period of time, to prevent undue strain on the motor and increase the brush pressure.

The brush is manufactured by setting bristles, wire, etc., into the drilled holes in the brush block. Most bristles are set vertically to the plane of the block. As the brush rotates, the bristles bend. Some bristles are set at an angle to the block so that in rotation the bristles bend to the vertical position, thus improving their action.

A wide variety of brush-filling materials are in use:

- Bassine fiber—A stiff fiber for heavy-duty scrubbing, dark red to black in color.
- Nylon—A very durable synthetic fiber which can be varied from very soft to very stiff. It is used for floor scrubbing and carpet shampooing, and is usually colored black.
- Palmetto fiber—A moderately stiff fiber used for medium to heavy scrubbing and also for rug shampooing. Light red or reddish brown in color.
- Palmyra fiber—A medium stiff fiber, somewhat stiffer than palmetto, generally used for scrubbing. It is mottled brown in color.
- Steel wire—Made from various grades of carbon steel wire, this brush is used for very heavy-duty scrubbing. A brush made from "butcher's wire" can be used for scarifying. Steel wire may be blended with fibers such as bassine. Steel wire brushes must never be used on any smooth finish floor susceptible to staining or scraping. These brushes must be dried carefully after each use to prevent rusting.
- Tampico fiber—A medium soft fiber for buffing, often blended with other fibers to provide more stiffness in the larger sizes. Tampico is generally white or tan in color.
- Union mixture—A fiber mixture, usually of palmyra and tampico, and therefore generally mixed brown and tan in color. This combination is medium stiff and is normally used for polishing.

When a solution tank is used with the floor machine, a specially drilled and channeled brush must be used to permit the flow of the solution through the brush and onto the floor. These channeled brushes may be made of wood (usually in the smaller sizes) or metal (usually in the larger sizes). Generally speaking, wood-backed brushes should be considered expendable, whereas metal-backed brushes should be refilled.

Here are a few guidelines for caring for brushes:

- Comb brushes frequently to remove foreign matter.
- Place brushes on the machine by hand.
- When not in use, hang brushes on a nail or rest them on the block.
- Air dry brushes. Drying by heat, such as on a radiator, may damage the block.
- Wash brushes occasionally, but take care not to overwet the wooden block.
- Never use brushes designed for dry cleaning in liquids.

Floor Machine Pads

Lambswool. Circular lambswool is used on a special holder or under a brush where buffing to an unusually high degree is desired. Lambswool is not recommended for normal buffing operations.

Nylon pads. These pads are made of nylon impregnated or treated with silicon carbide and are made in various grades for scrubbing or polishing. Some types of pads require a special holder, whereas others may be used under a regular floor-machine brush. The pads have the advantages of being rustproof, safe to handle, not "loading" quickly, not accumulating static electricity, and not spinning off water.

Steel wool pads. Woven steel wool is available in eight grades, #0000 to 4. For floor work, #0 is normally used for buffing, #1 for light scrubbing, and #2 for medium scrubbing. Grade #00 is sometimes used for fine polishing, while grade #3 may be used for heavy-duty scrubbing or for "shaving" cork floors. Steel wool should not be used on composition floors. When in doubt as to the proper grade of steel wool to use, it is safest to begin with the finer grades and work up to the grade which works most efficiently, yet still safely.

VACUUM CLEANERS

Vacuum cleaners are available in a large array of styles, types, and sizes. The equally wide range of attachments makes possible a combination for almost any conceivable need. Yet, the housekeeping department which operates without even a single wet vacuum, for example, is seen regularly. Space permits only a limited description of this versatile tool.

Selecting Vacuum Cleaners: 11 Guidelines

A number of factors affecting vacuum selection are identical with those for equipment in general. In addition, these items refer specifically to vacuums:

1. The vacuum should have a high suction rating. This is normally expressed in inches of water lift.
2. A motor should be of adequate horsepower to permit continuous operation without damage. The motor should be the by-pass type.
3. If the vacuum is to be used for wet pick-up, it must be provided with safeguards so that neither water nor suds is ever drawn into the motor.
4. The vacuum tank should be large enough so that frequent dumpings may be avoided.

5. A dump valve is desirable, eliminating the need to remove the vacuum head and turn the entire unit over when handling liquids.
6. The unit should be easy to transport and maneuver. Casters should be large and ruggedly mounted.
7. The vacuum should be adaptable to a wide range of attachments.
8. When in operation, the vacuum should be reasonably quiet.
9. The unit should be provided with white rubber guards to prevent marking or damage to walls and furniture.
10. Style, finish, and trim should be taken into consideration.
11. Stainless tanks should be selected for wet use.

Maintaining and Using Vacuum Cleaners

Vacuums will provide better service when properly used:

- Empty vacuums at the end of each day, no matter whether the unit was used for wet or dry service.
- On wet vacuums, check the automatic cut-off mechanism frequently to make sure that water cannot enter the motor.
- Wash tanks on wet vacuums at least monthly with a disinfectant solution to prevent the growth of bacteria and algae.
- Store the vacuum in an open position to allow entry of air.
- Remove strings, hair, or soil from vacuum attachments at the end of each day.
- Reverse the hose every few months to prolong its life because of the change in the wear pattern.

Types of Vacuums

Dry tank type. Some vacuums are designed for dry use only. They must be carefully marked for this purpose, as they are easily damaged by water. This is the most common type of vacuum. For carpet or general floor dry cleaning, this type of vacuum may be mounted or carried on a utility cart. A special class of this vacuum, of unusually rugged construction and high power, is used for collecting metal and plastic chips. Unless a dry vacuum can be restricted to use by a custodian who will never need a vacuum for wet cleaning, you should always provide wet-dry vacuums for greater flexibility and safety.

Wet-dry vacuum. This is the next most common and most versatile of the vacuum types. Constructed with a round or square tank ranging from 5 to 20 gallons in size, the wet-dry vacuum has a wide variety of uses when equipped with the proper attachments. Some of these vacuums have built-on carrying trays for these attachments. The use of this vacuum for wet cleaning, for example, produces a much cleaner and drier surface than can be obtained with a mop, even for removing soil and water in the joints between composition floor tiles. The use of this vacuum will also reduce the surface and airborne microorganism count, contributing to better industrial hygiene. No housekeeping department should be without at least one vacuum of this type.

Tile-type vacuum. Where floor drains are not available, a wet vacuum with a tilting tank, so as to dump into a commode, is very useful.

Drum adapter. A standard 55-gallon open head steel drum becomes a vacuum container when a regular industrial vacuum head is placed on it with a special adapter cover. Where this combination is to be used for wet use, the drum should be equipped with a dump valve near the bottom. Maneuverability is achieved by fixing casters onto the drum, or by placing the drum on a dolly. This arrangement is particularly useful in industrial operations where large quantities of dry or wet waste are to be collected, such as in cleaning sumps, removing sawdust from woodworking equipment, vacuuming overhead with extension tools, etc.

Hospital vacuum. Any of a wide variety of standard vacuums may be converted for hospital or similar critical use by providing a special filter system. These filter systems may remove dust particles and microorganisms as small as 0.5 micron. In this range, the hospital-type vacuum also becomes useful in the cleaning of a small "gray room," where the room is not of sufficient size, or the cleaning problem sufficiently acute, to warrant a special stationary system. You can determine the suitability of the hospital-type vacuum for cleaning areas requiring special attention by analyzing the exhaust air for dust count and particle size.

Pack type. This is a high-power dry vacuum worn on the back like a knapsack, or slung over the shoulder with a strap. The types having an internal filter bag seem to have the greatest acceptance and use. These units are most suitable where the worker will be cleaning in confined spaces or moving a great deal from place to place. It avoids the necessity and hampering effect of having to move a vacuum cleaner on casters to a location very near where the employee is working. The biggest drawback with the pack-type vacuum is the relatively small tank capacity.

Scoop type. This is a wet-dry vacuum having a built-on squeegee mounted either in front of or behind the tank. As the operator walks forward, pushing the tank, the squeegee scoops up the water. This arrangement eliminates the wand and hose and permits a very rapid cleaning operation. It may also be used for dry floor cleaning in a similar fashion. This type vacuum is limited in its maneuverability, and its use should be restricted to long wide aisles and open spaces. It also requires even, smooth floors for efficient operation.

Stationary system. Built-in systems have a central vacuum machine, normally located on the lowest floor, with pipes running to vacuum stations, permitting cleaning of any area of the structure. Stationary systems are particularly well suited for multistory buildings where a good deal of vacuum work will be needed, such as in precision assembly work. Stationary units are applicable to a large range of industrial uses and may actually connect a number of buildings. Open vacuum units may be stationed above inspection stations so as to keep these areas free of dust. The stationary system eliminates the need for a number of different vacuum units, but, on the other hand, if it becomes inoperative, no vacuuming can be performed.

Care must be exercised in using long hoses so that they are not dragged or bent, and so that they do not mark thresholds, furniture, door jambs, etc. The vacuum stations provide a specialized advantage: they may be used to remove dust and lint from dust mops at periodic intervals. The stationary system is almost a requirement for large superclean rooms, but is not desirable for general housekeeping.

Suction sweeper. This is a specialized style of scoop-type vacuum for removal of dry waste and litter. This unit is normally driven by a gasoline, propane, or electric motor. It has a very large collector bag and is particularly suitable for outside policing of trash and leaves, and internal collection of waste paper scraps and similar light materials.

Choosing Vacuum Attachments

A vacuum without attachments is as useful as a milling machine without cutters. You should select vacuums so that the attachments will fit all machines, thus improving flexibility and reducing the necessary stock of parts. Attachments should snap on and off easily, rather than having to be screwed on or off. Many of the attachments are available in various sizes and materials.

Basic components, without which the vacuum will not operate under the usual conditions, are:

- Filter bag—The filter bag is designed to permit the flow of air but to entrap most soil particles. A type of soft cloth known as "moleskin" is suitable, but the newer synthetic materials are preferable because they do not require removal for wet pick-up and may be more easily cleaned. The hospital-type vacuum has a special filter medium.

- Hoses—Vacuum hoses are available in the 1½-inch and 2-inch diameters, and the various tools must be sized to fit the hose, which in turn must be sized to fit the vacuum. Hoses are normally 10 feet in length, but shorter lengths are available for close work. These are available in several different materials, but the most useful hose is made of neoprene, as this can be used on solvents which would damage other materials.

- Wands—Wands are made of tubular metal and connect the hose to the various attachments, acting as a handle. Wands may be bent or straight, and should be light in weight. Extension wands permit the custodian to do overhead cleaning without leaving the floor.

Specialized tools are available for the following jobs:

- Cleaning carpets, wet or dry.
- Dry cleaning floors—Tools of this type use a sweeping action provided by bristles, a felt strip, or cotton yarn.
- Removing liquids from floors—This attachment is fitted with squeegee blades.
- Cleaning chalkboard erasers.
- Cleaning dust-mop heads.
- Dusting the tops of books while on library shelves.
- Cleaning venetian blinds while in place.
- Dusting walls.
- Removing loose soil from the tops of overhead piping—These tools are available for various pipe diameter ranges.
- Crevice cleaning, such as between radiator fins.
- Upholstery cleaning.
- Cleaning of bins and drawers—This tool is designed to remove dust but not heavier materials such as hardware fasteners.
- Boiler tube cleaning.
- Cleaning irregular surfaces, such as wire partitions.
- Some vacuums may be fitted with a spraying attachment on the exhaust side of the unit.
- A blower attachment may also be fitted on the exhaust side of the unit.

AUTOMATIC FLOOR MACHINES

The automatic floor machine, often also called the autoscrubber, is essentially a combination of the scoop-type vacuum with a floor machine. The autoscrubber dispenses a detergent solution from a storage tank onto the floor, the floor is scrubbed by single, double, or multiple brushes, and the solution is then vacuumed into a waste reservoir. The machine is also suitable for dry cleaning, in which case no cleaning solution is involved.

Such a machine is self-propelling. A sulky may be attached on which the operator may be seated, but this is only successful for very large operations. The need for a sulky might indicate the desirability of obtaining a rider-type machine, which resembles a street cleaner.

Automatic floor machines are powered by storage battery, electric cable, propane, or gasoline engines. The cable-powered unit is the most trouble-free, although its range of operation is limited by the length of the cable. The battery-operated models provide the greatest flexibility of use, but they require battery recharging and are unusually heavy.

The purchase of an automatic floor machine on principle alone can tie up a good deal of money with no return, if conditions are not suitable for its regular use. This machine can only be used in large exposed areas such as long corridors, expansive lobbies, wide aisleways, etc. An elevator or ramp is required to move the machine from one floor to another. Floors must be smooth and level to permit proper operation of the brushes and squeegee.

In a single pass, the autoscrubber picks up detergent solution almost immediately after it is placed on the floor. Thus, because there is no time for the detergent to provide a chemical cleaning action, a scrubbing on a single pass will perform only a medium cleaning job on a heavily soiled floor. For heavyduty cleaning, two or more passes will be required, and these steps must be taken into consideration when considering the efficiency of this machine. As with all equipment, manufacturers' claims for cleaning rates are often for an ideal situation over a carefully selected period of time.

In some cases, greater efficiency can be obtained if a second worker "services" the autoscrubber operator, thus permitting the machine almost uninterrupted use. The support worker removes obstructions, prepares cleaning solution and removes waste solution, mops corners which the machine cannot reach, removes water spots which it could not pick up, etc. In some cases, more than one machine can be serviced by a single custodian.

Under ideal conditions, the automatic floor machine can save a great deal of time. Under other conditions, it will be more trouble than it is worth.

POWER SWEEPERS

Power sweepers fall into a natural classification of types, on the basis of their small, medium, or large size:

Walking type. The operator walks behind the smallest of the power sweepers, which clean a path from about 20 to 32 inches wide. These sweepers may be manual or powered. It is only the mechanical type of sweeper which includes a built-in vacuum, which is desirable for keeping down the dust. A manual sweeper is tiring to operate, creates a good deal of dust, and should only be used for the most restricted situations.

Riding type. In the intermediate size of power sweeper, the operator rides either at the front or rear of the machine, which is self-propelling. The most popular sizes of this sweeper clean 36-inch and 48-inch paths.

Municipal type. This is the large power sweeper of the type often used in street cleaning, sweeping a 6-foot path. Even larger sweepers are available, such as an 8-foot sweeper used to sweep aprons and runways at airports. The larger machines are generally designed for outside sweeping and may include a water spray system for further controlling the dust.

When considering whether to buy a power sweeper, you should first analyze its dollar-saving potential, including such factors as:

- Type of surface to be cleaned.
- Degree and type of soiling and litter to be removed.
- Congestion of the area.
- Planned frequency of cleaning.
- Ease of operation, maneuverability, and maintenance.
- Efficiency of soil removal.

You should then consider the various methods of powering the sweeper, such as gasoline, propane, and storage battery; the method of dumping the waste, which may be either manual or mechanical; and the availability and cost of accessories such as:

- Engine-hour meter.
- Fire extinguisher.
- Oxy-catalyst exhaust filters to cut down on carbon monoxide.
- Side brushes for sweeping against curb or walls.
- Attachments so that the vacuum system may be utilized as a large vacuum cleaner.
- Warning lights and horns.
- Water-bath mufflers for use in fire hazard areas.
- Water-spray system for dampening the soil to keep the dust down.
- Enclosed seating area.

POWERED EQUIPMENT FOR SPECIALIZED CLEANING

Several types of mechanical equipment have been designed to perform specialized cleaning functions.

Can washers. These are automatic devices for the washing of garbage cans and waste receptacles. The can is inverted over the unit and held in position, while a foot pedal actuates the cleaning mechanism. A hot detergent solution is sprayed with considerable force against all inner surfaces of the container. The heat assists in drying the container. This type of equipment is also useful for the cleaning of small mixing kettles, milk cans, stock pots, and the like.

Grounds care equipment. The use of grounds equipment such as tillers, seeders, mowers, and trimmers is not normally a function of the housekeeping department, although common supervision is used at times.

Sprayers. Although sprayers are primarily designed to dispense insecticides, they are also useful for spraying herbicides on weeds, deodorizing, treating dust mops and cloths, and applying detergents and disinfectants. The types of power sprayers include:

- The electric vibrator type for spraying in small quantities.
- The electric aerosol type which disperses a fine mist—This machine can be turned off automatically with a timing attachment.
- Fogging guns which utilize compressed gas as the propellent.
- Fogging machines, either portable or vehicle-mounted, which utilize hot exhaust gases from a pulse-jet-type engine.

Sprayers and foggers, when handling flammable chemicals, should never be used in a closed or poorly ventilated area. When the machine is not in use, all of the insecticide or other material, as well as the propellent in the fogging type, should be removed and properly stored. Sprayers should be cleaned regularly, as the small orifices tend to clog. Insecticides should be used only by a licensed pest-control specialist.

Steam cleaners. Steam cleaners consist of a steam generator or steam source, a "gun" or nozzle attached to the machine by a flexible steam hose, and a solution tank for detergent. Steam cleaning is a very rapid method of removing heavy soils from certain types of equipment and floors, and it does this by the combined action of heat, turbulence, force, and chemical action.

Not all materials or surfaces may be safely steam cleaned, and care must be taken to prevent damage through the cleaning of fragile or delicate equipment, or the spoiling or damage of adjacent materials, equipment, or surfaces.

Steam-cleaning materials should be carefully selected. Operators should always wear goggles. Since the steam-cleaning action causes sprayback and floods the floor, the operator should wear a waterproof cap, apron, gloves, and rubber boots.

Tank cleaners. These are special devices which are lowered into tank cars, storage tanks, or vats, or similar enclosed vessels. Detergent spray is forced out of a revolving head to completely wet the interior of the vessel.

Transport cleaners. Special guns and spray systems are constructed for the cleaning of truck and trailer, aircraft, railway car, and ship exteriors.

Venetian blind cleaners. Special equipment is available either for cleaning individual venetian blind slats, or for immersing the entire blind in an agitated detergent solution or ultrasonic tank.

Wall washers. Wall-washing machines usually consist of two tanks, one for the cleaning solution and the other for rinse water. Air pressure is used to force the cleaning or rinsing liquids through flexible hoses to a trowel which is covered by a porous cleaning pad.

The wall-washing machine is only efficient for cleaning large expanses of smooth walls on a project basis. Where irregular walls or relatively small areas are to be cleaned, or where the work is subject to interruption, the walls should be cleaned manually using equipment such as sponge mops, dry mops, sponges, and cloths.

Waste-disposal equipment. The field of waste-collection equipment is extremely large. To treat it adequately is beyond the scope of this book. In general, the following standard and specially constructed equipment is available:

- Balers—for waste paper, corrugated cartons, and other waste materials.
- Slitters and grinders—to reduce corrugated boxes and other containers and waste material to shreds.
- Pulpers—to reduce cellulose-containing waste to a pumpable material.
- Conveyors of all types—to move waste material.
- Incinerators—for safely burning waste.
- Containers—for separating recyclable waste.

Special cleaning equipment is available or can be constructed for unusual needs. For example, some tunnels are cleaned by a fork truck which has been specially adapted with a washer unit.

23

Back to Basics: Making the Most of Manual Cleaning Equipment

Many of the common hand tools of the custodian have remained almost unchanged over the years. Improvements and innovations that have been made are not startling and have not contributed markedly to overall custodial efficiency, except in certain cases where specialized types of equipment have been developed.

Despite this, if you thoughtfully select and provide manual equipment, you can ease the burden of your custodial workers and enable them to work more productively. The carpenter, for example, may be equipped with a vast array of power equipment which will save a great deal of time and improve the quality of his work; yet, he selects his hammer, hand saw, and level carefully and treats them well.

An impressive number of manual equipment items are available for housekeeping use. Many of these are old standbys such as mops, brushes, and brooms. Others are more infrequently used, such as applicator pans, plow mops, and proportioners. This chapter provides a general survey of manual cleaning equipment.

BROOMS

"Straw brooms," or "corn brooms," are so named because of their manufacture from undyed corn-straw stock. Brooms have changed little over the centuries, except that the fibers are now arranged in a fan shape whereas in olden times they were wrapped in a circle around the handle, and plastic fibers are also now used.

For most uses, brooms have been supplanted by push brushes, dust mops, and vacuum sweeping tools. For certain situations, however, the broom still serves best, such as for the cleaning of ribbed rubber matting, removal of certain types of soil from rough concrete floors, for moving bulky litter, and the handling of lint and nondusting soil in congested places. Avoid using corn brooms where dust will be raised.

Brooms are classified by their weight per dozen. The 40-pound broom should be selected for use in industrial areas, and from 24 to 32 pounds in other areas. Smaller brooms are available for use in more congested areas or for light-duty purposes.

A special type of broom is the "toy broom," having a short handle, which is used in conjunction with a lobby dust pan for removing litter from public areas on a policing basis. Hand whisk brooms, which may be made of palmetto fiber or nylon bristles, are used for brushing fabric where vacuuming is impractical or unsuccessful.

Here are a few pointers regarding broom care:

- Brooms should be stored, to permit free circulation of air, and should be hung up rather than allowed to rest on the straws.
- Brooms may be cleaned by washing in a mild synthetic detergent solution, followed by a rinse with clear water. They then should be hung up to dry with the straws down.
- Wet corn brooms should not be used, and brooms should not be used on wet materials unless unavoidable.

BRUSHES

Thousands of years ago, a cave man rolled up an animal hide. He was about to make a painting on the wall of his cave. He used the brushing action of the hide's bristles to wipe the dust off the cave wall. This may have been the first brush, our oldest cleaning tool.

Composition of Brushes

The animal skin brush has developed into an enormous variety of brushes of all types, some hardly even recognizable as brushes. For example:

- Block material—About ten basic materials are involved, including hardwoods, plastics, aluminum, and various types of rubber.
- Block shape—About twenty basic shapes are available, including geometric forms such as round, rectangular, triangular, and oval as well as a number of special shapes such as wing tip, short-handle grip, long-handle grip, etc.

• Filling material—At least ten different varieties of each of five basic types:

— Animal—Bristle, horsehair, etc.

— Metallic—Brass, carbon steel, stainless steel, etc.

— Mixtures—Metallic and vegetable, plastic and vegetable, etc.

— Plastic—Nylon, PVC, saran, styrene, etc.

— Vegetable—Bass, cactus, palm, etc.

• Length of filling—Five basic possibilities: stub, short, medium, long, and extra long. Actually, length of filling material is measured in inches out from the block face. A single brush may contain more than one length of filling.

• Setting methods—There are about ten basic methods of setting brush fillings into the block, such as with the use of adhesives, nails, staples, woven wire, etc.

• Special designs—At least twenty different variations on basic designs are available for special applications:

— combination brushes, where one side of the block is filled for one purpose and the other side for another purpose.

— detachable handles.

— fountain brushes, where a cleaning solution flows through both the handle and the brush.

— bumpers, to protect the brush from marring or damaging a finished surface.

— gang brushes, which can be locked together to form a longer brush.

— angle-set brushes, with the bristles set at various angles to the block.

— shaped fillings, where the filling materials are of varying lengths.

You should be familiar with the different types of brush ingredients; however, when selecting a brush, consult the recommendations of a competent, experienced sales person.

Six Types of Brushes Used in Custodial Cleaning

The scope of this work permits only a general listing of the various types of brushes available to the custodian:

1. *Bowl brushes.* These are also known as toilet-bowl brushes, commode brushes, and sanitary brushes. The common bowl brush is made of stiff

bassine fibers fixed with wire to a hardwood handle about 24 inches long. Some of these brushes are equipped with "collars" for cleaning under the rim of bowls. Others of a special design include a tip of sponge. (Bowls are better cleaned by the use of cotton swabs or sponge applicators built onto bowl-cleaner containers.)

2. *Counter brushes.* These are made of horsehair, fiber, or nylon filling in a wooden block with a hard grip. This brush is most often used for sweeping soil into dust pans, or for reaching into corners. Its earlier use for dusting surfaces, such as desks and counters, has been replaced by the use of dusting cloths since brushes tend to stir up dust.

3. *Deck scrub brush.* This brush may be used by hand or with a long handle for scrubbing floors and other badly soiled, coarse surfaces. Scrub brushes may include a built-in squeegee for moving the scrubbing water. They are available in various block lengths. A special type of deck scrub brush is the baseboard brush. Triangular in shape, it is able to clean the baseboard and adjacent floor area.

4. *Hand scrub brush.* Also known as "scrubs," these brushes are used for the hand cleaning of equipment, fixtures, and small floor areas of moderate coarseness.

5. *Fountain brush.* This brush is designed to dispense a detergent or rinsing solution through the brush by means of a hollow handle. Fountain brushes are used for washing windows, vehicle bodies, and other surfaces such as badly soiled industrial walls. Long extension handles are now available so that windows may be cleaned as high as three or four stories. Both windows and vehicle bodies may be cleaned with brushes that are not of the fountain type. Vehicle body brushes should always be equipped with a rubber bumper to prevent damage to the finish.

6. *Special brushes.* The need to clean numerous types of surfaces has required the design of brushes for specialized applications. A special brush has been designed to clean each of the following:

Baseboards
Bed springs
Bottles
Drinking glasses
Fenders
Pots
Pipes
Radiators
Venetian blinds
Walls

When considering brushes, alternate cleaning methods should also be borne in mind. For dust removal, for example, both vacuuming and the use of a treated dust mop or dusting cloth will create less airborne dust than a brush.

Brushes should be cared for by combing or cleaning them occasionally and hanging up or resting them on the block when not in use. Brushes should be washed and dried carefully when soiled. Bowl brushes should be washed daily in a disinfectant solution.

Push Brushes for Cleaning Floors

Floor brushes are also known as push brushes, push brooms, and "sweeps." The floor brush, if properly used, will clean large areas of floors without raising a considerable amount of dust or overly tiring the worker. Floor brushes are available in a range of sizes and types to fit individual requirements. Because of lack of control, some industries arbitrarily purchase the cheapest push brush available in a rather small size and consider this item expendable. The implications in terms of efficiency and cost are obvious.

To use a push brush properly, a rough floor should be swept in areas which cannot be reached by a powered or push-type sweeping machine. Where dust is raised, a sweeping compound may be used sparingly. Push brushes should not ordinarily be used to clean resilient floors or any types of floors in administrative, medical, or dietary areas, as too much dust is raised. The proper tool for this job is the treated dust mop or vacuum sweeping tool.

When sweeping with a push brush, the custodian should stand with a straight back, with feet about 12 inches apart. The body weight should be swung on to the right foot with the left side of the body toward the clean floor. The weight will be swung to the left foot as the brush is extended. The stroke of the brush starts close to the body and should extend for about 6 feet. Shorter strokes, which are tiring, should only be used on congested areas, or where the dirt is composed of fine sand and soil. Dust "boiling" can be held down if the brush is not flipped at the end of the stroke.

Care and Maintenance. Push brushes require a little more care than most hand equipment:

- The brush handle should be rotated in the block frequently so as not to unduly wear one side.
- The handle should not be leaned on for resting.
- If the brush is permitted to rest on its fibers when not in use, it will bend them out of shape and make the brush useless.
- Push brushes should be used only for sweeping. They should never be used as a mop, squeegee, lever, or hammer.
- Selection of as large a push brush as practicable for a given job will get the job done quicker and cause less wear of the brush.
- The brush should be combed at least weekly to keep the filling material clean and flexible.

• Brushes with natural bristles can be attacked by moth larvae if in storage for some time. They should be protected with moth balls, paradichlorobenzene crystals, or insecticides.

• In storage, brushes should be rested on the block, hung up by the handle, or hung on special attachment rings.

• Very soiled brushes should be washed in warm neutral detergent solution and rinsed in clear warm water. The brush should be shaken to straighten the fibers, and should not be used until completely dry.

• Brushes of 24 inches in length and longer should be fitted with a handle brace to avoid the handle breaking at the block.

• Some brushes are fitted with scraper attachments. These should be used for light-duty scraping only. A special long-handled scraper should be used for heavy work.

CUSTODIAL CARTS

Carts are used to transport materials and equipment from one area to another. Their use eliminates frequent trips between work and storage areas and helps insure the use of the proper materials and equipment in the various cleaning operations.

Carts can be classified in three basic types: the waste cart, equipment cart, and utility cart.

Waste Cart

The waste cart is primarily used for waste collection and often also for waste transportation. In moving from one area to another, the custodian will empty wastebaskets and litter into the cart. When dumping wastebaskets, they should be held down into the cart at the level of the waste and upended. Dumping from above the cart will create dust and possibly strew bits of litter on the floor.

The most popular type of cart is made of a collapsible tubular frame from which is hung a fabric or plastic litter container. The frame is on casters for mobility. The litter bag normally holds about six bushels, and an extra bag should be kept on hand to be used while the soiled bag is being laundered, or while a torn bag is being repaired. For policing purposes, such a waste cart may be fitted with a small carrier, resembling a shoe bag in which can be placed tools and materials such as sponges, small brushes, bottles of detergent, scrapers, etc.

Waste carts may also be constructed of the rigid type such as a hamper, with a frame of metal or wood, with sides of cloth, plastic, fiber, wood, sheet metal, etc. An openhead fiber or metal drum may be used as a waste cart when placed on a dolly with casters. A drum or pasteboard box should never be rolled or slid on the floor for this purpose.

Equipment Cart

Equipment carts are used for the transportation of mopping buckets, floor machines, vacuums and accessories, and quantities of cleaning materials from the storage area to the work place. Special equipment carts are generally restricted to use by project workers, as they are not required by custodians performing daily cleaning. Equipment carts may be fitted with a small litter bag for waste that is collected during the performance of the project.

Utility Cart

Utility carts are designed to perform a combination of functions. The ideal utility cart contains a litter bag, has a tray for housekeeping chemicals, a storage shelf for items such as replacement paper towels or tissue, and also is able to transport manual equipment such as a mopping bucket, wringer, mops, and broom. Some utility carts for use by maids, or for special use by custodians, omit the storage shelf and may also omit the carriage of the equipment.

Utility carts should be cleaned completely each week, with the wheels being checked for strings and lubrication. If the cart is used in a dietary or medical area, it should be washed with a disinfectant solution twice weekly.

A well-selected and properly equipped and maintained cart can pay for itself in just a few months.

SOAP DISPENSERS

There are four basic types of soap dispensers:

1. Liquid soap dispensers, for dispensing liquid soap in liquid form.
2. Lather soap dispensers, for dispensing soap in lather form.
3. Powdered soap dispensers, for dispensing powdered soap.
4. Detergent dispensers, for dispensing detergents, hand lotions, or soap with a detergent content.

For specialized uses, there are also available paste-type cleaners and dispensers, impregnated cloths and papers, numerous types of bar soaps, protective creams and lotions, etc.

Generally speaking, soap dispensers are available for mounting as follows:

- Wall mounted—For mounting on the wall with screws or adhesive pads. Available for all four basic types.
- Basin mounted or for countertops—For mounting in the basin or countertop. This type of installation is generally not for use with powdered soap.
- Tank-type gravity feed systems—A battery of liquid or lather soap valves with a central soap supply.
- Recessed equipment—For flush wall mounting. Available for dispensing liquid, lather, or powdered soap. Recessed units may include other necessary washroom accessories, such as shelf, mirror, light, towel dispenser, etc.

To select the proper type of soap dispensing equipment, carefully study the needs and requirements of your particular installation, based on these considerations:

- Economy of operation:
  - — Powdered toilet soap provides approximately 250 hand washes per pound.
  - — Liquid soap dispensed in liquid form provides approximately 600 hand washes per gallon.
  - — Liquid soap dispensed in lather form provides approximately 2,000 hand washes per gallon.
- Economy of maintenance—Since labor costs account for most of the maintenance dollar, this becomes a basic consideration:
  - — The dispenser should be easy to fill and should have an unbreakable window to indicate to the custodian when it is time to refill.
  - — Depending upon the number of prospective users, the unit should have a large enough capacity to assure refilling only periodically, thereby allowing the custodian more time for other duties. In heavy traffic installations, the larger capacity dispensers have proven themselves to be far more economical to maintain than the customary 12-ounce size.
  - — The valve mechanism should be as trouble-free as possible. For instance, a high quality liquid or lather soap dispenser should avoid the use of packings or washers which require replacement.
- Design—The dispenser should employ as many tamperproof features as possible. It should be locked on the wall with a concealed wall plate. The container should be designed so as to be locked with a special key. The

unit should be constructed completely of metal (preferably stainless), and when translucent areas are required, they should be made of unbreakable plastics.

- Powdered soap dispensers should be used when there is a need for abrasive cleaning action. These features should be considered:
  - The unit should be designed to dispense any free-flowing soap powder and should not be designed to dispense one particular brand or type only.
  - The unit may incorporate an "adjustable-output" mechanism, thereby allowing proper adjustment, depending upon requirements.
- Paste-type dispensers should be used by mechanics, machine operators, and others whose hands become oily or greasy.
- Detergents (or soap containing a detergent content) cannot be used in regular soap dispensers, unless the dispenser is specifically designed for their use.
- In medical or dietary areas, the "hospital-type" dispenser should be used, which is operated by a foot pedal.

The following conditions may periodically occur:

- Liquid or lather soap dispensers:
  - An inoperative valve is usually caused by soap coagulation due to the use of a liquid soap with too high a soap solids content, the soap valve having been unused for a long period of time, or having corroded from the use of a detergent. Rinsing coagulated residue from internal parts is easily accomplished by filling the soap container with hot water, allowing it to stand for about ten minutes, then releasing it through the valve mechanism. Repeat if necessary.
  - A dripping valve is usually caused by packings which are worn with age and merely require replacing. It may also be caused by uneven coagulation in the valve.
- Lather soap dispensers: Poor lathering action in a lather soap dispenser is usually caused by the use of a soap solution having too low a soap solids content for the hardness of the water. In the case of a new lather soap dispenser, it is due to a special preservative compound containing oil, which kills the lather. In a new dispenser, work the soap through the valve several times, after which proper lather consistency will appear.
- Powdered soap dispensers: Some types of powdered soap dispensers have a strong tendency to clog, due to water splashing into the discharge opening, thereby causing a build-up or caking of the powdered soap. The dispenser mechanism should be checked for this periodically.

DUST MOPS

The ideal tool for sweeping smooth, lightly soiled, dry floors is the dust mop. When properly treated and cared for, the dust mop will remove loose soil from the floor quickly and efficiently, without leaving a residue.

Several types of dusting mops are available:

- The rectangular cotton dust mop. This is the most popular and useful of these mops. It consists of a mop head made of cotton yarn tied onto a wooden block or metal frame with handle. These mops are available in widths up to 60 inches.
- Dust mops are made in materials other than cotton, such as saran. The saran mop is normally factory treated, but once it is used and washed, it loses its dust-gathering efficiency and cannot be easily re-treated.
- U-shaped mop frames and mop heads are available which, while removing dust from the floor, collect litter in the enclosed area.
- Small dusting mops of a triangular shape are available for working in congested areas. These mops are sometimes used for dusting or even washing walls.
- The swivel-type mop, also for congested areas, permits the custodian to change the position of the mop head on the floor by twisting the handle.
- A large dust mop in the shape of a "V" is available for sweeping open areas such as wide corridors, gymnasiums, and lobbies. Sometimes these mops are equipped with double handles and are called "plow mops." A similar effect can be achieved by holding two rectangular mops together in a "V" position.
- Treated mopping towels and papers are available to be placed over the dust mop or over a push brush for dusting purposes. The appearance of these types is better than the dust mop, but they are less efficient since a smaller surface is exposed to collect dust and soil. As the yarn-type dust mop is utilized, new surfaces of the treated yarn are exposed, providing a continuing cleaning efficiency.

12 Ways to Extend the Life and Improve the Efficiency of a Dust Mop

1. Collect soiled dust mops (also dust mitts and wiping cloths) at the end of each working shift and clean them properly.

    - Shake out the mop carefully into a waste receptacle, dust bin, or in an area where the soil will not spread, or vacuum.

- Remove splinters or other foreign materials from the mop strands.
- Spray on 1 ounce of treatment for each 4 inches to 6 inches of dust mop length. Be careful not to overtreat the mop head.
- Roll up the mop head snugly and place in a closed metal container reserved and marked for this purpose.
- Allow at least twelve hours for the mop treatment to permeate the strands before reusing.

2. Launder the mop when it becomes soiled:

   - Place in a netted bag, or bind the strands in two places to prevent tangling while the mop is being washed.
   - Wash in a solution of warm water and detergent.
   - Dry the mop head carefully.
   - Comb the strands.
   - Treat the mop head (treating may be done during the rinsing cycle when washing, in some cases).

3. Store mops in a clean, dry area. Hang the mop so that it does not touch the walls and equipment and will not be brushed against by personnel. Mops should be hung with the mop heads down.
4. Keep the tie cords in proper repair and in place.
5. Do not allow the strands to become knotted or matted. Keep the strands free from splinters, metal particles, and keep them combed regularly.
6. When shaking out a mop, do not strike the handle or any part of the mop against a hard surface. Such action might damage the frame and weaken the handle.
7. Do not use a dust mop on wet or oily floors.
8. While mopping, keep the back as straight as possible and bend at the hips. Do not overreach. Following these rules of work posture will reduce fatigue and prevent strain.
9. Do not lift the mop from the floor unnecessarily; some of the gathered soil may drop from it.
10. Do not bear down on the mop; pressure is not necessary to proper mopping action.
11. If a vacuum is available, use it to remove loose soil from the mop frequently. If no vacuum is available, shake the mop head frequently into a large waste receptacle or a dust box to remove and collect loose surface soil.

12. When both the large and small dust mops are to be used by a custodian, carry the smaller dust mop on top of the larger one in a piggy-back fashion. Thus, when cleaning an administrative area, use a large dust mop for sweeping the corridors, while the smaller dust mop is carried on top to be used in the individual offices as they are passed.

For dust removal functions on surfaces other than floors, use dusting cloths and dusting mitts. These should be treated and laundered in the same way as dust mops.

MOPPING EQUIPMENT

The arsenal of mopping equipment includes buckets, mopping outfits, mops, squeegees, and wringers.

Buckets

Mopping buckets are usually constructed of galvanized steel, but they are also available in aluminum and stainless steel. Buckets may be either round or oval in shape. The former type has the advantage of being stronger and easier to clean, whereas the latter type takes less space and offers a wider opening to receive the mop.

Most mop buckets are equipped with "gliders," which are small beads of metal which permit the bucket to slide on the floor. Buckets are available with attached casters of various sizes, which protect the floor from such sliding marks. The 3-inch caster size is generally suitable. The smaller buckets, such as pails, are not so equipped.

Buckets should be selected on the basis of the metal gauge and the quality of the construction. Certainly, both the inside and the outside of the bucket should present a smooth surface that will not cut or scrape the hands of the user.

Buckets, as well as all types of metal mopping equipment, should be washed, rinsed, and dried after use. If the bucket develops a leak, it should be soldered or patched right away, and the patch spots filed smooth.

Liquids should not be permitted to remain in buckets overnight. Some liquids will cause odors, or, such as alkaline solutions, damage the bucket.

Mopping Outfits

For everything but the smallest mopping jobs, one of the two basic types of mopping outfits should be used.

The larger of the two types is the mopping tank, which is a compartmented tank on large casters, with a built-on wringer. These units have two or three compartments and usually range from a total of 30 to 60 gallons in size.

The smallest operation will require two buckets, one for the detergent solution and the other for rinsing. These two buckets can be mounted on a special carrier or truck and fitted with casters and a pulling handle. A wringer is placed in one of the buckets to complete the outfit. Systems consisting of three buckets are also available, and are very effective in some cases.

Mops

The mop is the basic hand tool for wet cleaning.

Mops may be either permanently attached to a wooden handle or may be attached and removed from a special frame having a wooden or metal handle. The detachable mop head has the considerable advantages of permitting laundering in a machine, placing the entire strand length in contact with the floor, ease of cleaning and drying, and reusability of the handle. The old-fashioned stick mop should not be considered for industrial use.

The mop head itself may be constructed of several different materials:

- The cotton mop—This type is most extensively used. It is inexpensive and easy to launder. It is also well suited for the application of wax by trained workers. The cotton mops, as well as other mops, are available in various sizes and are usually classified by weight. A 16- or 20-ounce mop should be utilized by a smaller or handicapped custodian, while a 24- or 32-ounce mop should be used by the stronger housekeeping worker.
- Mops made of rayon—These are lighter in weight than cotton and more absorbent. They are not as suitable as cotton for applying wax, and where mops are to be used for both wet cleaning and wax application, standardization should be made of the cotton mops.
- Cellulose sponge mop heads—Although these are considerably more expensive than cotton or rayon, they are much more absorbent and durable when used on smooth floors.
- The sponge mop with a built-on levered wringer—This is a special type of mop that is useful for policing operations and stair tread cleaning, as well as for certain types of wall washing.

16 Tips for Caring for Mops. Most mops do not wear out through use, but rather are rendered useless through misuse. Although mop damage can sometimes be attributed to mopping equipment, such as a poor choice of wringer, it is most often caused by abuse to the mop itself. Here are some pointers on mop use and care:

1. The basic purpose of the wet mop is to transfer liquid to and from floors. The use of mops for other purposes should be discouraged.

2. Soak a new mop in warm water for at least twenty minutes before use, in order to remove excess oils and expel entrapped air, thus providing better absorbency.

3. Change mopping water frequently while mopping to prevent the water from becoming overly dirty. Rinse the mop each time the water is changed. Dirty water will cause a redeposit of soil during mopping, leaving soil streaks.

4. While mopping, make sure the mop stays on the floor and is not flung about.

5. Turn the mop from one side to the other frequently while in use in order to expose more clean, moist strands.

6. Squeeze mops or press them carefully to remove water. Twisting the mop will tear or weaken the strands.

7. Use mops for scrubbing small areas and spots by folding the mop over on itself so that the tips of the strands are in direct contact with the floor, under pressure. Never scrub with the part of the mop closest to the holder because this will tear the strands.

8. Avoid the use of mops with lye, caustics, or strong undiluted cleaning solutions because these materials attack the mop strands.

9. If a rough surface is to be mopped (such as splintered wood), mop very carefully and try to mop in a direction which will avoid snagging. It is best to reserve a special mop for use on such a surface.

10. Do not continue to use worn-out mops—they should be discarded.

11. Always rinse a mop carefully after use and squeeze it dry.

12. Store mops in a warm dry area where air circulates freely.

13. Hang a mop in storage with the yarn away from the wall, with the strands down.

14. Do not let wet mops touch each other or come in contact with walls or equipment.

15. If mop strands become loosened, remove or cut them off with scissors in order to prevent snagging and splattering.

16. In addition to rinsing the mop whenever the mopping water is changed, carefully wash the mop periodically (daily if possible). The frequency of washing will depend upon the use to which the mop is put, but mops in regular use should be washed carefully at least weekly, preferably before the weekend. Where a number of mops are in regular use, you can use a mop-washing machine which exposes the mop to a high-pressure clear water bath.

Suitable training in mopping techniques enables a custodian to obtain rapid coverage with a large mop without fatigue. Unsound mopping procedures will give backaches, even with the use of smaller mops in restricted areas.

Squeegees

Squeegees, through the wiping action of their rubber blades, are used to move large quantities of water on a floor. Thus, the squeegee can collect water in a confined area for pickup with a vacuum, or can move the water to a floor drain.

Squeegees are straight or curved. The curved squeegee has the advantage of moving more water without spillage around the sides, but it is a little more difficult to handle and cannot be pulled back as the straight squeegee can.

Water may also be picked up from a floor with a squeegee and a special squeegee pan, or pick-up pan, which traps the water for later disposal. Pick-up pans have a liquid capacity of 3 to 5 gallons and should be used in conjunction with the straight type of squeegee.

When using squeegees, follow these procedures:

- After each use, wipe the rubber blades clean and dry.
- When the squeegee is not in use, store it in a cool place away from liquids.
- In storage, make sure the squeegee does not rest on the rubber blades, which are easily damaged. It is best to hang up a squeegee by its handle.
- Never hang up the squeegee to dry outside: the rubber blades are damaged by sunlight.
- Keep a spare set of squeegee blades on hand. Squeegee blades should be replaced when efficiency is impaired.

Wringers

The name "wringer" should be a misnomer, since the wringing of a mop to remove water loosens and damages the strands. The proper drying action for a mop is squeezing, which is the design of most "wringers."

This squeezing is accomplished either by moving the mop between rollers, like an old-fashioned clothes wringer, or squeezing the mop between the perforated metal plates of a box-type wringer. The squeezer-type action of the latter is more efficient and less destructive to the mop. Squeezer-type wringers are either of the horizontal or downward pressure types. When choosing a wringer, consider these factors:

- The horizontal type can take more abuse without failure.
- The downward-pressure wringer is more efficient.
- A mop wringer needs to be rather closely sized to the mop to be used.
- The mop wringer of plastic construction has proven to be quiet, non-rusting, light weight, and durable.

From a construction standpoint, the roller type seems to be more suited for the compartmented mopping tank because of the height of the tank.

For the proper care of wringers, here are 6 tips:

1. Keep working parts properly oiled.
2. Keep screws and bolts tightened.
3. Clean the wringer after each day's use.
4. Do not lengthen handle or press too hard on the foot lever, because too much force on a wringer can break it.
5. Make sure the wringer is large enough for the mop that is being used.
6. Remove loose mop strands and other articles caught in the wringer.

WASTE RECEPTACLES

In addition to the waste and utility carts previously mentioned, a wide range of waste receptacles is in use:

• The "torpedo" receptacle—This type is named for its shape, as it is cylindrical with a round top. Refuse is placed in the container through a single swinging door. The waste is collected in a separate inner container which may be made of fiber, galvanized iron, plastic, or other materials. The torpedo receptacle is an attractive and durable container for small to medium quantities of waste, such as in administrative areas.

• Rectangular containers—These metal receptacles, in various colors, are usually fitted with a swinging top, through which waste can be deposited from either side. The inner container may be a burlap or duck bag, metal, or plastic. The smallest of this type of container is classed as a sanitary napkin disposal unit, while the largest has a volume of about 60 gallons.

• Special types of rectangular containers, either with open or self-closing tops—These are fitted with brackets to support paper towel dispensers, mirrors, and the like.

• Another variation of the rectangular container that includes a tamping mechanism—This mechanism compresses the waste to a fraction of its original crumpled size, which reduces frequency of waste disposal.

• Wall-hung receptacles—Although these are limited in the amount of waste they can accommodate, they are very handsome in appearance and do not have to be removed for floor maintenance.

• Flush-type receptacles—These are built in during construction or renovation and can be had in combination with soap dispensers, mirrors, and towel dispensers. They are usually fabricated of stainless steel.

• The common galvanized garbage can—This type is still in widespread industrial use, particularly for handling wet waste and food waste. The sizes in most common use are of approximately 20- and 32-gallon capacity.

• Fiber waste cans—These are also available, but they create odors and deteriorate rapidly if misused for wet waste.

• Plastic garbage cans—These have the advantages of not denting, rusting, absorbing liquids, or being noisy. The newer plastic materials can be steam cleaned successfully. Some plastics have the disadvantage of supporting fires.

• Drum covers—These are available to convert an openhead drum to a waste receptacle. The cover has a torpedolike head with a self-closing door.

• Waste receptacles for classified material—These are constructed of heavy-gauge metal and cannot be opened for disposal without a key. Waste is inserted through a small opening so that the materials cannot be easily pulled out.

• Containers for oily waste—These are also made of heavy-gauge metal. They have a special lid and are properly vented underneath to avoid spontaneous combustion.

• Woven-wire receptacles for paper litter—These are inexpensive, but rather unsightly. Some people will use these receptacles, however, while paying no attention to the closed units.

• Wastebaskets—These are generally used in office areas, although they may be used in the plant as well where small amounts of dry waste are created. Empty buckets and pails which may have contained raw materials or chemicals are usually substituted for this purpose in the plant. If this is done, they should be painted. Wastebaskets are available in metal, fiber, and plastic.

• Special metal cans for edible waste—These cans are not affected by steam cleaning, and they have no edge or lips which might interfere with cleaning.

• Cigarette urns—These are a special class of waste receptacle (as is the ash tray). Cigarette urns may rest on the floor, or may be mounted on the wall. The wall-mounted type is certainly preferable in administrative areas, as floors may be cleaned without their removal. It is also preferable in plant areas where malicious mischief is not a problem. If cigarette urns are not located wherever cigarette smoking is permitted, the natural result is that cigarette ends will be thrown to the floor.

OTHER SPECIALIZED MANUAL EQUIPMENT

There are a number of specialized hand tools available for the custodian. Here are some of them:

• Applicators of wax and seals—These are made of lambskin, fitted to a wooden block and handle. They should be used with an applicator pan which places the proper amount of liquid onto the applicator. Unless lambswool applicators are carefully cleaned after each use, they harden and become worthless. The same material, often called chamois, is available for hand cleaning and drying, as well as polishing.

• Deodorant cabinets—These are mounted in rest rooms and other areas. They are vented and disperse deodorants from a solid or liquid state.

• Detergent guns—These dispense cleaning solutions through water or air pressure, or a combination of both.

• Drum accessories—These include faucets, pumps, level indicators, racks, and wrenches.

• Dust boxes—These can be used to shake out dust mops and push brushes, although other methods are generally preferable.

• Dust pans—These are fitted with a hand grip for removing chips and granular waste, or fitted with a short handle for lobby policing.

• Feather dusters—These are still being used, although other tools have made them obsolete.

• Foot-bath trays—These are used to prevent athlete's foot in shower areas, through the immersion of the feet in a suitable disinfectant. Mechanical foot baths are also available which spray the feet.

• Funnels and measuring cups —These are used for pouring and measuring small quantities of liquids.

• Gloves, aprons, hard hats, goggles, and face shields—You should issue these permanently to custodians who must work with hazardous materials.

• Ladders, scaffolds, and hoists—These are required for cleaning walls and overhead areas. Their use should be restricted to specified individuals. Any work requiring this type of equipment should be inspected frequently.

• Proportioners—These mix a measured quantity of detergent or other chemical water by an aspirating action.

• Safety signs—These provide warning when floors are wet and slippery.

• Scrapers—These are used to remove chewing gum and other sticky substances from floors. The common putty knife should be carried by a custodian in office areas, whereas the long-handled floor scraper may be used in plant areas.

• Sprayers—These are used to dispense chemicals in a fine fog, mist, or spray. Sprayers are used for lacquers, insecticides, deodorants, dust-mop treatments, and detergents. Some larger sprayers are carried by a strap over the shoulder, whereas most small sprayers are carried by hand.

• Steel wool—This is best used by hand for metal cleaning. Where rust is to be avoided, stainless steel wool is preferred.

• Tool holders—These permit the hanging and the storage of brooms, dust mops, squeegees, and wet mops by their handles, rather than resting them on the floor.

• Wiping cloths—These have replaced sponges, which are now considered unsanitary. Obtain cloths that do not produce much lint, such as diaper cloth.

24

Choosing the Best Cleaning Chemicals for Your Needs

In any given building service department, housekeeping chemicals can be a morale builder or a morale destroyer, a legitimate expense or a financial burden, a minor or time-absorbing part of the manager's work. The typical custodial department devotes from 3%–10% of its budget for cleaning chemicals. The figure will be higher or lower depending on the quantity of chemicals purchased, and thus price reductions based on that quantity; on custodial wages, since as they go higher they tend to reduce the percentage cost for chemicals; on consumption control devices that prevent overuse; on the prevalence or absence of pilferage; and on the actual amount of cleaning being performed. An improved cleaning operation will result in a more frequent performance of certain operations, which will utilize more chemicals. Conversely, minimizing chemical use may limit the amount of cleaning work that can actually be done.

Selecting appropriate cleaning chemicals is important. For example, if you select an inadequate floor finish, one that does not level well or powders or shows scuff marks or water marks easily, then your workers will conclude that the difficult and time-consuming job of finish stripping and reapplication will not yield good results. They will understand that certain activities will be required more often because of the poor chemical quality than would have been necessary had a better material been selected.

When an organization needs to cut costs, a finger is often pointed at custodial chemicals as a logical place to make savings. Be very careful in this regard, because of the problems that can arise in work performance when lesser-quality chemicals are introduced. A change in chemicals can create a change in procedure and thus create a different time allowance for cleaning. The paradoxical result might be that instead of saving work, you are actually creating more difficult work.

Although chemical products represent only a small part of the housekeeping budget, they can have an effect on personnel and their time utilization far beyond this figure if these products are not properly selected for the work, and the personnel trained in their use.

This chapter explores the most common cleaning chemicals found in housekeeping today and offers practical suggestions for on-the-job use.

THE TOP FIVE CLEANING CHEMICAL PRODUCTS USED BY CUSTODIANS

The typical housekeeping department will have a storeroom containing a couple of dozen varieties—in many cases still more—of cleaning chemicals. Most of these, however, are used either infrequently, or in small quantity. But in terms of dollar value and importance to the custodial effort, five products dominate the picture:

- Floor finish
- Finish stripper
- General detergent
- Disinfectant
- Hand cleaner

Floor Finish

There are a few types of flooring that under special conditions can be utilized without finish, such as special pedestal floors for data processing areas, and pure vinyl floors under lightly trafficked and soiled conditions. But in all other cases, something must take the wear, and the choice simply lies between the flooring itself and a protective coating, also called a finish or "wax."

Organizations whose floors are primarily vinyl asbestos, asphalt tile, or terrazzo find that the floor finish is the most important single product when the factors of unit cost, total cost, impact on custodians, and the effect on the public is considered. Selecting an acceptable floor finish is tedious and expensive; once a suitable product is found, a change should not be made unless a very significant cost improvement is effected. Certainly do not consider changing floor finishes if only a moderate per gallon saving will be realized over an existing good performer.

Tips on Applying Eight Types of Floor Finishes. These are the varieties of floor finishes now available, and their applications:

1. *Solvent or spirit waxes.* These are yellowish-colored solutions of natural wax, such as paraffin and beeswax in solvents, such as mineral spirits. Their principal application is for wood, which would otherwise be damaged

by the application of water and water systems. Solvent waxes are also usable on cork, which is a form of wood. Since wooden floors can also be protected by using penetrating and surface sealers, thus eliminating the need for waxes, the use of the solvent waxes, either in liquid or paste form, is minor.

2. *Water emulsion waxes.* When composition tiles, such as asphalt tile, were first developed, it was necessary to find some protective surface other than the solvent waxes which would soften and damage the tile. Carnauba wax, a natural wax scraped from the leaves of a plant found in South America, had formerly been used in solvents and could also be utilized in an emulsion form in water, much as milk is an emulsion of fatty materials in water. When applied to composition tile or other surfaces, the water evaporated and the wax particles coalesced into a thin surface. Various additives were added to the carnauba or other natural waxes to control their leveling, slip-resistance, softness, sheen, and other such factors.

These water emulsion waxes serve their purpose fairly well, but drawbacks were noted, such as darkening, slipperiness and the requirement for dry buffing to produce a good gloss. These materials have been superseded in their use by the polymers mentioned below, except in cases where gritty or sandy soil will scratch the harder polymers, and the softer characteristics of the water emulsion waxes provide an advantage in healing the worn spots through their mobility while buffing.

3. *Polymer finishes.* The most successful floor finishes in use are the synthetic polymers which, like the water emulsion waxes, are suspensions of particles in a water system, but rather than being natural waxes, they are the products of chemical invention. (Many people, however, still call polymer finishes "waxes.") Various chemicals go into making these products, the basic item being acrylic plastic, but other plastics are utilized as well, such as styrene and ethylene. The advantages of the polymer finishes over the water emulsion waxes are that polymer finishes:

- Form into a single film on application, rather like a sheet of cellophane, and thus dry to an initial gloss without buffing.
- Are more durable and tougher than the waxes.
- Discolor much more slowly.
- Can be spray-buffed most successfully (as elsewhere described), a technique that is most efficient for floor care.

4. *Metal-link polymers.* A special form of polymer floor finishes is the metal-link material, a chemical invention that makes polymers much less susceptible to removal from the floor by detergents. This has been accomplished by making the polymer molecule much heavier through the interlocking of metallic ions, typically zinc or zirconium. These products generally have all the advantages of the polymer finishes, including spray-buffability, so they represent an excellent investment.

5. *Solution finishes.* It is now possible to produce a polymer finish in a water solution (involving alkali soluble resins and coalescing agents) rather than as a water emulsion. The pros and cons of polymer finish and solution are not nearly as definitive as the differences between solvent waxes and water emulsion waxes, and the solutions are still in the state of development. A compromise exists with translucent finishes that incorporate polyethylene for better wear.

6. *Water emulsion pastes.* Both waxes and polymers can be prepared in a pasty form. This represents not so much a change in formulation, but a change in consistency and in the method of application; these products can be smeared around a floor under a floor machine pad, which acts as an absorbent to remove some of the older material and soil, somewhat like spray-buffing. A moderately good job can be done with this product and technique, although there are often complaints of slipperiness in certain formulations. Using these pastes should be most successful in conditions of light traffic and soil, as a floor so prepared tends to become increasingly dingy in appearance.

7. *"Lock and key" finishes.* Another special development of the polymer finishes was one which required an acidic stripper, thus being resistant to detergents which are primarily alkaline in their action and composition. Now that the metal interlocks are available, this type of product has become obsolete, especially since the use of acidic materials can become injurious to people and surfaces alike.

8. *Thermoplastic finishes.* Used with an ultra high-speed floor machine, these finishes melt under the heat of the pad, then coalesce into a continuous film.

Various Floor Finishes. Figure 24-1 illustrates a floor finish testing record. Use this form to evaluate various floor finishes, by following this procedure:

1. Select an area, such as a corridor, where several test sections can be given nearly equal traffic and exposure.

2. Strip entire test area completely of any old finish and rinse thoroughly (at least twice).

3. Set up test "bands" the width of the corridor and preferably ten feet or more in the long dimension of the corridor. By some convenient method—such as plastic tape or masking tape attached to the baseboards—mark each test strip.

4. Deliver each finish to be tested to the person responsible for testing in a clean container identified only by a letter or a number so that opportunity for prejudice will be eliminated.

5. Make sure all custodians uniformly apply all products tested. After the floor has dried, apply two light coats successively, allowing the first coat to dry thoroughly before application of the second, and noting any special

Figure 24-1. Floor Finish Testing Record

Product Code ____________

Area Identity ____________

Date Applied ____________

Time of Evaluation After Original Application	Qualities Rated (1)											Effects of Maintenance: Buffing		Effects of Maintenance: Spray-Buff		Total Score at Each Evaluation
	Good Levelling	Dries 30-60 Minutes	No Cloudiness	No Discoloration	Gloss	No Slipperiness	No Powdering	Little Scratching	Little Scuffing	Little Black-Marking	Strips Readily (2)	Resultant Gloss	Removal of Scuffs & Marks	Resultant Gloss	Removal of Scuff & Black Marks	
As Soon as Dry							x	x	x	x	x	x	x	x	x	
2nd Day	x	x	x	x							x					
1 Week	x	x	x	x							x					
	x	x	x	x												
	x	x	x	x												
	x	x	x	x												
	x	x	x	x												
	x	x	x	x												
	x	x	x	x												
	x	x	x	x												
	x	x	x	x												
TOTAL RATING (3)																

<u>NOTES AND COMMENTS:</u>

(1) Rating values: 2 = Good; 1 = Acceptable; 0 = Poor

(2) Strippability: See Instruction 9

(3) Highest Total Rating <u>usually</u> will indicate best performance; but, product may be disqualified for unsatisfactory performance in one important category, such as "No Slipperiness" or "No Powdering"

drying characteristics. (First coat should be fully dry in 45 minutes unless inside humidity is high, or temperature excessively low.) Enter on a Floor Finish Testing Record the product code and the exact area where test product is applied, along with date of application.

6. Using the rating scale indicated on the Floor Finish Testing Record, rate each product individually for the first six quality characteristics shown in the table.

7. Do not buff the product before the second day after application. At that time, rate all characteristics except strippability and spray-buffing. (Spray-buffing should not be needed so soon after application, but if it is, because a very heavy traffic area has been selected, those characteristics should be rated on all products beginning at the same time.)

8. At the end of one week, rate all characteristics except the first four and strippability.

9. Continue ratings about once a week for whatever period deemed necessary to reveal a meaningful difference among the products tested. With careful spray-buffing, stripping should not become necessary within any reasonable time. However, at the end of the period arbitrarily chosen, it will be desirable to strip all products and rate them as to strippability.

10. Remember that the highest total rating will usually indicate the best performance. However, products may and should be disqualified for unsatisfactory performance in any one important category, such as slipperiness, powdering, or inadaptability to spray-buffing.

11. If bad weather and heavy soil necessitate an unusual amount of wet cleaning, do not let this affect the validity of the test if all test surfaces are subjected to comparable procedures.

Finish Stripper

Periodically floor finishes must be removed, although the time between removal and refinishing can be greatly lengthened by trapping soil before it reaches the floor (such as with the use of gratings, mattings and runners), and through techniques such as spray-buffing. Some floors which receive great traffic and soiling and have a poor finish may be stripped every month or so; others may be stripped only after a number of years if properly protected and cared for. Minimization is the rule, since the stripping activity is injurious to floors, as it involves a harsh, alkaline chemical and a great deal of water.

The stripper works by attacking the surface of the finish, much as paint is attacked by paint removers.

The formulation of a stripper is a chemist's job, and should not be done by using a higher concentration of a general detergent, or simply by adding boosters to other types of chemicals; simply adding ammonia to a general detergent may strip the finish, but it may damage the floor as well.

Ideally, the floor finish stripper should be provided by the manufacturer of the wax or finish being used, as the two would be formulated as companion products. Each finish formulation should require a specific stripper formulation; for example, the metal interlocks require aminos or other chemicals to break the bond, whereas other finishes may not require this.

The stripper is used in far less quantity and with far less frequency than the floor finish, but as can be seen is an important custodial chemical.

General Detergent

The original and fundamental cleaning material is a detergent: water. Since a detergent is a material that provides a cleaning function, water fits this bill admirably, not so much because it is abundant, but because it dissolves more substances than other chemical materials.

The job of the chemist is to make water a still better detergent through the addition of "boosters" of various types (for example, television commercials may claim that a given product "makes water wetter"—and it may). These additional builder chemicals are only required in a very small quantity to do an effective job. In a use-solution of a detergent diluted to two ounces per gallon, the active ingredients present may be no more than one fourth of one percent.

Detergents can be formulated to provide various functions, depending upon the nature of the soil, the cleaning process, and the characteristic of the water. There is no such thing as an "all-purpose detergent"—detergents are formulated to perform specific jobs. A floor finish stripper and a hand soap are both detergents—no one would think of reversing their roles or trying to have one product for both jobs.

The housekeeping department needs a general detergent which can be used for such jobs as mopping floors, washing fixtures, and other general cleaning functions. Just as we have progressed from natural spirit waxes to polymer floor finishes, so have we progressed from natural soaps made from animal and vegetable fatty material, to synthetic detergents produced in the laboratory. Such detergents are not generally difficult to select; you must simply be aware of their performance and the effect on the user.

Since we are basically talking about water in any discussion of detergents, the formulation of the product must reflect the condition of the water being used. The primary factors are water hardness, which is controlled at the place of use, and the pH of the formulated detergent.

Water is hard or soft depending upon the quantity of metallic salts in solution, principally formed from the elements calcium and magnesium. The measurement is usually taken as parts per million (ppm) when calculated as calcium carbonate. Soft water is considered to have less than 200ppm, moderately hard water from 200–400ppm, and hard water over 400ppm. Occasionally one sees water hardnesses of 2,000ppm, but fortunately this is a rare condition. Water hardness may also be calculated in grains, with one grain equal to approximately 17ppm of calcium carbonate. Where hard water is used, detergents must be formulated to sequester this hardness, or to precipitate it if the precipitate is not objectionable.

The acidity or alkalinity of a solution is identified as its pH, which is shorthand used by chemists to indicate the presence of hydrogen ions. An arbitrary pH scale has been developed from 0–14, ranging from the strongest acid at 0 to the strongest caustic at 14, with 7 being neutral. See Figure 24-2. The scale is complicated by the fact that it is logarithmic—that is, the pH of 5 is ten times as acidic as a pH of 6. It is up to the chemist, not the custodian, to determine pH requirements for various uses. While a bowl descaler would be acidic and thus of a lower pH, a finish stripper would be alkaline and of a higher pH, while a hand soap should be nearly neutral.

Once the detergent has been properly formulated, other factors affecting or creating cleaning are:

- Time—Increased contact time provides greater chemical action.
- Temperature—The higher the temperature, the quicker the chemical reaction.
- Mechanical action—Some cleaning is done by rubbing, scratching, and scraping.

Be sure that custodians are not putting additives into an otherwise acceptable product. Sometimes workers throw in handfuls of scouring powder, which can scratch a floor or other surface due to its abrasive particles—or put in some TSP (trisodium phosphate), which can leave a white film on a floor due to the precipitation of metallic salts, etc. One custodian was seen to stir in broken-up urinal discs, which are completely insoluble in water, and have no detergent action whatsoever.

When selecting a general detergent, consider the following:

- The pH should normally range from 7.0 to 9.5.
- Excessive foaming is undesirable, as it requires a secondary rinsing job, or leaves a film that attracts soil and looks unsightly.
- Avoid natural soap content, and stick to synthetic detergents, which do not leave a precipitate (such as the "ring around the bathtub").
- For general purposes, use-dilutions should be no higher than two ounces per gallon of water.

Figure 24-2. pH of Some Cleaning Chemicals

	pH	
↑	14	
	13	SODIUM METASILICATE
	12	TRI-SODIUM PHOSPHATE SODIUM CARBONATE
	11	AMMONIA
INCREASING ALKALINITY	10	POLYPHOSPHATES
	9	SOAP BORAX
	8	SODIUM BICARBONATE SODIUM TETRA PHOSPHATE
NEUTRALITY	7	DISTILLED WATER
		SODIUM HEXAMETA PHOSPHATE
	6	BORIC ACID
	5	
	4	
INCREASING ACIDITY	3	ORGANIC ("WEAK") ACIDS
	2	
	1	MINERAL ("STRONG") ACIDS
↓	0	

• The formulation should avoid ammonia, free alkali, solvents, abrasives and free oils or fats.

Disinfectant

Germ killing is one of the most important aspects of housekeeping in a hospital—yet it is also one of the most important activities in all types of buildings, since the control of disease-causing bacteria is necessary in rest rooms, locker rooms, food service and vending areas, and even on telephone mouthpieces.

In the past, disinfection was performed by first cleaning with a soap (and later a neutral synthetic detergent), and then a second step of disinfection was performed. Now it is feasible to perform both cleaning and disinfection steps in one—these products are typically called germicidal detergents or cleaner-disinfectants. In order to gain the advantages of standardization, and to assure that germ killing will always occur during cleaning, a number of health care organizations have eliminated the general detergent and have decided to use a combined product for all cleaning.

To kill micro-organisms, various degrees of control are possible:

• *Antisepsis* is the killing of germs on living tissue, such as with mercurochrome, or tincture of iodine.

• *Sanitization* involves reduction of bacteria levels to numbers which are generally considered safe—a 50% reduction in bacteria count qualifies one to use this term, which is often applied to food service surfaces, such as counters, and glassware.

• *Disinfection* is the term that housekeeping personnel are most concerned with, as it relates to killing bacterial organisms that are potentially harmful. The strict definition of the term varies but is now generally taken to include the killing of most (but not necessarily all) disease-causing bacteria and fungi, but does not necessarily include the killing of spores (germs which encase themselves in a hard shell) or viruses (the simplest living things, resembling a complex molecule). Disinfectants as well as other materials that kill are controlled by the United States Department of Agriculture—a chemical can only be classified as a disinfectant if it kills at least 99,999 out of 100,000 organisms under controlled conditions. Thus, complete kill is not necessary (or even possible under most use conditions). Bacteria are usually found in such enormous numbers that to leave only a single organism alive out of 100,000 might still leave millions alive on a given surface.

• *Sterilization* signifies the complete killing of all organisms which may be present, and is generally achieved in an autoclave through the use of steam or lethal chemicals or perhaps by irradiation.

Disinfection is the primary concern of the housekeeping department, in whatever type of building. Custodians should be familiar with a few other terms generally associated with this type of activity:

- *Bactericides* kill all bacteria, those which are beneficial as well as those which are harmful. One should not forget that the body cannot survive without the beneficial bacteria which live in and on it.
- *Fungicides* kill fungus growths, such as that which causes athlete's foot.
- *Germicides* kill those bacteria—or germs—which cause disease.
- *Sporicides* kill spores, which are the protective form of certain bacteria which build a shell around themselves, difficult to penetrate; these are the hardest of the bacteria to kill.
- *Virucides* kill viral organisms, such as those which cause viral pneumonia and influenza. Some forms of viruses are very difficult to kill.

Since the control of disease-causing micro-organisms is so important, it is not surprising that the federal government rigidly controls the marketing of disinfectants, a term generally synonymous with germicides. Since the government cannot exercise this control except over products shipped in interstate commerce, it is most desirable to purchase such products from companies doing business in more than one state, so you can be assured of controlled manufacture and testing. The Department of Agriculture requires all pertinent information to be shown on the label in such cases.

To test disinfectants, chemists rely on the Use Dilution Confirmation Test, as developed by the Association of Official Agricultural Chemists, and called the AOAC Test. This laboratory test simulates field conditions and measures the kill of the given chemical at control dilution rates over specific periods of time for individual test organisms.

The AOAC Test does not qualify a chemical to kill all bacteria, but just those against which it was tested. In selecting disinfectants, you may wish to require the following:

- At normal use-dilution (this is usually two ounces or less per gallon) the product should kill Staphylococcus aureus and Pseudomonas aeruginosa, as well as Salmonella typhosa and usual test organisms.
- The product should kill fungi, such as those causing athlete's foot, and should inactivate many of the viruses that are pathogenic to man.
- In a general hospital environment, the product should also kill the tuberculosis organism.
- Bearing other requirements in mind, the product should be selected for minimum skin irritation.

- The germicidal activity should not be affected by the available water supply.

Several types of chemicals can be used for germicidal purposes, and despite the advantages of standardization, the housekeeping department may be required to use more than one variety. Generally, however, you should basically rely on either a phenolic type disinfectant or a quaternary ammonium type for most usages, with the use of an iodine disinfectant for very special cases under hospital conditions.

- The earliest disinfectants were strong, odorous products such as simple phenol, cresol, and pine oil. These products led to the feeling still prevalent, that for a germicide to be effective it must necessarily smell strong! All of these materials are obsolete, even though great amounts of money are spent promoting them. They do not kill, for example, Staphylococcus aureus, typically the greatest culprit in cross-infection.
- Synthetic, or modified phenols, have been developed to overcome the disadvantages of the earlier materials, and are quite effective in doing this. They are essentially odorless, and kill a wide range or spectrum of organisms, significantly including tuberculosis germs. They have a fairly low degree of skin irritation, although some people develop a rash with continued use—just as some people can be irritated by pure water. The synthetic phenols—now simply called phenols—are popular disinfectants.
- In order to overcome the limitations of phenols, the chemists have developed quaternary ammonium chemicals, the "quats," but at the same time have introduced limitations that the phenols do not have. The "quats" have the advantage of good odor control, very low skin sensitivity and a high dilution ratio—but they do have the disadvantages of not being able to kill the tuberculosis organism, and being inactivated by a smaller quantity of soil than the phenolics. The "quats" are non-staining, odorless and non-toxic. Their characteristics make them ideal for use where food is processed or served, and are very useful in hospitals where tuberculosis is not considered a problem. Many hospital workers prefer to use phenolics because of the legal implication of using a material that will not control TB.
- Sodium hypochlorite chemicals, being corrosive, should be restricted to specialized use in certain processing operations, and should not be generally used by a housekeeping department.
- Iodine is a powerful disinfectant, and can be prepared for general use in the form of "Iodophors," a solution based on a non-ionic syndet (synthetic detergent). When in use, these materials release active iodine which will kill a wider range of pathogens than the phenols or "quats." Their acidic nature, however, limits them to specialized use, such as for "dirty cases" in hospitals where one would fear that the phenol or "quat" would not be sufficient. One would use the Iodophor, for example, for a gas gangrene or spinal meningitis

case. General use of this type of material is not recommended and where biological laboratories exist, their use should be carefully avoided, since vapors from the chemical can interfere with such tests as protein-bound Iodine measurement.

Both synthetic detergents and soaps, when they are dissolved in water, break up into particles which are charged positively or negatively. Charged particles are called ions, and in a solution the positive charges counterbalance the negative charges. In a solution of soap or synthetic detergent, however, there are large and small particles or ions, the large ones being the active constituents, and these materials are classified according to the charge on this larger ion—thus, ions are negatively charged, and cations are positively charged. Where there is no charge at all, these are called nonions. Ionic classification is important in the manufacture of chemicals, because a cationic material can neutralize an anionic material when they are mixed together. Thus, the housekeeping department is once again urged not to do its own formulation, but to leave this to the skilled chemists.

This is even more true when you consider the numerous factors involved in germicidal activity:

- The presence of soil and organic matter.
- Water hardness.
- The presence of other chemicals.
- The pH of the solution.
- Contact time (which should be a minimum of two minutes for adequate kill).
- The temperature of the solution and the surface being cleaned.

Hand Cleaners

Although hand cleaners are not a material used by the housekeeping department in its service of environmental control, it is in the category of paper towels or toilet tissue: a dispensed product for use by other employees and the public in rest rooms. Hand soaps are such a personal product that many people judge the function of the entire housekeeping department on the basis of the type of hand cleaners being offered. It is not unusual to see a housekeeping supply room with as many as a half dozen or more varieties of hand cleaners available. In general, two products from the following list should be sufficient:

1. a hand cleaner for lighter soils, such as is used by administrative or clerical people (the liquid hand soap dispensed in lather form being a good example)

2. a heavier-duty material for greasy or oily soils (the lotion-type cleaner with solvent content being a good choice)

Here are some of the varieties of hand cleaning materials available:

- Bar soaps are the oldest of the hand cleaners, made from natural fatty materials. Bar soaps are inexpensive, but have the disadvantage of being a "community" product which many people wish to avoid. It is possible that the use of bar soaps can lead to cross-infection, and even if this is unlikely, people still give it thought. The use of bar soap also complicates the cleaning of the lavatory surface.
- A dispenser is available which will grind up bar soap into a powdered or chipped form, thus removing the community aspect of the material. Most people do not like the slippery, sticky feel of the resulting product, and the dispenser tends to become clogged and soiled.
- Synthetic detergent bars are also available, having the same advantages over soap bars as neutral synthetic detergents have over liquid soaps. For example, when used in a shower room, a natural soap film or curd is not formed on the walls. But these bars have the disadvantages of defatting the skin and thus causing irritation, and also they decompose rapidly in the presence of water.
- Soap can also be dispensed in leaf form, from either disposable or permanent dispensing devices. Again, the user complains of a sticky, slippery film in the hands which is difficult to rinse away, and when such leaves fall to the floor they cause cleaning problems.
- Soap towelettes are also available: soap impregnated into rectangles of paper which are rubbed onto the skin when wet and then rinsed off. The disadvantages include a feel which is unpleasant to some people, the litter which the discarded towelette may cause, as well as the time involvement in its use.
- Powdered soaps, in numerous formulations, can be dispensed from a device by lifting up on a plunger, turning a crank, or otherwise. The shortcomings include clogging of the dispenser, dispenser rusting if it is made of steel, powder falling on the lavatory or floor requiring further cleaning, and a tendency toward overuse.
- Similarly, granular products are available in dispenser form for heavy duty cleaning, and these products may contain such abrasive materials as wood flour (a finely ground sawdust), cornmeal, pumice, etc. Their rough nature can be injurious to the hands, not to mention the face.
- Pasty hand cleaners may simply be scooped out of a large open-mouthed container, or dispensed by extrusion, much as is toothpaste from a tube. These materials can be useful for cleaning grease and oils from the

skin, as they may contain a solvent component, although this tends to de-fat the skin and lead to irritation or chapping.

- Lotion-type cleaners generally have the heavy-duty cleaning advantages of the paste-type materials, but are milder, and thus more generally accepted. Some of these contain emollients to protect the skin.

- Liquid soaps are most frequently used in industrial and institutional situations, and by the public in airline terminals, office buildings, and the like. Glass dispensing globes should be scrupulously avoided from the safety hazard standpoint, and plastic used instead—or a central tank provided for remote dispensing. Care must also be taken that a concentrated soap is not introduced into a tank system or the individual dispensers, as they will clog up. It is best to have such materials diluted immediately upon arrival in the storeroom, and then marked "ready for use." From the standpoint of public relations, a higher quality liquid hand soap is certainly desirable.

- Lather dispensers are a cut above liquid dispensers, as they take the same liquid soap, but a special valve mixes air and produces the material in a foam or lather. Not only is this form of dispensing more acceptable to the public, but it also uses less than half as much material as the liquid soap dispenser. Some liquid soap dispensers can be converted to the lather type by the purchase of a special valve. The lather dispenser also eliminates the problems of selection by color, as the color of the original material is not visible—but the odor problem remains!

SPECIALIZED CLEANING PRODUCTS

An imposing, if not confusing, array of specialized products is available for use by the housekeeping department. Whether they are of major, minor, or no consequence depends upon the surface to be cleaned and the activities which are the responsibility of custodial workers.

Anti-Static Materials—In climates where low humidity causes static problems with carpet, an uncomfortable situation for tenants and visitors, it is possible to spray the carpet with a material which reduces this static. Such chemicals may be based on quaternary ammonium bromides, in the same chemical family as the quats which are used for disinfectants. Where carpets are purchased with conductive fibers built in them, there is no need for such chemical agents. Sometimes the anti-static agents tend to cause faster re-soiling—for this reason they may be combined with anti-soiling agents.

Carpet Cleaners—A number of chemicals are available in this category, as discussed in Chapter 27. Carpet care products include:

- Spot cleaners (often available in kit form)
- Powdered or granular cleaners

- Surface brighteners
- Foam-producing cleaners

Chalkboard Cleaner—In general, products of this type do more harm than good, as they tend to fill the "tooth" or pores of the surface with an oily or clay-like material, that eventually leaves the board too smooth and too slick for good writing and reading. For the most part, chalkboard care should be dry, without even washing, except on some of the newest synthetic boards, and then only as recommended by the manufacturer.

Chelating Agent—These are special materials formulated to remove scum and other metallic deposits, through the ability of the chemical to encapsulate these materials. They are often marketed under the name of "terrazzo scum remover," etc. When available in a jellied form, they may be used on vertical surfaces.

Cleanser—The cleanser, by convention, is a scouring material—that is, one which cleans primarily by abrasion, but also containing detergent components to remove the loosened soil. The ingredient that performs the scratching may be silica dust, volcanic pumice, or granular minerals. Cleansers are available in three physical forms:

- Cleansing powders—These are most familiar in terms of products advertised on television and used by general consumers. But they also contain the largest particle size, and therefore are the most injurious to surfaces. Cleansing powder regularly used on plated restroom fixtures will completely remove the metallic plating, and will scratch the finish on the fixture itself so that it is much more readily soiled and becomes progressively more difficult to clean—a vicious circle indeed.

- Paste cleansers—This type material has a lesser particle size than powdered cleansers, and also avoids the problem of overuse that the powdered materials have, since they are often shaken into a sink and they clog the drain. The material should be put onto a damp sponge or cloth if used at all.

- Lotion-type cleanser—These have the smallest particle size of all and should be first choice for general use. Graduation to paste and then to powdered cleansers should only be on a basis of unavoidability because of the nature of the surface or the soil.

Concrete Curer—Products of this type are often marketed by janitor supply companies and sometimes applied by housekeeping departments, but basically it is a material for use in concrete construction, to harden and cure concrete floors. The material forms a latex-like structure that permits the migration of water molecules to the point where the seal will not lift, but retains enough moisture over a long enough time to allow good curing. Actually, such material should be applied by a construction contractor.

Deodorant—The housekeeping department is regularly called upon to control odors, and this can be done through several approaches, all of which should be used simultaneously where possible:

- Providing impermeable surfaces through design or through sealing that will not permit the harboring and growth of bacteria.
- Ventilation to remove transient odors.
- Killing the bacteria from which most odors derive by disinfection and sanitation before the use of odor-controlling chemicals.

The last item provides us a number of possibilities.

Deodorant blocks are made of paradichlorobenzene, and provide a masking odor—that is, one which is perhaps more pungent than the odor that is objectionable. With rare exception, there is no worthwhile application of this product, and its use often indicates a failure to clean and disinfect.

Drip fluid is another masking chemical, made by adding a perfume (typically oil of eucalyptus) to a paraffin mineral oil. The installation of unsightly cabinets, the damage to ceramic walls and the overpowering odor created when fixtures have not been used for some time, are some of the reasons that this material should never be used.

Jellied deodorants are available, with the active component generally being chlorophyll, which can actually combine with the particles in the atmosphere which cause odors for their direct elimination. When available in containers where the amount exposed can be varied, this product performs well.

Wick deodorants are dispensed in liquid form through a cotton wick, the amount dispensed depending upon the length the wick is pulled from the bottle (and of course the amount of air movement). These materials may also be based on chlorophyll, and some have a formaldehyde content which literally anesthetizes the nasal membranes, so that the nose cannot smell! They also may contain a perfume. These products are generally good performers.

Aerosol deodorants may contain some of the same ingredients as the jelly or wick deodorants, but all too often they rely on perfume, and thus may not actually remove odors. Further, any chemical dispensed in aerosol form, whether manually or on a timer, will be expensive.

Descalers—Acid products used for removing mineral deposits, such as are found in toilet bowls and urinals, attack minerals directly, so that they form new compounds with the acid which was washed away. It is desirable that descalers contain less than 10% equivalent hydrochloric acid, so that they do not need to carry the poison label, and can be used with less hazard to personnel—some of these products are marketed with the acid content as high as 20% or more. Also, their formulation with organic acids tends to make them less hazardous than when they use mineral acids, such as muriatic (which is a commercial grade of hydrochloric acid). Descalers should be used only periodically and not on a regular basis for cleaning, and their use should be by designated personnel only, wearing proper protective clothing and face shields.

Drain Opener—Although many housekeeping departments are issued products for unstopping drains, this is really a plumber's function. A number of accidents have been caused by "blow back," that is, the formation of gasses in the piping which create enough pressure to blow back in the face of the user—some of these accidents have resulted in blindness. Again, drain opening by chemical means should be performed by a qualified plumber, rather than a custodian.

Dust Mop Treatment—Materials are available for impregnating the yarn of the dust mops so that they will pick up and hold dry soils more easily. These treatments are of two types: oil-base and wax-base. The hazard with oily materials is that over-treatment (especially if the mop is used before the material has penetrated the fibers—24 hours should be allowed for this purpose) may leave a film on the floor which can cause falls or deterioration of a resilient floor through actually dissolving the material. For best results, the housekeeping department should either rent treated dust mops which are carefully controlled by the supplier (but not all suppliers provide this careful control), or have the treatment done, perhaps in one's own laundry machine, by a designated person for all dust mops (and perhaps dust cloths as well). Individual custodial workers should not perform their own dust mop treatment.

Foam Depressant—Sometimes it is necessary to put an additive into a detergent so that too much foam will not be formed—this would especially be important where the detergent will be removed by a wet vacuum, such as in an automatic scrubbing machine—too much foam can damage the equipment. Foam depressants are typically based on silicone type chemicals. A properly formulated detergent for use in automatic scrubbing machines, however, will eliminate the need for this special product.

Glass Cleaner—Where glass is to be washed in small quantities, aerosol cleaners or spray-bottle cleaners often containing quantities of isopropanol to improve evaporation can be useful. For larger surfaces, a neutral synthetic detergent can be prepared or purchased for the job, generally in much more diluted form than would be used for, say, floor mopping. Window cleaning detergents also may contain additives for attacking greasy soils or metallic soils. Be certain not to use glass cleaners that contain any abrasive materials, such as pumice.

Insecticides—Although the control of insects is not strictly a cleaning activity, it is still the responsibility of many building service departments, especially on a limited basis. Where insect control becomes an important aspect of environmental control, such as in food processing plants, large institutions, and the like, this is very often handled on a contracted basis with specialists in this field. In the housekeeping department, insect control may range from having an aerosol container on each custodial cart, simply in order to kill any insects which may come into view, to a regular fogging or misting of fairly large areas.

Insecticides typically consist of a small quantity of active ingredient in a solvent carrier. The buyer is therefore concerned not only with the properties of the active insecticide ingredients, but with the properties of the carrier as well. In general, it is safest to use insecticides, especially in the hands of workers with limited specialized training in this field, that are non-toxic, and these are primarily the Pyrethrins and their derivatives. Especially where insecticides are to be used around food stuffs, the Pyrethrins should be used.

Another class of insecticides are the chlorinated hydrocarbons, with such names as chlordane, DDT, dieldrin, and lindane—these have the advantage of residual activity, but the disadvantages of toxicity. For example, a great deal of publicity was developed concerning the problem that DDT caused in the contamination of growing foods and injury to animal life. The United States Department of Agriculture controls insecticides by classing them as "economic poisons," just as it controls germicides, weed killers, and the like. Thus, in selecting insecticides, it is best to deal with an organization in interstate commerce whose products fall within the purview of the Department of Agriculture.

It is also best to deal with specialists in this field if you do not have a qualified expert on your staff.

Metal Polish—A number of metal polishes are available for cleaning and shining such metals as copper, brass, and bronze. They contain both abrasives and chemical cleaning components. Most polished metals are being "designed out" of new construction, as the cost of maintaining them is excessive. Attempts to avoid metal polishing by the use of clear lacquer usually fail except in very lightly trafficked areas.

Oil Absorbent—These materials are made of porous clays or minerals of various types—they are able to absorb almost twice their weight in oil or water, and thus are useful around machine shops and processing equipment to prevent hazardous, slippery floors. When purchasing oil absorbents, consider both the mesh of the clay—too fine a mesh can create dust—and its efficiency of absorption.

Seal Stripper—Water emulsion sealers are removed with the same finish strippers used on polymer floor finishes, but the solvent sealers must be removed with special chemicals typically composed of methanol and methylene chloride—these are the basic constituents of paint removers, and the operation is quite similar. The chemicals attack the surface, blister it, and thereby make it possible for it to be lifted. This removal is facilitated in practice through the introduction of some granular material to which it becomes attached, such as sawdust or cornmeal. The surface must be carefully rinsed with solvent as well. The whole procedure, then, is tricky, messy and to a certain extent, hazardous.

Again, you should call on experts for help, either in the recommendation of chemicals and method, or in the actual performance of the work.

Sealers—A seal is a material used to protect a floor surface (as defined for a building service department) over a considerably longer period of time than a floor finish. These materials are called penetrating seals when they are thin enough to sink into the pores of the structure, and surface seals when viscous enough to remain primarily on top of the surface. There are two basic types of seals: water emulsion sealers and solvent sealers.

Water emulsion sealers are very much like water emulsion polymer finishes, except that the solids content is higher and other provisions made for a more permanent use. Such sealers are useful on terrazzo floors in order to provide a base upon which a surface can be developed. They are also useful on resilient floors, such as asphalt tile and vinyl-asbestos, which have become roughened, dried out, or damaged so that their pores need filling—otherwise, a regular finish would sink into these floors and leave no surface for wear or shine. A water emulsion seal is not normally needed on a resilient floor in sound condition.

Solvent sealers, as the name implies, are solutions in solvent and have much longer and better wear characteristics than the water emulsion seals—but they are much more difficult to remove, as well (see seal strippers above). The most popular solvent sealers are the phenolics, epoxies and polyurethanes, each of which has two sub-varieties. These seals may variously be used on concrete, wood, terrazzo, stone and other surfaces. The only one which can be used in the presence of moisture, with any degree of success, is the moisture cured polyurethane, which was developed for the sealing of wooden floors in textile areas where the atmosphere is normally humidified to prevent the breaking of yarn fibers.

The application of seals of various types is a complex subject, the variables involved including the nature of the surface, soiling, moisture, removability, hardness, slipperiness, ease of application, toxicity, etc.

Stainless Steel Polish—Stainless steel is a metal that belies its name—the multitude of microscopic parallel scratches introduced in the metal in its forming hold oily soils that cause unsightly stains. In many cases, the chief culprit is finger marks. The proper solution to this problem is to first clean the stainless steel, then dry it carefully, then fill these scratches and grooves with a product made of silicone material or some oily material, so that the metal will no longer become stained.

Sweeping Compound—Sweeping compounds are mixtures of oily or waxy materials to attract and hold particles of dust and soil, along with granular materials such as sawdust and sand to provide a suitable carrier and surface for the oil or wax. In the past, sweeping compounds were utilized on almost all types of floors since no better cleaning method was available, but with the advent of vacuum sweeping, treated dust mops and cloths, automatic scrubbing machines and better floor finishes, sweeping compounds are now used primarily on unfinished floors, such as concrete and stone—their use on finished floors would be where a great deal of soil has been introduced and no other means of removal is available.

Sweeping compounds can be quite injurious to resilient floors if the wrong type is selected. Sweeping compounds are primarily of the waxy type on a sawdust base, designed for use on floors which are easily damaged, and the oily type on a sand base for rougher floors. If the latter is used on a finished floor, the sand will cause scratching, while the oil will soften the resilient material.

Further, sweeping compounds are typically overused and thus may create soils of their own by raising a dust cloud. The material should not be spread liberally over an entire area to be swept, but rather formed into a line of compound which is then moved forward with a pushbroom.

Solvent Cleaner—Solvent cleaners are needed when the soils consist of oily or greasy materials, or tars, which may not yield to cleaning by soaps or synthetic detergents. The purpose of the solvent is to dissolve the soil, then to suspend it so that synthetic detergents can then emulsify the soil and flush it away. This dissolving action may be provided by kerosene, pine oil and certain alcohols, for example. The products may or may not be formulated in association with water.

One problem associated with solvent cleaners is the effect on the skin, since the solvent will tend to de-fat the skin and cause irritation. Where such irritation is present or suspected, protective skin cream or gloves should be provided.

Another problem deals with flammability, but fortunately safety solvents have been developed which reduce the flammability hazard to a reasonable minimum and also overcome noxious vapors associated with some solvents. Highly toxic materials such as carbon tetrachloride should never be made available in a housekeeping department.

Solvent cleaners should be carefully used around resilient flooring, as their action can soften and damage the tile. Special equipment has been designed to utilize solvent cleaners and their vapors—these cleaners are especially adaptable for use in ultrasonic equipment, where a very high frequency of vibration separates soil particles from the microscopic cracks and crevices in the surface to which they are attached.

Wood Polish—Special polishes are made for wood—they are also called furniture polishes—and some of these can be beneficial in healing scarred portions of wood. Of course, such products are also useful for natural wood paneling and feature strips. The amount of natural wood in design is decreasing as plasticized surfaces come more and more into use, in order to minimize cleaning labor time.

25

Floor Care Guidelines

The care of floors represents the average housekeeping department's greatest single expense. The International Sanitary Supply Association indicates that floor care consumes 40% of the housekeeping dollar. The cleaning of the next largest category, equipment, requires only 24%. A survey conducted by one maintenance industry magazine showed that floor-care items brought sanitary chemicals suppliers 70% of their business and 80% of their profits.

Thus, floor care is important to both the housekeeping department and to its suppliers. It is important to others as well because the condition of the floors is the first thing noticed when a person enters a building. Often the level of maintenance of an entire facility can be judged by the condition of the floors.

CATEGORIZING FLOORS AS "HARD" OR "SOFT"

Floors may be categorized as "soft" or "hard." Soft floors include composition materials such as the various resilient tiles and cork. Hard floors include concrete, wood, terrazzo, and stone.

SELECTING AND INSTALLING FLOORS

A great many problems in floor care can be avoided through proper selection of easily maintainable types. Sometimes the surface is the problem. Manufacturers of resilient floors are now preparing tiles that imitate travertine marble, and as mentioned above create an insoluble cleaning problem unless the cracks are filled, just as with natural "rotten marble." Embossed tiles are also very difficult to maintain; they cannot be spray-buffed well, and the raised portions cause any finish—and the tile as well—to wear and discolor more quickly.

Coloration is a constant problem—colors that are too light or too dark show marks and scuffing too easily in the first case, and show soil too easily in the latter.

The installation and repair of all floor types is a job for professionals. The skill with which a floor is first installed has a considerable bearing on the appearance of the finished floor, as well as on the effort necessary to keep it properly maintained. This is not only true of the final floor surface itself, but also is true for subfloors. An irregular concrete or wooden subfloor will show up very quickly in any composition covering, which immediately assumes the contours of its support. This is particularly true of vinyl. Attempts to properly maintain a resilient floor with an irregular subfloor are frustrating. The depressions fill with wax and soil and take on a darker tone than the balance of the floor. Floor-machine brushes span these depressions so that they cannot be scrubbed or buffed properly. The peaks of the floor receive almost the entire wear so that they thin out quickly, exposing color and texture variations. If such a floor is not properly protected by wax, it will wear through to the subfloor very quickly. On the other hand, proper wax protection under such conditions calls for inordinately frequent waxing.

Providing a good subfloor and quality floor covering material, both installed by skilled workers, represents a worthwhile investment.

THE ADVANTAGES AND DISADVANTAGES OF SIX MAJOR FLOOR TYPES

The following listing indicates the variety of major floor types now in use. The consideration or selection of flooring types is somewhat simplified by the fact that all floors can be put into one of the follow categories:

1. Composition (asphalt or vinyl tile, rubber, cork, linoleum)
2. Carpet
3. Cemented (such as concrete, ceramic, terrazzo)
4. Wood
5. Stone
6. Metal

1. Composition Flooring

Asphalt Tile

Description: A resilient floor tile, made of asphaltum, asbestos fiber, lime rock, colored pigments, and sometimes wood flour. It is the asphaltum which binds the materials together. Asphalt tile is usually cemented directly to a smooth-troweled (preferably machine-troweled) concrete floor, or to a felt underlayment over a wood subfloor. Where asphalt tile is to be laid over wooden planking, an intermediate layer of plywood, $^3/_8$ inch or thicker, should be used to obtain a smooth surface. Although asphalt tile is no longer manufactured, it is still found in many buildings.

Grades: Normally supplied in 9-inch square tiles (occasionally 12-inch square tiles) in $^1/_8$- or $^3/_{16}$-inch thickness. Colors are graded from A to D, with A being the lowest and cheapest color grading. The colors are generally mottled or striated. A special coloration gives asphalt tile the appearance of cork. A special grade of greaseproof asphalt tile is available for use in kitchens or in precision assembly areas where greases or lubricants may be spilled.

Characteristics: Solvents and oils attack asphalt tiles, causing them to soften and break down. Strong alkalis are also injurious, and the Asphalt Tile Institute recommends that cleaners be limited to a pH of 10, although general resistance to alkalis is considered good. Resistance to most acids is also good. They are deformed easily by heavy loads (limit 25 pounds per square foot), and are one of the noisiest of floors. They have fair light reflection.

They are inexpensive, available in a good range of colors, can be used on or below grade; easily and quickly installed; and long-wearing when properly maintained.

Maintenance: Most natural soaps contain materials that form a film over asphalt tile. Therefore, neutral synthetic detergents should be used for wet cleaning this material. For dry cleaning and dust removal, care should be used in treating dust mops. A wax-based treatment is the safest, although most users still prefer the oil-based treatment. Since oil is injurious to asphalt tile, it must be carefully applied. Asphalt tile should always be protected with natural water-emulsion waxes or polymeric floor coatings. Never dry buff or strip.

Cautions: Avoid solvents and oils. Remove factory coating before applying wax. Do not use any liquids within two weeks of installation. Always use water sparingly. Avoid harsh alkalis. Use rubber or plastic cups under heavy furniture to avoid indentation. Temperature extremes are injurious. Sweeping compounds containing sand will scratch. Mopping will dull the finish, requiring rebuffing or new wax application.

Cork

Description: Ground cork bark is molded, compressed, and heat-treated, and made into sheets or tiles. Tiles are usually 1/8- to 5/16-inch thick, with either plain or tongued and grooved edges. Different shades, from tan to brown, are obtained by changing the temperature levels of the heat-treating process. Normally available in 9-inch square tiles.

Characteristics: Cork is an excellent acoustical insulator. Its thermal conductivity is very low. Cork is the "springiest" of resilient floors but does not have quick and permanent recovery from static loads. However, its maximum static load without permanent indentation is three times that of asphalt tile.

Maintenance: A neutral cleaner should be used to clean cork floors, preferably a synthetic detergent, which cleans well in soft and hard water and leaves no scum. Solvent cleaners may also be used to clean cork tile floors. Factory finishes are available, or cork may be sealed with a penetrating sealer. Use solvent or water-emulsion waxes. To remove stains or discolorations, buff with an abrasive pad or dry #0 steel wool pad. Sanding may even be necessary for badly discolored areas. Sanding is best accomplished with #00 grade sandpaper under a regular floor machine. Do not attempt to sand a cork floor with a drum sander unless you are an expert with this type of equipment. Sealing is definitely required after such abrasive treatment of cork floors.

Cautions: Avoid greases and excessive water. Avoid alkaline cleaners. Do not wax unless the cork is factory finished or well sealed. Do not drag heavy or sharp objects over cork. Cork naturally tends to fade and discolor.

Linoleum

Description: Linoleum's primary ingredients are linseed oil and a filler. Originally the filler was cork, but now manufacturers have largely replaced cork with wood flour, whiting, and other inert materials. This mixture is compressed onto a backing, and the linoleum is then heat-treated to form the finished product.

Linoleum is produced in sheets and tiles. An excellent range of colors and designs is available.

Characteristics: Static load capacity of both tile and sheet is good (75 pounds per square inch). There is a hard linoleum tile available with excellent static load characteristics (200 pounds per square inch). Linoleum's grease resistance is excellent. Its resilience, quietness, and thermal conductivity is good, and it is available in a wide variety of colors and patterns.

Maintenance: Mild soaps or mild detergents that are not alkaline, or only very mildly alkaline, should be used for cleaning. Avoid ammonia-type cleaners as they may cause discoloration. Linoleum should be rinsed thoroughly after cleaning. Do not flood the floor as linoleum will deteriorate after repeated use of excess water. Linoleum floors should always be waxed after cleaning with a good water-emulsion wax, polymeric floor coating, or solvent wax. Keep cleaning frequencies to a minimum, as overcleaning reduces the life of linoleum. Abrasive-type cleaners may be used for spot cleaning stains, but are not recommended for general use.

Cautions: Avoid alkaline cleaners. Avoid excessive use of water.

Rubber

Description: Natural, synthetic, and reclaimed rubber is made into tiles and sheets. Inert fillers and color pigments are added, and the mixture is heated and rolled out under pressure. Most rubber flooring of today is synthetic rubber. Rubber is one of the most resilient flooring materials, and one of the safest of the resilient-type floors.

Characteristics: The high resilience of rubber makes it one of the most comfortable floors for foot traffic. It is a very quiet floor and one of the most durable of the resilient-type floors. There are many colors available in either a solid, variegated, or mottled design.

Maintenance: Rubber floors should be mopped daily with a synthetic detergent. Soaps are not recommended for rubber floors. It is recommended that rubber floors be waxed with a polymer water-emulsion type finish. Occasionally, the color pigment will come loose on the surface of the floor and will appear to "bleed." This should be scrubbed up with a good wax stripper and rewaxed or sealed with a water-emulsion type of seal.

Cautions: Avoid oils, grease, turpentine, petroleum solvents, and carbon tetrachloride. Do not use oily dust mops or sweeping compounds. Rubber floors should not be laid in bright, direct sunlight. Rubber floors are expensive and difficult to maintain.

Vinyl Asbestos Tile

Description: Vinyl asbestos tile is similar to asphalt tile. The binder is a vinyl-type resin. Other ingredients are asbestos, color pigments, and inert fillers. The tile is generally 9 inches square, sometimes 12 inches square. The asbestos content has made this floor obsolete.

Maintenance: The floor should be dust mopped or swept daily. It can be mopped as often as necessary with almost any soap or synthetic detergent. It should be waxed regularly with a natural wax or polymer-type emulsion finish. Older floors should be sealed with a water-emulsion seal to fill the pores of the tile and to provide a new, smooth surface to facilitate cleaning. Waxed floors should be stripped at least after every six waxings or semiannually, then sealed (if necessary), and rewaxed.

Cautions: Avoid abrasive cleaners on unwaxed floors. Use floor cups on thin table and chair legs to avoid permanent indentations. Get professional advice on the asbestos hazard—never dry clean, dry strip, or buff. Remove tiles only under flooded conditions.

Vinyl Tile—Homogeneous or Plastic

Description: This vinyl, sometimes referred to as "pure" or "100 percent vinyl," is similar to rubber tile. However, it does not have as much inert filler and pigment, and the binder is a vinyl-type resin. The tile is generally sold in 9-inch squares, sometimes 12-inch squares, and it is also available in sheet form.

Characteristics: This homogeneous, or plastic vinyl tile, is very expensive. Its resistance to permanent indentation is very high (200 pounds per square inch). This tile, like vinyl asbestos tile, is available in many colors and designs and is very resistant to acids, alkalies, oils, greases, soil, and solvents.

Maintenance: Homogeneous, or plastic vinyl tile, should be maintained in the same manner as vinyl asbestos tile, above.

Cautions: Avoid abrasive cleaners on unwaxed floors. Dirt and grit will cause scratching and it quickly shows subfloor irregularities.

2. Carpets and Rugs

Description: Carpets and rugs are fibrous floor coverings made by various weaving and back-covering processes. Wool and cotton once were the most common type of carpet fiber, but have been superseded by synthetics, such as nylon. These synthetic fibers are as great a boon to durability, appearance, ease of maintenance, and cost in carpeting as they have been in their other fields of use.

Some of the more popular types of weaves employed by carpet makers are:

- Wilton—This weave contains buried yarn beneath the pile surface. This gives the carpet a sturdier and longer life.
- Velvet—All the pile yarns in this weave are on the surface. It is a less expensive version of the Wilton weave.
- Axminster—Most of the pile yarn in this weave is on the surface. This is the most versatile weave for color, pattern, and design; it is also inexpensive, but not too durable.
- Tufted—The pile is pushed up between a grid-pattern backing. This weave dominates the carpet market.

Wilton and velvet weaves have loop or uncut pile counterparts. Wilton loop and velvet loop are very similar to them in characteristics, except the loop-type is more durable and harder to clean. Other types of carpeting are the coated-back carpets which sometimes are made of a normal weave with a rubber or plastic coating on the back, or simply with tufts set in a rubber or plastic cement solution.

Grades: The various grades of carpet are determined by these interrelated factors:

- The number of rows of pile per inch lengthwise.
- The number of rows of pile widthwise.
- Pile weight.
- Pile height.
- The type of mat construction—Generally, the firmer the backing, the more durable the carpet.
- The nature of the fiber, whether natural or synthetic.

Characteristics: In sound absorption, carpeting is in a class with acoustical tile and is far superior to any other floor covering in this respect. Carpeting is very slip-resistant but it tends to fray and unravel, and this can create a tripping hazard, especially on the edges. The prestige and dignity carpeting lends to the decor of an office, lobby, reception room, conference room, etc., are generally recognized. Carpets may also be permanently laid or merely placed without any physical attachment to the building. Carpets are available in a practically unlimited range of colors and designs. Daily maintenance is inexpensive.

Maintenance: Carpets should be vacuumed at least once a day in trafficked areas. Spot cleaning should be performed as necessary. Dry cleaning, which can be accomplished fairly easily and quickly, is recommended at intervals of approximately four to eight weeks, depending upon soiling conditions. Wet cleaning, or shampooing, is a more difficult method of cleaning and should be performed once or twice a year by competent personnel trained in this technique. Chapter 27 discusses the care of carpet in more detail.

Cautions: Avoid spilling liquids of any kind. Wet cleaning or shampooing should only be done by trained personnel. Frayed and unraveled edges are a tripping hazard. Watch for undue wearing and soiling in heavily trafficked and exposed areas.

3. Cemented Floors

Ceramic Tile

Description: A hard floor tile is made of clay which is kiln-baked to the hardness of stone. The two types of ceramic tile are glazed and unglazed. The grout, or binding cement, is available in acid-resistant, stain-resistant, and standard types. Ceramic tile is generally found in areas that are subject to many spillages and where rapid and easy cleaning is desired, such as restrooms, kitchens, cafeterias, laboratories, and workshops, as well as heavily trafficked areas.

Characteristics: Various coloring processes make a great many colors available. Glazed tile has an impervious, glassy finish, much the same as glazed china, and should not be used on floors. The unglazed tile has no surface finish and is the same throughout. This tile is durable, long lasting, and very easy to maintain.

Maintenance: Clean with a nonalkaline synthetic detergent. Soap tends to build up films which are dust catching and slippery. Ceramic and quarry tile floors are most commonly installed with Portland cement grout. While the tile itself is non-porous and highly resistant to the penetration of any type of soil, the grout areas present the same problems as those encountered with Portland cement floors. Floors made up of small non-glazed ceramic tiles can be successfully maintained with water emulsion sealers and finish. This approach is seldom successful with quarry tile because of the fact that such floors are normally installed where they are rather continuously exposed to moisture or to very heavy traffic. The most desirable approach in this case is to use organic binders in place of Portland cement for installation of the tile and then to avoid any surface coating. In some instances, good results have been obtained by using a penetrating sealer to close the pores of the grout, but at the same time buffing the floor vigorously to remove any sealer from the tile surface itself.

Cautions: Alkaline cleaners cause disintegration of the grout. Grout is adversely affected by acids (however, acid-resistant grouts are available). Grit, abrasive powdered cleaners, and other cleaners sometimes cause scratching.

Concrete

Description: Concrete consists of a chemically bonded mixture of cement, sand, and gravel. It can be of ordinary hardness or compacted for extra heavy-duty use. Concrete is usually gray to white in color, but it may be pigmented to obtain various colors.

Grades: Concrete grades are determined by the installation. This entails such factors as thickness, percentage of cement to sand and gravel, type of finishing process (machine trowel, broom finish, float finish, etc.), and location.

Characteristics: Concrete is the hardest and most durable floor surface, if properly cared for. It is also inexpensive to install. One of the inherent disadvantages of concrete floors, perhaps with the exception of good portland cement concrete properly installed, is its dusting or blooming.

Maintenance: New concrete should be properly cured. There are several ways to promote the proper curing of newly laid concrete. In the past, it was often done by covering the concrete floor with about one inch of water or by covering it with strips of wet burlap bags or kraft paper.

A better method of curing uses a polymeric concrete curing aid that is sprayed or mopped directly on a new concrete surface an hour after troweling.

Curing: One or two coats of this curing aid is sufficient to promote the complete cure of the floor. The polymer forms a thin membrane over the floor and promotes more efficient chemical reaction in the setting of the concrete, thus effecting a more complete cure. Simultaneously, the curing aid helps to seal and dustproof the floor. If a sprayer is not available, the combination curing aid and seal may be applied with a mop or lambswool applicator. This can be done about four hours following the final troweling or as soon as the surface is hard enough to permit use of an applicator.

The action of the curing aid is such that curing goes on thereafter for many weeks and produces the maximum structural strength. This seal should be nonskid and must easily accept other surface finishes and filling materials over itself after the cure is complete. The seal will prevent dusting or blooming. It also prevents the penetration of greases into concrete, or the adhesion of paint, soil, or plaster that may fall from walls or be tracked in during construction.

The cost of maintaining a floor that has been covered with a polymeric curing aid an hour after trowelling is tremendously reduced during the years that follow.

About sixty days after the curing aid has been applied, the floor should be cleaned and one coat of a permanent concrete seal applied. On old or properly cured concrete floors, it is important that the floor be carefully cleaned before the seal is applied. You should wet the floor with hose or mop, then apply a synthetic powdered concrete cleaner or a liquid detergent to the surface. Let it stand, then scrub the floor with a floor machine or deck scrub brush. Pick up the dirty solution with a wet vacuum. Rinse, then vacuum again.

It is important that the floor be thoroughly dry before the new seal is applied. If no wet vacuum is available, use a squeegee to remove the water from the floor. If the floor has not been evenly laid, or if it was not designed to drain properly, extra care has to be taken to see that the squeegee removes all water. Let the floor dry overnight with no traffic before applying the new coat of seal.

Sealing: Don't apply seals on a rainy day or a day with very high humidity. Humidity directly affects the drying time of the seal. A sealing operation is also aided if the area being sealed is well heated. Spread the seal evenly with a lambswool applicator or mop, being careful to "feather" overlapping strokes to avoid lap marks or an uneven appearance.

If for some reason a new concrete floor is to be sealed, but no combination curing aid and seal used, the floor should be neutralized or "etched" with 10% muriatic acid solution before sealing. Mop on the solution and allow it to remain for 5 to 15 minutes, pick up, and rinse *thoroughly* with clear water.

After the neutralizing or "etching" is completed, a penetrating type of seal should be used as a primer coat at this state. After this primer coat has dried, a permanent seal should be applied. Application procedure is the same as described above.

In the sealing of old concrete floors, the first step is to remove the soil. Apply detergent that will chemically, as well as mechanically, clean the floor surface. After the detergent has been applied to the floor, scrub the floor with a floor machine, using a wire brush. If you do not have a floor machine—and any building with a large area of flooring, whether concrete or any other type, should have one—you may use a heavy deck scrub brush for this cleaning.

If you are cleaning an old concrete floor where there is a heavy crust of grease, as in a machine shop, use a strong detergent. In severe cases it may be necessary to scrape the floor first to get the top crust off. Then proceed with the cleaning operation just described. *Be sure to remove all grease deposits.* In many cases it is wise to clean the floor, then wait several days to see if oil comes to the surface of the floor by capillary action. If it does, do the cleaning operation a second time. After cleaning, the seal is applied as above.

Neutralizing is not usually necessary with old floors, although it certainly can do no harm. Sometimes it is considered a good safety precaution. If the floor is smooth troweled, it *must* be neutralized to open the pores and provide a good bind for the seal.

The most commonly used surface sealers for concrete are the phenolics, epoxies and polyurethanes, The phenolics are the oldest formulation of this type of sealer and have the disadvantage of considerable initial color and of darkening on exposure to oxidation and the effect of ultraviolet rays. The epoxies and polyurethanes are much more color stable, as well as tougher and more resistant to wear, than the phenolics. They—and in particular the epoxies—have the disadvantage of being much more difficult to remove than the phenolics, should that be necessary.

In the application of a surface sealer, it is important that the first coat be thin enough to allow good penetration of the pores in the concrete. This can usually be provided for by cutting the first coat about one part of solvent to three or four parts of the sealer; the solvent used for this purpose should be the one recommended by the seal manufacturer. After the first coat has dried thoroughly, one or two successive coats are normally used to protect the floor against traffic. Under warehouse or manufacturing conditions, floor finishes, such as those used on resilient floors, are seldom employed. In some situations, their use is practical if a high level of appearance is desired, since they will protect the base coats of sealer and reduce the frequency of resealing.

A routine maintenance program should be utilized for concrete floors, sealed or unsealed. If the concrete floor is subject to heavy soiling or heavy traffic, it will have to be cleaned more frequently, with a heavy-duty cleaner. Set up a schedule for doing this cleaning job at regular intervals. Some plants clean their concrete floors once a month to remove grease and provide maximum cleanliness. Others clean much less often, depending on the kind of use the floor gets and the level of maintenance desired. On the other hand, in most food-processing plants, schools, hospitals, and in some factories, cleaning is done daily.

Cleaning: To clean the floor, move as many items of furniture out of the way as is practicable. Dust mop the floor with a treated dust mop. This prevents frequent changing of cleaning solution and rinse water. Then apply the cleaning solution to the floor with a mop. Let it stand for four to five minutes. Scrub to remove soil. Pick up the solution with a wet vacuum, a squeegee, or a mop. Rinse the floor with a hose or wet mop. If you use a mop, change the rinse water frequently.

Waxing: Where high gloss and added protection is desired, a sealed concrete floor may be waxed regularly. (Unsealed floors should normally not be waxed, but should be sealed if more gloss and protection is desired.) This is especially recommended in personnel areas such as lunch rooms, locker rooms, etc. Contrary to popular opinion, waxing does not create a slip problem on sealed concrete floors if a slip-resistant wax is used.

The waxing procedure for sealed concrete floors is the same as for composition floors. If certain spots or lanes are subject to very heavy traffic, you may want to clean and wax them more often, taking great care to feather out the wax so that it joins the old coat smoothly. For this purpose, it is permissible to dilute a polymeric floor wax somewhat with water and spread a thin coat over the trafficked area every few days.

Painting: Concrete is alkaline, and when new the lime salts are very active. Usually for a period of six months the presence of these salts will prevent proper drying and adhesion of the conventional type of floor paints. There are many pros and cons about the painting of concrete. Generally a clear seal is recommended instead of a colored seal or paint. However, many colored seals are available if you especially desire to make a floor blend into a decorative color scheme. Colored seals may be applied over cured concrete, using two coats of the colored seal. Or, if the floor has not been cured and is very smooth or glazed, you should clean and etch the floor and apply two or three coats of colored seal. The first coat may be thinned with turpentine for better adhesion. Paints and colored seals are surface finishes and should *always* be protected with wax if heavily trafficked.

Patching: Severe crazing and cracking of an old concrete floor may require grinding off the top surface. This is then replaced with a new topping. However, this is seldom necessary if good patching is done. Use a regular surface mixture of concrete as recommended by your concrete dealer. Clean out all holes and cracks. Square off the bottom edges of the hole so that new cement poured into it is actually a plug set down into a chiseled-out hole, rather than just a surface layer. The hole and the surrounding area should be saturated with water for several hours before patching.

If the concrete floor is sealed, patching should be done at a time just before you plan to strip off the old seal and reseal the floors.

Caution: Avoid strong acid and alkalis. Seal to prevent dusting.

Terrazzo

Description: Terrazzo is a mixture of marble chips and portland cement. It is often separated by metal strips. The mixture is poured and troweled out and allowed to dry for five or six days. It is then ground and polished with a stone grinder. The appearance of terrazzo varies with the color of the marble used. Epoxy Terrazzo provides a matrix that is much denser and less porous than the portland cement.

Characteristics: Terrazzo is a very beautiful floor with marble, in its variegated colors, as the major portion of the mixture. It is very easy to maintain and, if installed and maintained properly, it is very long lasting. Terrazzo is a relatively inexpensive hard floor, and is very durable.

Maintenance: While using water-emulsion sealers and finishes, along with the spray-buff technique, is generally applicable to maintaining terrazzo floors as well as resilient floors, conditions sometimes dictate other approaches. If a terrazzo floor is to be treated with any kind of finish, it is especially important that walk-off matting be provided at entryways where water might otherwise be tracked in and where slippery floors might result. If this precaution is taken, slip hazards seem to be encountered only under exceptional conditions.

In practice, it is generally impossible to maintain uniformly good appearance on terrazzo floors without the use of either finishes or sealers. If the procedures already described for the use of the water-emulsion products are not followed, there remains a choice among three other frequently used systems of maintenance. In the past, many terrazzo floors were maintained by the application of a special soap, often based on linseed or similar oils, which left a hard water residue that was then buffed into the surface of the terrazzo. Today few authorities consider this a satisfactory approach.

The most popular type of sealer among terrazzo floor contractors is a solvent sealer based on synthetic waxes which do not discolor with oxidation and sunlight as do the carnauba and other natural waxes. This type of sealer is very easy to use and, since it can be stripped at least off the surface of the terrazzo, can seldom result in any serious difficulty. For large areas of terrazzo exposed to heavy traffic where the emphasis must be on economy in time more than on extremely high levels of appearance, this approach may be the choice; in fact, quite good appearance can be maintained with regular cleaning and buffing of the surface.

Other solvent sealers are available to provide surface coats which are relatively color stable and which can be patched with reasonable success as wear occurs. Generally thee sealers are used without additional floor finish, and when traffic aisles occur which can no longer be satisfactorily spot-finished, the sealer must be removed with some type of organic solvent, sometimes a methylene chloride stripper. Stripping this type of product is a rather complicated operation requiring well-trained and supervised personnel; thus, this is probably the most expensive procedure for the maintenance of a high level appearance on terrazzo.

Cautions: Avoid alkalies and acids. Avoid inorganic and crystallizing salts. Avoid oily sweeping compounds and treatments. Do not use steel wool for scrubbing or buffing as small pieces of steel left on the floor will rust and discolor the terrazzo.

Granolithic

Granolithic floors are similar to terrazzo, except that granite chips are used instead of marble chips, which are used in terrazzo floors. These floors use no metal dividers, although they have expansion spaces which run end to end. For maintenance, see the section on terrazzo floors.

Magnesite

Description: A hard floor very similar to terrazzo. The binder is magnesium chloride with fillers such as cork, asbestos, wood flour, sawdust, and sand. Grades and cost of magnesite vary according to the filler used.

Characteristics: The great variety of fillers that may be used to manufacture magnesite offers a wide range of appearance, durability, and resilience. The harder the filler, the more durable and less resilient the floor.

Maintenance: The floor should be dusted, swept, and mopped regularly. Because of the nature of most of the fillers, magnesite floors are sensitive to too much water. Magnesite floors should be sealed to prevent blooming or dusting. Regular waxing is also recommended.

Cautions: Avoid excessive water. Avoid acids and alkalies.

4. Wood

Description:

(1) Oak—The most extensively used wood floor. There are two main groups of oak, white oak (nine species) and red oak (eleven species). There is little difference in quality between these two groups. Oak floors owe their beauty to the open cellular construction, which gives the grain its unique character.

(2) Maple—The next in popularity of wood floors is maple. There are two species used for flooring, sugar maple (also called sugar tree) and black maple (also called rock maple). The wood is close-grained with small and evenly distributed pores. It is very resistant to abrasion and very hard. For this reason, it is an exceptionally good floor for heavily trafficked and heavy-wear areas.

(3) Birch and Beech—These two woods are very similar. They are very hard and abrasive-resistant. Like maple, these two are close-grained with very small and evenly distributed pores.

(4) Pecan—The hardest of all wood flooring is the pecan wood. A very durable wood, it is also close-grained.

(5) Other species—Walnut, cherry, ash, hickory, teak, and cypress. These woods are not too popular because of their scarcity and high cost. The use of soft woods for the sake of initial economy should be discouraged.

Maintenance: The correct maintenance of wood floors is of paramount importance, if their beauty and life are to be preserved. For old wood that is in poor or fair condition, clean with a good solvent-type stripper, using a floor machine with a #3 steel wool pad or abrading-type pad. For oak or open-grain wood, apply a filler, then a penetrating sealer. Wax old wood floors with a solvent-based wax, or, on properly sealed floors, with a water-emulsion wax. New floors should be sealed and waxed as above. All wood floors should be dust mopped, swept, or dry vacuumed daily. Sealed and waxed floors should be damp mopped to remove excess soil.

Cautions: Avoid excessive water as the wood will tend to warp and cup after saturation. Avoid oils and greases as they will stain and streak the wood.

Wood Block

Description: Wood block floors are wood floors with the grain end or cross section on the surface. As we might expect, this cross section would absorb a great deal of water and moisture, so wood block floors are creosoted (or similarly treated) to prevent warping and buckling. Wood block is almost exclusively an industrial floor.

Maintenance: Wood block floors should be swept daily. Sweeping compound is recommended for dust control in manual sweeping. Scrubbing should be done when the floor becomes heavily soiled. The scrubbing solution should be vacuumed up immediately after scrubbing to prevent the saturation of the wood block floor. Mopping is recommended for spillages. This floor, because of its block construction, is very easy to patch by removing the old or broken block and replacing it with a new block.

Cautions: Avoid excessive water and liquids. Near composition floors, wood block floors should be kept very clean. Entrances to composition floor areas should be covered by mats and runners to prevent trackage of the creosote or other similar treatment.

5. Stone

Stone floors are generally of granite, slate, or flagstone (crab orchard stone) and are used in lobbies, patios, courtyards, stairs, and other outside areas. These types of stone are very durable and also very hard. Stone surfaces, which are not directly exposed to the elements, should be sealed with a penetrating type seal to prevent "blooming" or dusting. On inside areas, stone may be sealed with a penetrating type seal as a base coat and then covered with a surface-type of seal.

Marble

Description: Marble is a naturally occurring rock composed of carbonate of lime. The rock is cut, polished, and laid to form a smooth surface. Most of the finer marbles are used for walls, window sills, and other decorative purposes. The well-known travertine marble and honed marble are used for floors. The impurities in marble cause the variety of colors. Marble without these impurities is white. The various decorative schemes create the different grades and costs of marble.

Characteristics: Marble is found in solid colors such as black, red, and green. There are variegated colors with unusual swirls and wisps that make very beautiful patterns and designs. Marble in all its color forms is one of the most beautiful floor surfaces. The travertine and honed marbles are more rough and porous on the surface and, hence, are safe for foot traffic. They also have less tendency to scratch and wear than other types of marble.

Maintenance: Marble is very expensive and very easily damaged by improper care. Therefore, the maintenance of marble floors must be excellent. Regular dust mopping with a wax-treated dust mop should be done at least daily. For removing surface dirt, mopping with a synthetic detergent is recommended. Floors should be sealed and waxed with a slip-resistant wax. For stubborn soil, a mild abrasive may be used.

Cautions: Avoid acids of all types, as the calcium carbonate, of which marble is composed, will react with an acid. Even fruit acids will ruin the finish on marble floors. Avoid staining with oils, fats, and greases. Avoid harsh abrasives, especially those containing ammonia and dyes.

6. Metal

Description: Metal floors are fairly unusual and are generally considered as special-purpose floors. They are almost always ferrous in nature as most other metals would be too expensive or lacking in the properties that are so valuable in steel and iron. The main types are sheet, plate, and grating. Some of the special purposes for this type of floor would be for exceptionally heavy machinery or to cover old floors that have been worn beyond the state of acceptable use. Grating is used where there are excessive spillages. You may also find metal used for temporary floors or where readjustment or redesign of the basic floor plan is desired.

Maintenance: Plate or sheet-type metal floors can be swept or scrubbed. Steam cleaning is probably the best method for cleaning metal floors.

Cautions: Metal floors are especially dangerous where there is a possibility of electrical shortages. Metal floors can be very slippery, especially where spillages of oil and greases occur.

Conductive Floors

Description: Conductive floors are used in areas where explosive vapors are present. Sparks resulting from the accumulation of static electricity constitute a very real hazard in these areas.

Grades: The resistance of a conductive floor should be less than 1 million ohms and more than 25,000 ohms. The National Fire Protective Association recommends a standard special tester for measuring the conductivity of floors.

Conductive floors are available in ceramic tile, terrazzo, and oxychlor in hard floors, and rubber and various types of vinyl tile in soft floors.

Characteristics: Conductive floors have the same characteristics as floor surfaces in which no conductive properties are incorporated, plus the desired quality of conductivity.

Maintenance: Conductive floors should be mopped each day or after any spillage or use of the floor. Soap is *not* recommended for such cleaning as it leaves a film on the floor. A synthetic detergent is the correct material for cleaning conductive floors. *Conductive floors should not be sealed or waxed.* Although conductive waxes and seals are available, the risk involved in the floor retaining its conductive properties by incorrect application or the erroneous substitution of a nonconductive seal or wax is neither worth the savings in floor wear nor the improved appearance.

Cautions: Metal floors, although completely conductive, present another hazard, as they can cause a severe shock from faulty electrical equipment or wiring. Conductive floors must be cleaned regularly. Avoid soap and other film-forming products.

HOW TO CLEAN FINISHED FLOORS

The most important types of finished flooring are linoleum, rubber, terrazzo, and vinyl. It is generally desirable to protect these floors with some type of finish; this includes vinyl flooring, despite the claims of its manufacturers when this type of material first became available. Floor finishes protect the floor from damage by traffic and environmental elements, improve sanitation, and help in maintaining more readily cleaned surfaces and improved appearance.

The use of protective finishes is *not enough.* For adequate and economical protection, walk-off matting should be installed at all entrances where any soil is likely to be carried in. This includes not only entrances from out-of-doors but also entrances from manufacturing, lab, or shop areas. Further, it is very important to consider the need for supplementing normally-used walk-off matting at times when new construction or building renovation may be in progress. Floors should be protected from the mechanical damage which may result from dragging desks and heavy furniture from place to place by occupying personnel. Special equipment is available which makes a simple and easy job of moving desks, file cabinets and other heavy furniture.

Another consideration in protecting floors from unnecessary damage is preventing rusting of metal surfaces in direct contact with the floor. Often this type of damage results when water is allowed to run under the bottom of a file cabinet and the sheet metal rusts, quickly producing permanent stains in the pores of the floor. Use plastic corners or angle strips installed on or under the cabinet so that the sheet metal does not touch the floor.

Dry Floor Cleaning

The major factors in floor care are the removal of heavy soil by some form of sweeping or vacuuming, wet cleaning of the surface at the proper intervals and with necessary accompanying mechanical processes, and the application and maintenance of floor finish.

The most commonly used and generally most suitable tool for dry floor care is the treated *dust mop.* There are many variations to the basic tool, which usually consists of strands of yarn some five inches in length assembled into a mop which may vary from a small wedge shape up to lengths of six or more feet. For dust mops of three feet or greater length, it is generally desirable that the handle be rigidly fixed. For narrower dust mops used in congested areas, the mounting should be designed to allow the mop to swivel through 360°.

The dust mop should be treated with a compound which will cause dust to adhere to the strands of the mop and thus minimize the tendency for dust to be distributed into the air during the mopping operation. Mop-treatment compounds consist of formulations of light petroleum fractions. The mop treatment selected should not be too oily or too solvent in nature, for this can damage finishes and surfaces or can produce slippery floors. Mop-treatment formulations sometimes contain quaternary ammonium chlorides or other bacteriostatic chemicals.

Variations on the classic dust mop include *treated cloths* of a patented form which are applied to blocks and handles so that they are used in the same manner as dust mops. The contention of the manufacturers of this type of equipment is that better sanitation and comparable costs are obtained through folding these cloths in different ways to present clean surfaces as those already used become loaded with dust. Another substitute for the dust mop is a device which uses a disposable so-called non-woven cloth for the dust-collecting surface. Both of the devices appear to do a better job of collecting very fine dust than will the string dust mop; however their use becomes less advantageous when trash, paper scraps and larger sized particulate matter are present.

For dense coarse particulate soil, such as sand encountered in large quantities, the use of a *push broom* and *sweeping compound* may be advisable in place of dust mopping. Proper selection and proper use of sweeping compound are both important to avoid serious damage to resilient tile surfaces. The dust control component in the compound should be predominantly waxy, rather than oily. Oils can seriously soften and damage tile, especially that composed of asphalt or rubber. Custodians frequently make the mistake of scattering sweeping compound over the entire area which is to be swept. This is wasteful of both time and compound and is likely to result in staining of the floor if the compound is walked on. Just a few handfuls should be put down along a line the width of the area to be swept; then this line of compound is moved forward with each stroke of the broom in the sweeping operation. Working in this fashion, the custodian does not accumulate a large volume of sweeping compound to be picked up with the dust pan and discarded. If he finds that the compound is becoming overloaded with dust before he reaches the end of the area which he is sweeping, a small amount of new compound can be added, or the material already accumulated picked up and all new compound put down.

The type of broom fiber used should be selected in accordance with the type of soil likely to be encountered. For instance, for a mixture of coarse and fine dirt, the best broom would probably be one with a Tampico center and a casing of horsehair; or a mixture of horsehair and bristles may be used.

In an industrial clean-room or critical areas of a hospital, it may be necessary to avoid dry sweeping and use a special *vacuum cleaner*. This is a much slower procedure than sweeping. Two suitable types of vacuum cleaners are available: those which have a dust separation and collection device located remotely from the area being cleaned (central systems) and those which use special filters capable of removing better than 99% of all particles larger than about 0.5 of a micron. Each system has its own advantages and disadvantages, and these should be weighed carefully in deciding which to select. Central systems generally have the disadvantage of requiring the custodian to work with rather long sections of hose; also careful design is required to insure a sufficient number of risers and connections and a system of piping which does not result in too much loss of air flow efficiency.

Wet Floor Cleaning

Daily wet cleaning (possibly with a germicidal detergent) is normally the rule in hospitals or other areas where rigid microbiological control is required. Frequent wet cleaning may be required during periods of unusual soiling. The frequency of wet cleaning is normally kept to a minimum. Well-designed battery-operated auto scrubbers are now available to apply clean solution to the floor, scrub with a mechanically driven brush, and pick up the soiled solution, all in one operation.

Purchasing Considerations. If at least two of the following conditions exist, then the purchase and use of an auto scrubber is probably desirable:

- If large, open-floor areas exist without obstacles or congestion.
- If frequent wet cleaning is necessary or desirable.
- If the inevitable removal of a significant amount of the floor finish layer with each use of the equipment can be tolerated.

Select the largest equipment readily maneuverable in the open space, even though there will be some supplemental manual wet cleaning in certain congested areas.

Manual wet cleaning methods should be varied according to the amount of soil encountered. If the soil is light, damp mopping or mopping with a single bucket with a fairly light solution of neutral synthetic detergent may be entirely satisfactory. With heavier soil, it may be desirable to put down more solution and pick it up, wringing it into an empty bucket to avoid soiling the clean solution. Under conditions of very heavy soil load, rinsing will be essential; then it will be desirable to use a three-bucket system with detergent solution in one bucket, rinse water in another bucket and a third bucket into which both the mop for applying detergent and the mop for applying rinse water may be wrung.

If only cleaning of the floor is necessary, without accompanying disinfection, a neutral synthetic cleaner will normally be used. Such a product formulation is usually based primarily on wetting agents with moderate emulsifying powers which do not produce scums with hard water, as do soaps. The pH of a neutral cleaner should generally be less than 10.

The two classes of cleaner-disinfectants most commonly used are modified phenolics and the quaternary ammonium chlorides. The selection of the product should be based on the types of organisms which must be controlled and the cleaning requirements.

The remaining type of cleaning compound most frequently used in floor work is the stripper for removal of wax or finish. Generally strippers are formulations with high pH (up to about 12.0) provided by either ammonia or organic amines; sometimes organic solvents are used, but these must be carefully selected so that they do not result in deterioration of resilient tile. The so-called "metallized" floor finishes normally require the presence of ammonia in the stripper. Some companies have developed floor finishes formulated with acid systems, so that an acid-type stripper must be used for their removal; however these have not proven widely popular.

Formulations of cleaners used in autoscrubbers usually vary from those used in manual operations in that the wetting agents employed do not produce as much foam. High foamers will fill the receiving tank of the machine too rapidly, causing trouble with the vacuum system as a result of foam and moisture getting into the bearings of the fan shaft or motor bearings.

Figure 25-1 is a useful chart for troubleshooting floor care problems.

Figure 25-1. How to Troubleshoot Common Floor Care Problems

To help solve some of the most common floor care problems, use this chart as a troubleshooting guide or a preventive checklist.

POOR GLOSS	
Most Common Causes	**Solution**
• Insufficient coats applied or recoat schedule off	• Scrub, rinse, and apply additional coats. (Maintain 4 to 6 coats.)
• Thin coats of finish	• Apply in medium to full coats; avoid wrung-out mops.
• Wrong scrubbing or buffing pads/brushes used (usually too aggressive)	• Use red or green pads for routine scrubbing, blue or black pads for deep scrubbing, and approved pads or brushes for buffing. On uneven floors, an appropriate brush is recommended.
• Excessive amount of sand and grit on floor	• Use approved mats, dust mop frequently, and remove grit outside doors.
• Floor not properly rinsed before recoat	• Strip, rinse well, and apply new finish.
STREAKS IN FLOOR FINISH	
• Recoating too soon before prior coat has dried properly	• Strip, rinse, and reapply. With most finishes, allow at least 15 minutes after finish is dry to the touch before recoating. Do not recoat if mop drags.
• Floor finish applied over factory finish on new tile	• Strip thoroughly, rinse, and re-apply finish.
• Dirty mop or equipment used	• Insure equipment is clean before applying finish.
• Floor finish frozen or stored in extreme heat	• Replace damaged product, strip, rinse, and re-apply.
• Contaminated finish put back in container	• Do not save leftover finish in container for future use. Dispose of it.

DISCOLORED FLOOR FINISH	
• Finish applied with new cotton mops before soaking/cleaning them	• Soak and clean new mops thoroughly. Use rayon mops to apply finish.
• Incorrect concentration of cleaner	• Follow recommended dilution rates.
• Dirt coated into floor finish	• Strip, rinse, and re-apply finish. Deep scrub finish properly before recoating.
• Outside dirt, sand, and grit ground into finish	• Use approved mats and runners.
• Ineffective daily cleaning (particularly with UHS program prior to burnishing)	• Damp mopping may not be sufficient. Use auto-scrubber daily with proper cleaner and pads.

POWDERING/POOR ADHESION OF FLOOR FINISH	
• Factory finish not stripped from new tile before finishing	• Thoroughly strip, rinse, and re-apply finish.
• Applying coats too thinly	• Apply in medium to full coats. (Avoid wrung-out mops.)
• Wrong buffing pads/brushes used (too aggressive)	• Use recommended pads or brushes.
• Burnishing a finished designed for low-speed buffing or spray buffing	• Match the finish to the maintenance procedure.
• Floor not thoroughly cleaned or rinsed before finishing	• Before applying finish, floor must be thoroughly cleaned and rinsed.

SCUFFING AND SCRATCHING OF FINISH	
• Wrong scrubbing or buffing pads/brushes used	• Use recommended pad or brush for each procedure.
• Excessive dirt and grit on floor	• Use approved mats and dust mop frequently.
• Floor finish film is too thich from excessive recoating	• Deep scrub with a black pad before recoating. (Maintain 4 to 6 coats.)
• Not scrubbing or buffing often enough	• Separate main from secondary traffic areas and schedule appropriate maintenance.

MAINTAINING THE FLOOR FINISH

Using Waxes, Finishes and Sealers

Waxes in organic solvent systems have been used for many years to coat floor surfaces, as have water emulsion systems containing finely divided wax particles. Various natural and synthetic waxes have been used in such formulations, but the greatest emphasis has generally been placed on the content of carnauba wax, a natural derivative from the leaves of a palm plant. Usually the total solvents in a good product range between 12 and 18%, and only a portion of these may be made up of carnauba or other natural wax.

Because waxes tend to produce slippery surfaces, finely divided silica may be introduced into these products, under the trade name "Ludox." Practically all of the natural waxes used on this type of floor finish tend to darken with exposure to light and oxygen, with the result that the floor coatings eventually become discolored and must be stripped and replaced for the sake of appearance.

About the end of World War II, complex polymers, some of which had been developed in efforts to produce substitutes for natural rubber, began to be substituted in emulsion form for the waxes. The most important of these products coming into early use were the styrenes and polystyrenes. The major disadvantage of these products was the tendency toward hardness, brittleness, and powdering. To distinguish them from the water emulsion waxes these products were referred to as "finishes," "polymers" or "polishes." Because of their hardness, the styrenes did not lend themselves to buffing and consequently were referred to as "non-buffable floor finish."

Acrylics or polyacrylics and polyethylenes are not as hard and brittle as the styrenes and possess greater toughness; they are also buffable to some degree. The acrylics generally have the disadvantage of being very difficult to remove with floor strippers. Often, combinations of the styrenes and acrylics, perhaps along with microcrystalline waxes of petroleum derivation were formulated in an effort to combine the advantage of each of the individual components. Certain types of molecular combinations are referred to as co-polymers. The trend in formulations has been away from the use of styrenes and to a greater use of the polyethylenes and the acrylics.

Polymers: The Favored Choice for Commercial Use

Under average commercial conditions, the choice between the water emulsion waxes and the polymers or synthetic finishes usually favors the latter. Polymers form continuous surface films upon drying, rather than layers of discrete particles (as in the case of the waxes). There is less tendency toward slipperiness, the surfaces are more water resistant, and less buffing and repair are needed. An exception to this rule may be found in the case where floors are unavoidably exposed to unusually large amounts of sand and grit; under those conditions, it may be easier and more economical to buff the water emulsion wax frequently rather than to restore a polymer subjected to excessive scratching.

In the instance of terrazzo floors or of highly porous resilient floors, such as old asphalt, it is sometimes desirable to use a water-emulsion sealer. These sealers, whose solids are usually predominantly acrylics, have a greater ability to bridge floor porosity than do the finishes themselves; however they usually will not withstand the friction of traffic well and thus are used only as a base for the coats of finish which actually receive the wear. Some persons responsible for floor maintenance prefer to avoid the sealers because most of them are more difficult to strip than are the finishes; these persons simply use additional coats of the floor finish to obtain desired appearance.

The Three-Step Approach to Applying Floor Finish

When applying water-emulsion sealers and finishes, one floor finish should never be applied over another unless it has been previously ascertained that the products are compatible.

Step 1: Strip the floor completely. This is accomplished with the chemical stripper used in concentrations recommended by the manufacturer, with the aid of a floor machine with suitable stripping pads and a wet vacuum and other necessary tools for picking up the emulsified floor finish and soil. If there is a heavy buildup of old finish on the floor, it is quite probable that recommended concentrations of stripper will not remove the entire coating in one application. The temptation to increase concentration beyond that recommended by the manufacturer in an effort to remove all old finish in one operation should be resisted, for to do so may result in "burning" or damaging the floor surface by excessive removal of plasticizers and pigments. The safe procedure is to re-strip those limited areas where the first application may not complete the job.

Step 2: Thoroughly rinse the floor to remove all old finish and stripper solution. This is essential to a good job of refinishing. Unless this step is properly carried out, the new finish may not bond properly or it may have a dull appearance. There seems to be no need for the "neutralization" of floors with vinegar or other acids if the floor is completely rinsed.

Step 3: Apply the finish. Generally the fastest and most satisfactory way of applying floor finish is with a mop. However, it is often desirable to use a 16-ounce rather than a larger size mop, and most workers prefer thin cotton strands (string mops) rather than heavy cord or synthetic fiber mops. The application of finish in thin coats is most important to avoid the hazards of tackiness, dullness, and slow drying.

Under reasonable conditions of temperature and humidity, a good polymer floor finish should dry in approximately 30 to 45 minutes. It should be completely dry to the palm of the hand before the second coat is applied.

While either the first or second coat should be applied up to (but not on) the baseboard, all other coats should be kept about one nine-inch floor tile width away from the floor edge, since the absence of traffic in that area may otherwise permit the development of buildup. Unless the floor is excessively porous or the total solids in the finish fall below the desirable level of 14 to 18%, two or three coats of finish will normally give an excellent base.

Spray-Buffing

In the past, the primary method of floor finish maintenance has been regular cleaning, periodic buffing, patching of heavily trafficked areas by blending in a partial wax coat with the mop, and total stripping and refinishing of the floor when indicated by the development of mottled appearance or the buildup of finish in the more lightly used areas.

Today, custodial managers often depend upon spray-buffing as the major approach to floor maintenance. Spray-buffing is a technique wherein the floor is first cleaned and then those surface portions damaged by traffic are restored by the application of a light spray of a dilute solution of the original floor finish (or a compatible substitute) and the buffing of this spray with a pad before it has dried. This technique will result in partial emulsification of the very top layer of finish, the repair of scratches and scuffs, and the removal of black marks, as well as the production of a tough surface with very high gloss. A necessary final step is dust mopping the treated area to remove residues of old finish which have been loosened by the spray-buff pad and left as a coarse residue on the surface.

Spray-buffing is not a technique for coating large areas of a floor surface which actually needs the application of a full layer of finish, nor should it be depended upon to clean a dirty floor. The operator must understand the precautions necessary to avoid soiling walls and furniture with the spray and to prevent build-up in areas where the floor receives a little traffic; spray should be applied only where it is needed.

Provide your custodians with the proper equipment: good open mesh spray-buff pads, a good machine and suitable pad-driving block, and a convenient spray device. The proper use of the spray-buff technique can extend the time between floor stripping many fold. The advantage to management is the economy in time—as well as the less important economy in materials—which results.

Specialized Cleaning Approaches

Resinous Poured Floors. Several organic binders are suitable for the production of floors similar in many characteristics to classic terrazzo or, with the substitution of granite chips for marble chips, to granolithic floors.

Since these binders are much less porous than portland cement, many of the binder manufacturers and floor contractors advocate that they can be used without surface finishes. In practice, continued high-quality appearance seems to require the use of sealers or sealer and finish combinations. Under some conditions, a significant amount of powdering has been experienced with some of these floor types. Marble chips are characterized by variable but significant amounts of porosity, and sealing may be desirable simply for protection of this aggregate, although discoloration or staining of the chips may not be noticeable if chips of mixed color are used.

One of the important advantages of the resinous poured floors, whether used with or without aggregates such as marble chips, is the reduction in weight both from the lower specific gravity of the organic binder and from the fact that these toppings can be poured much thinner than can Portland cement or classic terrazzo because of flexibility and other physical characteristics. There are also numerous significant variations from one to another. These include variations in resistance to physical and chemical damage as well as variations in problems of application and maintenance. Consider carefully the relative advantages of individual binders before making a decision on an installation of this type. After that, it is extremely important that the flooring contractor be experienced and qualified in the type of installation, since otherwise unsatisfactory results are practically a certainty.

Some architects and installers of the resinous floors contend that long-range savings are achieved by avoiding floor finishes and by applying a new glaze or surface coat of the binder material itself when undesirable traffic patterns appear in the surface. It is not always made clear to the user of the flooring that many problems may be involved in this resurfacing. For instance, volatile solvents used in the binder which dissipate during the curing stage may be highly toxic, so that repair may require either evacuation of the building or rather complicated arrangements for safe ventilation. It may be necessary to restrict traffic from the areas under repair for a period of 48 hours or even more.

Although flooring manufacturers claim that resinous poured floors require no waxing and may be cleaned only (using floor-care detergents), in practice, many users put a finish on the floor to prevent powdering and traffic patterns. Thus, the floor is spray-buffed, stripped and rewaxed as if it were a terrazzo floor. This requirement eliminates most if not all of the benefits of this type of floor.

Concrete Floors. Almost any non-acidic cleaning agent can be used on concrete floors without fear of damage to the flooring surface. The major problem in concrete maintenance stems from the fact that Portland cement is highly porous, so that grease and soils are readily absorbed; it is also subject to powdering and to the action of acids. Most of these problems can be eliminated through the proper use of sealers.

26

How to Solve Floor-Care Problems

This chapter describes common floor-care problems and offers recommendations on how to prevent and solve them. Obviously, we cannot provide a complete list, as there are literally countless problems that custodial workers encounter in floor maintenance. A special case, for example, would be the staining of resilient floors from the black solvent tracked off a wooden block floor. This can be controlled by sealing the approach area of the wood block floor with polyurethane seal, using plenty of matting and keeping it policed, and stripping the resilient floor more frequently than would normally be the case.

Problems listed in one category may have application in another. For example, conditions relating to the finishing of resilient floors will also be applicable to the finishing of terrazzo and other types of floors. The causes of a powdering floor finish applied on terrazzo will typically be the same as for a finish applied on a resilient floor.

RESILIENT FLOORS

Indentations

Permanent depression of the surface may be caused by several factors: the type of floor material, the furniture and equipment placed on these floors, and employee traffic. To prevent and correct this problem:

1. Move the furniture or equipment an inch or so occasionally, such as when the floor is being stripped. Or, use plates or gliders to spread the weight over a larger area of the floor.

2. Use proper equipment when moving furniture and equipment to prevent grooves. Desk-moving and file-moving equipment, for example, should be on hand in most organizations.

3. Women's shoe heels of the stiletto type can cause severe damage to floors. Some organizations have actually prohibited the use of such heels, while others suggest the use of larger heels or even provide heel caps which tend to enlarge their size.

4. The type of resilient material is an important factor. For example, pure vinyl has a poorer recovery rate from indentation than does ceramic-impregnated vinyl.

5. When cleaning resilient tiles, keep in mind that they are softened or even dissolved by solvents and oil. Therefore, use the oily type of dust mop treatment sparingly so as not to soften the floor and cause numerous indentations.

Damaged Tiles

A great deal of damage to resilient floor tiles is caused by chemical agents, such as strong caustic or alkaline solutions or by brittle materials such as varnish or shellac. Avoid using any of these materials, since they will cause the tiles to curl or pucker, crack, shrink, or round off at the corners. Use neutral synthetic detergents for general cleaning, and when stripping is required, use specially formulated chemicals, properly diluted. Where floors have been damaged to a considerable degree, it may be necessary to replace them or to consider covering the area with carpet or other material.

Loose Tiles

The principal cause of loosened tiles is first their warping or curling because custodians use excessive water and/or improper chemicals as described above, then because the mastic deteriorates progressively by additional water running under the tile. Do not flood floors for general repetitive cleaning, and remove the solution as quickly as possible. If a floor shows some signs of damage, you can extend the life of the floor by using wet vacuums, which will remove a good deal more of the water than a mop and wringer.

Colors Bleed

Do not use excessive oils or solvents, as in some types of dust mop treatments. This softens the tile to such an extent that the colors run, especially in asphalt and rubber tiles. This condition may also be caused if you use a solvent-type wax. This type of wax, normally yellow in color, and of viscous or pasty consistency, should not be used at all on tile floors; instead, use it on wooden or cork floors.

Excessive Porosity

You may damage the floor surface microscopically by using improper chemicals, such as harsh alkalies, or even normal detergents which are used in concentrations which are too high. The result is a "drinking up" of the floor wax or finish, so that it becomes difficult to achieve good gloss or surface protection. A similar effect is created if you use floor pads, either steel wool or synthetic, which are too abrasive. In this case, use a water-emulsion sealer to fill the pores after each stripping.

Tacky Surface

This condition is caused if you select or apply the wax or finish improperly. Poor-quality waxes may be tacky because of the raw materials in the formula. When wax films are applied that are too thick, they may become tacky. A residual soap film, remaining because of improper rinsing, may also feel tacky. Improper cleaning may leave chemicals, dirt, and other foreign materials which can cause stickiness.

Slipperiness

So called "fast floors" may be caused if you use an improper wax formulation or apply overly-thick films of wax. A fine film of paper or chemical dust can also make a floor slippery, as can the presence of water or other liquids, either in droplet or film form. You can eliminate these problems by protecting these areas with matting and by good cleaning techniques.

Fuzzy Surface

Tiny particles of paper dust, fabric lint, chemicals, or other foreign materials may sometimes cling to a floor and present a fuzzy or woolly appearance and feel. This can be caused by a finish that is too tacky or soft. You can solve this problem by selecting and applying the finish properly. Another cause is static electricity, and sometimes this may be cured by a change in the humidity. Or, you can apply an anti-static agent to the cleaning solution.

If these fail, then the area will simply need much more regular cleaning than before, preferably with the use of a wet vacuum or automatic scrubbing machine.

Darkening

Darkening of the floor is usually caused by a build-up of successive layers of wax or finish, mixed with soil; this would be especially prevalent around baseboards and under furniture. It is the result of improper floor care techniques, principally because custodial workers place the wax and finish in areas where it will not be worn off. A general darkening over the entire floor area may be caused by a redistribution of soil where the spray-buffing technique is mistakenly used for cleaning purposes, or where the resin-emulsion system is used; the dinginess is simply caused by successive layers of dirt and finish.

Soiled Baseboards

The soiling of baseboards indicates poor mopping and waxing techniques. The untrained operator splashes water, wax, and chemicals on the baseboard, where they build up into an unsightly condition. Prevention through training is the best approach to this problem, since correction requires a good deal of labor: the custodian must use a baseboard-scrubbing attachment to a floor machine, or a good deal of hand work.

Blanching

The over-use of stripping solution can actually leach the color out of resilient tile, leaving a bleached or whitened appearance. This might be especially noticeable along baseboards, perhaps where a custodian has used stripper in excessively high dilution—or even straight from a container—in an attempt to dissolve a heavy build-up. The proper approach is to make successive scrubbings with stripper of the proper dilution. A general washed-out appearance can be caused by excessive application of detergents on the basis of frequency, concentration, or formulation.

Traffic Lanes

Patterns worn in the floor by foot traffic may be caused by inferior finishes, which powder under pressure; or by improperly-maintained floors where dry buffing or spray-buffing is not provided to heal damaged surface areas. Where spray-buffing is not used, you should patch-wax the worn area in thin coats and feather it along the edges. Later, dry buff the floor.

Fading

Color fading may be caused by excessive sunlight. This might be controlled with shades or blinds. It may also be caused by the use of strong cleaners, the same cause of bleeding and streaking; make sure you use the proper concentration.

Powdering

Excessive powdering represents a breakdown of the surface finish. You can eliminate this by selecting a quality finish. Low temperature and humidity can aggravate this condition. Also, the powdering may be tracked in, and granular powdering is naturally caused by spray-buffing. Remove this by dust mopping. Powdering may also be caused by applying a finish on a floor that has not been properly cleaned or rinsed, or by putting a hard finish on top of one that is soft (such as putting a polymer finish over a natural wax).

Water Spotting

White splotches caused by water typically occur on floors which have been finished with an inferior type material. You can reduce the problem greatly by using higher-quality finishes and by buffing the floors regularly.

Dark circular splotches may be caused if the custodian drips wax from a bucket or mop during waxing. The custodian might also spray wax in droplets that are too coarse or that are permitted to dry before being reached with a machine pad in the spray-buffing technique.

Dullness

A flat, dull appearance may be caused by a porous floor absorbing the finish into the structure of the tile and leaving none on top to form a gloss, in which case a water emulsion sealer or additional coats of finish are required. If the floor has not been properly cleaned and rinsed, or if a poor grade of finish has been purchased, dullness can also occur.

Splotchiness

A floor showing irregular areas of varying reflectivity is generally caused by an improper stripping technique, where either all the old material was not removed, or some was allowed to redeposit. The reflection of light from the surface with the old material beneath it would be different, naturally, from a surface completely stripped and refinished.

Swirl Marks

Swirl marks may be caused by poor troweling of the concrete sub-floor, or by over-application of the mastic used for cementing the tile to the concrete. These two problems cannot be solved without replacing the floor, or covering it with another surface such as carpet. Swirl marks may appear in the floor when the finish is of inferior quality and does not level properly, or when it is applied too thickly.

Worn Tiles

Resilient surfaces may be completely worn through, and even the concrete sub-floor dished out, if soil accumulation is not controlled and if you do not provide a protective surface. Such wear areas typically occur at the bottom of a stair well, where a great deal of force is applied and also a turning motion is given, thus grinding the soil into the floor. You can protect such wear areas with matting where possible, and give special attention when cleaning and refinishing. Where necessary, repair the concrete floor and put in new tiles.

TERRAZZO FLOORS

"Blooming"

As the cement grout ages, some dust will be formed in the crystallization and wearing process that percolates up through the pores of the floor, ending up as a white powder on the surface. Make sure you seal floors properly with a membrane or polyurethane seal to prevent this occurrence.

Staining

Concrete grout is very porous, and acts almost like a sponge. The same sealing procedure that prevents blooming from the bottom up will prevent staining from the top down.

Pitting

Pits in the terrazzo surface may be caused by the loosening of the marble chips or mechanical damage to the grout. In either case, sealing the floor properly will toughen its surface and tend to limit this damage. Where the damage has occurred, you can repair the floor by "densifying," or you can fill in the holes with clear epoxy resin. You should bring in specialists for such work.

Yellowing

An amber or yellow cast to the floor may indicate the use of an improper type of seal; for example, the phenolic seals are not transparent as are the polyurethane seals, and darken the floor. This darkness may also be caused by excessive coats of finish, the use of waxes of an inferior grade, or by improper cleaning.

Blackened Strips

Black streaks alongside the metal divider strips often indicate that the strips are at a higher level than the surrounding terrazzo. To correct this, the terrazzo contractor should re-grind the strips to a flush condition.

Rust Spots

Ferrous metals should not be placed in contact with terrazzo, as the presence of moisture causes rusting which can badly stain the grout. Steel wool is one of the worst offenders, as the fine bits of metal can get caught in cracks; never use steel wool pads on terrazzo floors. During construction, any nails, screws, or other metal objects should be removed daily to prevent not only rust stains but mechanical floor damage as well.

CERAMIC FLOORS

Wear Areas

Where ceramic floors are sealed or finished, wear will take place on the square or rectangular surface of the tiles, but not in the rounded area of the tile or the grout which lies beneath, thus causing regular but unsightly wear patterns. When sealing ceramic floors, be careful to rub the seal into or off of the surface of the tile, concentrating it in the grout. Also, where finish will be applied to ceramic surfaces, custodians should finish floors regularly so that wear spots do not become apparent.

Loosening Tile

Grouting between the tiles can become damaged by alkaline detergents, or by imbedded soap. Thus use only neutral synthetic cleaners in proper dilution. If the tile does become loosened, however, you should repair or replace it.

CONCRETE FLOORS

Oil Spots

Because of the porous structure of concrete, oils and other materials easily penetrate the surface and can actually stain the concrete through its entire thickness. Even a surface removal of the stain will not prevent the stain from reappearing as it re-enters the surface area through capillary action. Just as for terrazzo, which after all is made of marble chips and a cement matrix, the best action is to prevent oil spots by sealing the floor.

Cracking

There are many causes of cracks in concrete, dealing with the design of the floor and the stresses to which it is exposed. The important thing to remember is that once a crack has appeared, you should patch it as quickly as possible with any one of a number of commercial crack-filling agents. Remember, a damaged floor is very apt to become still further damaged since it is no longer a continuous surface.

Traffic Patterns

Traffic patterns may appear in concrete as a result of the surface being abraded to form a slightly dished-out condition. Use matting where soil is introduced to such areas, and perform patch-sealing in these wear areas more frequently. Finally, give special attention to these areas with both dry and wet cleaning.

MARBLE FLOORS

Slipperiness

Marble, being a highly polished stone, is an inherently slippery material. You can reduce its slipperiness by applying finishes, rather than leaving the floor unprotected! Of course, the causes of slipperiness to finished floors, as described above, also apply here.

Soil Retention

Travertine, or "rotten marble," is an extremely difficult floor to care for. The numerous cracks and crevices trap soil of all kinds which becomes almost impossible to remove. A preventive approach is to fill all the pores with a clear epoxy resin. If you don't do this, use a wet vacuum to help remove a good deal of the soil, but the floor will never actually take on a clean appearance. Travertine is an exceedingly poor choice for a floor surface, although it is favored by some architects because of its unusual appearance.

WOOD FLOORS

Warping

Warped, cupped, or twisted wood floors are typically caused by the over-use of water. If you use any water at all on wood floors, you should first seal the wood properly; then use water sparingly and remove it as quickly as possible. However, it is preferable to set up a floor care technique that avoids the use of water altogether; instead, use treated dusting mops, solvent cleaners, solvent sealers, and the like.

Peeling

When a varnish or seal is applied to floors, it may peel if the coating is too thick, or if the floor has not been roughened to permit proper adhesion. When applying sealers, apply a dilute penetrating seal at first, which improves the adhesion characteristics of the surface sealer which then follows after suitable drying time.

Raised Grain

As in warping, this condition is caused by failure to protect the floor through sealing, and by the overuse of water.

Brittle Cork

Cork, being a form of wood, is susceptible to the same damage by overuse of water and improper detergents as is a planked floor. You should maintain a cork floor in the same way as a wood gymnasium floor, for example.

TIPS ON REMOVING NINE TYPES OF STUBBORN STAINS

Use all chemical agents with extra care to prevent fire, skin irritation or injury, and damage to surfaces. Try to destain a small inconspicuous "test spot" before beginning a larger operation. Many variables (surfaces, dilutions, application, supervision, etc.) affect the results obtainable.

1. BLOOD
   - Asphalt Tile, Linoleum, Vinyl, Wood: Rub with cloth dampened in clear cold water. If stain persists, dampen cloth with ammonia.
   - Concrete, Marble, Oxychloride Cement, Terrazzo: Rub with cloth dampened in clear cold water. Bleach with peroxide if stain persists.
2. CHEWING GUM
   - Concrete, Linoleum, Marble, Oxychloride Cement, Terrazzo, Wood: Remove gum with putty knife. Apply alcohol, rub with clean cloth, or freeze with compressed nitrogen.
   - Asphalt tile: Remove gum with putty knife. (The use of dry ice will make the gum easier to remove.) Do not use alcohol on asphalt tile.
   - Vinyl: Remove as much as possible with putty knife. Rub with #0 steel wool dipped in all-purpose synthetic detergent solution.
3. GREASE OR OIL
   - Asphalt Tile, Linoleum: Scrub with warm all-purpose synthetic detergent solution. Rinse with clear water.
   - Concrete: Pour alcohol on spot. Rub on spot. Rub with clean cloth.

- Marble, Terrazzo, Oxychloride Cement: Pour solvent on spot, cover with fuller's earth and let stand for several hours. Repeat if necessary.
- Vinyl, Wood: Pour kerosene on spot. Permit to soak for a short time. Wipe dry with a clean cloth. Wash with all-purpose synthetic detergent solution, rinse dry.

4. INK
   - Asphalt Tile, Concrete, Linoleum, Marble, Oxychloride Cement, Terrazzo: Use warm all-purpose synthetic detergent solution. If stain persists, mix 2 tablespoonfuls sodium perborate in pint of hot water. Mix whiting to form paste. Apply to spot and let dry.
   - Wood: Apply solution 1 part oxalic acid crystals to 9 parts of warm water. Permit to stand until dry. Mop with clear water.
5. IODINE OR MERCUROCHROME
   - Asphalt Tile: Use warm neutral soap solution.
   - Concrete, Linoleum, Wood: Apply alcohol and rub with clean cloth.
   - Marble, Oxychloride Cement, Terrazzo: Apply alcohol and cover with fuller's earth.
   - Vinyl: Wash with all-purpose synthetic detergent, rinse, then dry. If stain persists, scrub with scrubbing powder and warm water.
6. PAINT
   - Asphalt Tile: Rub with steel wool and all-purpose synthetic detergent solution. If area is large, use steel wool on buffing machine.
   - Concrete: Scrub with one pound trisodium phosphate in one gallon of hot water; rinse with clear water.
   - Linoleum: Rub with # 0 steel wool dipped in turpentine. Wash with all-purpose synthetic detergent solution and rinse.
   - Marble, Oxychloride Cement, Terrazzo: Rub with # 0 steel wool dipped in turpentine.
   - Vinyl: Rub with # 0 steel wool dipped in kerosene.
   - Wood: Use oxalic acid solution or one pound trisodium phosphate in 1 gallon of warm water.
7. RUST
   - Asphalt Tile: Rub with # 0 steel wool and all-purpose synthetic detergent solution.
   - Concrete: Dissolve 1 part sodium citrate in 6 parts water. Mix with equal parts of glycerin. Make a paste with whiting and apply to stain. For bad stains, wash with sodium citrate solution 1 to 6 parts,

add pad soaked in sodium hydrosulfite for 10 to 15 minutes. Wash thoroughly with water.

- Linoleum, Vinyl: Apply solution 1 part oxalic acid to 9 parts warm water. Let dry. Rinse thoroughly with clear water.
- Marble, Oxychloride Cement, Terrazzo: HORIZONTAL surfaces; to $^{3}/_{4}$ gallon water add 1.9 pounds sodium citrate and 1 pound sodium hydrosulfite. Add enough water to make a gallon solution. Cover stain with solution and let stand $^{1}/_{2}$ hour. Absorb with cloth by rubbing. Rinse with clear water. VERTICAL surfaces; make paste with whiting and 3 ounces sodium citrate and 3 ounces sodium hydrosulfite. Apply with putty knife and allow to remain 1 hour. Wash with sodium citrate.
- Wood: Wash with all-purpose synthetic detergent. Rub with # 0 steel wool if necessary.

8. SHOE MARKS
   - Asphalt Tile, Linoleum, Wood. Rub with # 0 steel wool or wash with all-purpose synthetic detergent solution.
   - Concrete, Marble, Oxychloride Cement, Terrazzo: Wash with all-purpose synthetic detergent solution and rinse.
   - Vinyl: Rub with # 0 steel wool dipped in all-purpose synthetic detergent solution.
9. TAR
   - Asphalt Tile: Remove surplus with putty knife. Do not put kerosene on asphalt tile. Wash with warm, all-purpose synthetic detergent solution.
   - Concrete, Linoleum, Vinyl, Wood: Remove tar with putty knife. Soak with kerosene. Rub with clean cloth. Wash with all-purpose synthetic detergent solution.
   - Marble, Oxychloride Cement, Terrazzo: Remove surplus with putty knife. Soak with alcohol and cover with fuller's earth.

27

Carpet Care and Cleaning

You should give careful thought to the desirability of carpeting an area before you make a selection. Much too frequently, this decision is related to cost alone, and even then the cost basis is often not an objective one. Some of the important considerations in deciding to carpet or not to carpet are acoustics, aesthetics, bacterial control, comfort and prestige.

While carpeting may eliminate slip hazards associated with some floor finishes, it must be installed properly to insure against tripping hazards. Excessive pile depth may present difficulties for enfeebled or elderly users; also, deep pile or an excessively spongy base may interfere with the movement of wheel chairs and carts.

CARPET VERSUS TILE: COST AND USAGE CONSIDERATIONS

Associations having vested interests in each type of flooring material attempt to present the maintenance cost for their product in the most favorable way. This explains how parallel articles in the same issue of a given publication can show economic advantages for carpeting versus resilient tile, and cost savings for resilient tile versus carpeting! Similarly biased figures can be found for installation and maintenance costs of wooden floors, terrazzo and ceramic tile.

Comparisons favorable to carpeting are criticized both by building managers and contract cleaners for exaggerating the probable useful life of carpeting and minimizing the useful life of resilient tile. The comparative cost of amortizing and maintaining carpeted floors versus resilient tile floors indicates that they are far closer than partisans from either field concede. If the comparison is being made strictly on the basis of maintenance cost, without consideration of installation and amortization, then the cost advantage will depend a great deal on the usage of the facility.

Carpeting maintenance costs are generally advantageous in office areas where soiling and traffic are moderate. In a facility such as an airport, where traffic is extremely heavy, with extensive trackage of soil, studies have shown carpet maintenance costs to be substantially higher than those for terrazzo. One of the reasons for the higher carpeting maintenance cost is found in the present technological limitations on the productive capacity of equipment used for carpet care.

It is seldom desirable to decide upon the use of carpeting on the basis of economic factors alone. Before the investment is made, you should read and listen to the available information about various types and should look at installations.

HOW TO CHOOSE THE RIGHT CARPETING FOR YOUR FACILITY

Fiber type, weave, pile (height, weight and density), color and pattern all will affect the economics and maintainability of carpeting.

Rayon and cotton are not considered suitable fibers for commercial carpets. Of the remaining types available, nylon is probably the best for heavy commercial use, with wool being suitable for prestige applications. The other synthetics currently seem to be rated as follows: polyesters, acrylics, olefins, and modacrylics.

For heavy wear, Wilton and velvet weaves are satisfactory. Axminster and knitted carpeting are generally somewhat inferior to the foregoing for commercial applications. High quality tufted carpeting is generally quite satisfactory. For heavy commercial applications, continuous filament nylon woven with a tight loop, backed with sponge rubber and constructed so that moisture will not penetrate this rubber backing, seems to provide the best wear and permit easiest maintenance.

Generally speaking, the greater the pile weight, the better the wearing quality. This weight is generally stated in terms of numbers of ounces of pile yarn per square yard of carpeting. For good wear, nylon pile weight should equal or exceed 48 ounces. Pile height is the distance from the top of the backing to the top of the pile. In commercial carpeting, close weave is more important than pile height, but ordinarily a height of .25 inches would be desirable for nylon. It is important that cut staple fibers be avoided in nylon, since these are likely to pill or fuzz and almost certainly will give poor wear. Pick or row (referring to the number of loops per inch of carpet length) and pitch (referring to the number of pile ends in each 27 inches of width) are both measures of pile density.

Color and pattern are very significant in carpet maintenance. Extremely dark or extremely light colors, as might be anticipated, are difficult to maintain; desired general color effects can often be obtained through the use of tweedy mixtures, while retaining high resistance to evidence of soiling.

ROUTINE CARPET CARE AND PROTECTION

Probably the greatest hazard to carpeting is the grit which is allowed to become embedded in the pile and which then cuts the carpet fibers, resulting in their ultimate failure (or at least causing changes in light reflectance, with apparent color variations). A fundamental step in carpet care is to protect it from the trackage of soil through the use of walk-off matting and other devices for soil entrapment. (Chapter 18 provides suggestions for preventing soil accumulation.)

Carpeting should also be protected from the permanent depression of piling which can result from any furniture being allowed to remain in one position at all times. You can help alleviate this problem through the use of furniture leg cups or casters specially designed to transmit the load to the carpet backing or floor without allowing it to rest uniformly upon the pile itself. The furniture leg cups are designed with teeth which support the weight of the furniture and which also permit a major part of the piling to remain erect. The casters have a honeycombed surface which results in only a small proportion of the pile fibers being packed down.

Since carpeting used in institutional and commercial applications is subjected to much heavier soiling than under household conditions, cleaning methods must be adjusted accordingly. Carpet sweepers and soft bristle brooms have little value in the maintenance of carpeting under commercial conditions except for policing to remove trash. The major tool for cleaning carpeting is the vacuum cleaner.

Assuming that the carpet is of suitable color and texture, it is often possible to omit vacuuming for a day or so without this becoming evident except where coarse soil is present. This is one characteristic of carpeting which gives somewhat more flexibility in the maintenance program than is enjoyed with resilient or other smooth floor surfaces. Nevertheless, omission of daily vacuuming in traffic areas introduces the hazard of excessive damage to the carpet pile.

A distinction is commonly made between the daily vacuuming of trafficked areas and a less frequent—often weekly—vacuuming of non-trafficked areas. Some of those persons responsible for programming maintenance have set one time standard for spot vacuuming and another for overall vacuuming. Actually the major time difference lies in the percentage of the total area covered in the operation, although there is likely to be some factor of lost time resulting from travel about the area in the spotting operation, as compared to continuous vacuuming of the total area.

Types of Vacuum Cleaners. The time requirement will also be affected both by the type of vacuum cleaner used and by the nature of the soil being removed. If trashy or linty soil is encountered, an upright with a driven brush will be much more efficient than will the simple wand-type vacuum; on the other hand, more difficulty is involved in reaching under low furniture with the upright type vacuum cleaner. The upright vacuum should be selected in most cases for the major daily cleaning operation, while the wand type vacuum, especially that adapted for wet pickup, may be very useful for policing heavily used areas during the day. In some instances, a carpet litter vacuum may be a needed policing tool where large amounts of litter accumulate, as perhaps in the corridor of a school building.

Whether the upright vacuum with driven brush or the wand type vacuum (either portable or central system) is used for daily cleaning of trafficked areas, it is essential to use a vacuum with driven brush for the less frequent overall cleaning of the area. In fact, most conditions dictate the use of a moderate duty pile lifter vacuum; this equipment actually has two motors, one driving the turbine to produce the vacuum, and the other driving the brush which raises the pile and loosens any embedded dirt and grit.

SPOT CLEANING

There is a great difference in the extent to which greasy or wet soil will penetrate the various carpet fibers. Continuous filament nylon normally has a maximum moisture absorption of about six percent. This is probably most resistant to actual staining, although various characteristics including its tendency to build up static electricity will sometimes cause the fiber to pick up surface soil. There is more opportunity for stains to penetrate wool fiber and the acrylics than there is with nylon.

Regardless of the carpet fiber, any spills which could possibly spot the carpeting should be picked up as quickly as possible. When spots are allowed to remain, it is frequently a matter of guess work to determine their nature and to decide on a cleaning approach.

Most soils can be classified under two general headings: those which are water soluble and those which are oil soluble. This fact has resulted in the frequent use of the so-called "two-bottle spot cleaning kit." This is a convenient kit which can be prepared by the housekeeping department and which consists of one bottle of solution for removal of water-soluble soil and another for the removal of oily soil. Water-soluble soils can usually be removed by a diluted solution of neutral synthetic detergent with a slight amount (perhaps one teaspoon to a pint of solution) of white vinegar. The other unit in the kit consists of a dry cleaning solvent, which of course should not be highly flammable; nor should it consist of toxic solvents such as carbon tetrachloride.

Commercially assembled spot cleaning kits are available from several manufacturers. They normally contain a set of instructions showing which material should be used for removal of a given type of soil and just how it should be employed. Make a preliminary test of each material on a swatch of the carpeting or in an area hidden from normal view to determine in advance that there is no tendency for the cleaning material to change the color of the carpet. Such a preliminary test will guard against serious discoloration of the carpeting, which often can be more noticeable than the spot being treated.

SURFACE BRIGHTENING

Frequently the traffic flow in a carpeted area is such that limited portions of the carpeting becomes quickly soiled. Often these areas are too large for spot cleaning, but small enough that you would like to avoid the usual "wet shampooing" approach.

One approach to this situation is dry shampooing. Dry shampoo usually consists of an absorbent material, such as fuller's earth or a fine hardwood sawdust, impregnated with a volatile hydrocarbon solvent. Some of these materials traditionally used carbon tetrachloride as the solvent and consequently were dangerous to the workers; pressure from health and industrial hygiene authorities has resulted in those products now on the market being safe to the user.

To apply these absorbent materials, sprinkle them over the soiled areas and then brush them into the carpet piling, either manually or with specially designed electrically driven brushes. Then, allow the dry shampoo to remain on the carpeting until the solvent has volatilized. The evaporation of the solvent tends to draw up any soils which are soluble in it—probably by a wicking or capillary action—from the pile into the absorbent carrier of the dry shampoo. This dry material, along with the absorbed soil, is then picked up, and herein may lie a problem, since these materials are difficult to remove from the carpet pile. Generally, a pile lifter, or at least an upright vacuum with a good motor-driven brush, is necessary for complete removal of the dry shampoo.

Certain formulations are available for "mop-on" cleaning of carpets. These may be wetting agents with emulsifiers and solvents which are effective in dispersion and suspension of soil. Normally these materials are diluted about one part to four or five parts of water and then used in the following manner. Place the cleaners in a shallow open pan or basin of suitable size; then quickly dip a small, short-strand mop (similar to a small standard dust mop but with shorter strands) into the cleaner with the object of wetting only the ends of the strands. Shake the mop to remove excess solution and use with a brushing action to go over the soiled area of the carpeting. Very good results have been reported for this type of carpet maintenance under certain conditions. Some of the manufacturers profess to include anti-soiling agents in their formulations; however it is important not to leave excessive cleaner residue in the carpet pile, for it apparently will accelerate the re-soiling of the carpeting if this is done. Some manufacturers advocate that the cleaner be used one day, and on the next day the same procedure be followed using clear water without cleaner.

A type of dry-foam shampooing machine has been developed wherein the detergent is fed through special pumps so that foam is developed before it is fed into the brush of the machine. In this type of equipment, the brushes are usually mounted on a horizontal axis rather than on a vertical axis, as in the case of the conventional buffer-shampooer. The advantage of this type of equipment is that much less moisture is introduced into the carpet pile so that it is easier to train an operator and much more unlikely that the carpet will be over-wetted and damaged. This type of equipment cannot clean the carpet pile as deeply as water extraction or the single disc floor machine type equipment, but it is so much easier to use that it is entirely practical to shampoo more frequently, which in the long run should result in better carpet care than with the other equipment.

In any shampooing operation, the initial step should be the use of a good pile lifter to remove all dry soil and to position the pile for best shampooing. After the foam has been picked up following the shampooing and the carpet has dried, a pile lifter should again be used to insure uniform lay of the pile and thus uniform light reflectance and appearance. Actually, with the foam type shampooer, many operators find it unnecessary to use a wet vacuum after the shampooing operation if this work is done frequently enough to prevent heavy soiling. In that case, the foam is simply allowed to dry on the surface of the pile and any dry soil thus left is then picked up with the pile lifter.

WET SHAMPOOING

For area rugs or carpets, there are advantages in removing them to a special location for shampooing. It then becomes possible to turn the rugs pile side down over a grating and to beat the back of the carpeting to loosen and remove any gritty material which is not taken out by the vacuum cleaner or pile lifter. Otherwise, the wet shampooing procedure is normally carried out in the usual manner, whether it is done in a special location or on the floor.

While the procedure is referred to as wet shampooing, the actual objective is to shampoo the carpet piling with a detergent foam, so that the foam serves as the emulsifying and suspending agent for the soil. This equipment normally consists of a conventional single-disc floor machine, with a detergent solution tank mounted either on the handle or above the motor, a solution release control, and a shower feed brush through which the detergent solution is fed into the fibers of the brush. Originally, soft natural fibers, such as Tampico, were used for the brush fill. These have generally been replaced by nylon bristles, with better control of the actual bristle stiffness.

In the procedure, a high foaming detergent, often with an organic solvent, is fed into the fibers of the brush so that enough of it is present to produce a high volume of foam when the machine is turned on and friction is produced between the brush fibers and the carpet pile. A properly trained operator can maintain a fairly consistent amount of wetting of the carpet pile by controlling the rate of feed and the intervals of feeding as the machine is moved from side to side on the carpeting.

Carpet shampooing is a good team job; one operator will usually start picking up the foam with a wet vacuum cleaner after approximately twenty square feet of carpeting have been shampooed by the other operator. In this way, good operators can keep from wetting the carpet backing excessively. Over-wetting has many disadvantages: it requires excessive drying time; it can cause shrinkage of certain types of carpet backing even to the extent that the carpeting will pull away from the fastenings; or it can cause mildewing of the carpeting when poor drying conditions exist. Even with good technique and suitable temperature and humidity, this type of shampooing will normally require a drying time of four to twelve hours before the carpet can be subjected to traffic. Also, precautions should be taken to prevent rust spots when furniture with metal gliders is moved back onto the carpet, in case it is not completely dry.

WATER EXTRACTION CLEANING

The water-extraction system (mistakenly called "steam cleaning") uses equipment that injects warm water into the carpet, and removes it by vacuum in the same motion, using a wand attached by hose to a machine. It provides the best cleaning of all in-place systems, but is rather slow (and thus expensive), and requires a longer drying time for the carpet. Further, wool and the mastic used in glue-down carpets can be damaged if the water is too hot. For these reasons, the water-extraction system should be used only periodically, such as after two or three surface brightenings. For smaller quantities of carpet or rush jobs, contractors should be considered.

Where the injector head is fitted with powered rotating brushes, the effectiveness of cleaning is even further improved.

A water extraction system can be truck-mounted and equipped with long hoses, so that it can be used for a series of small buildings, such as bank branches or apartments.

SPECIAL CHEMICAL TREATMENTS

Chemicals are now available for application to carpets and other textiles which significantly retard the penetration of soils. Some of these materials have proven highly effective and well worth the cost of purchase and application under certain conditions.

Anti-static agents which remain effective for reasonable periods of time (though the time will vary according to conditions of humidity and soiling) are now available. If static build-up is a significant problem, as it frequently is during winter in low humidity areas, the application of the materials may give adequate relief. A tendency to build up static is one of the major disadvantages of nylon carpeting. One application may not last the entire winter, the season when static is most noticed, but most reports indicate effectiveness for a period of two or three months.

At least one carpet manufacturer will supply carpeting containing conductive metal fibers woven in at proper intervals to insure against the build-up of static electricity. The addition of this conductive fiber does not cost much per yard, but the purchase of a full mill run may be necessary to obtain this feature.

HOW TO REMOVE 13 COMMON TYPES OF STAINS

The removal of stains from carpets should be done carefully and cautiously. Try to clean a small inconspicuous "test" spot before beginning the stain removal of a larger area. By all means try to determine the cause of the stain before attempting removal. Supervise the work carefully. Many variables (surfaces, dilutions, application, supervision, etc.) offset the results obtainable.

In general, work from the outside to the inside of the stain area to prevent spreading and formation of rings. For fresh stains, blot up as much of the staining material as possible before beginning the stain removal.

1. ACIDS: Use ordinary household ammonia to neutralize the acid but allow the ammonia to stay on only a moment. Rinse thoroughly with a cloth and cold water.
2. CANDLE WAX: Remove as much as possible with a dull knife. Cover with a clean, dry blotter and then press with a warm iron.

Sponge with Chlorothene (toxic chemical—handle carefully, do not inhale fumes, provide adequate ventilation).

3. CANDY: Scrape off with a dull knife. Apply Chlorothene. Sponge with hot water. (For chocolate, use lukewarm water, then follow with pepsin powder rubbed in, then sponged out after 30 minutes.)
4. CHEWING GUM: Scrape off with a dull knife. Apply Chlorothene. Or, freeze with compressed nitrogen.
5. COFFEE: Lather on a paste of egg yolk and warm water. Add a few drops of alcohol for an old stain. Rinse with clear water. (If creamed coffee, treat for cream.)
6. CREAM, ICE CREAM, OR MILK: Place a pan under the rug at point of stain. Pour on hot water, allowing it to soak for a while. Remove traces with hot water and cloth. Finish with Chlorothene.
7. ENAMEL, LACQUER, PAINT, VARNISH: Apply turpentine, then wash with strong, lukewarm soapsuds. Rinse with clear water and let dry. For a dry stain, apply half benzene-half denatured alcohol, working out the softened paint with a dull knife. Repeat as necessary for complete removal.
8. GREASE, OIL, OR TAR: Apply Chlorothene with a cloth, using a circular motion and working from the edge of the stain to the center to avoid causing a ring.
9. INK: Place a pan under the rug at point of stain. Pour on milk, allowing it to soak for a while. Use Chlorothene to remove grease left by milk.
10. MUD: Allow to dry thoroughly, then brush off and vacuum.
11. SHELLAC: Place paper under rug to protect floor, then rub with alcohol.
12. SOOT: Cover with dry salt. Sweep both up together and vacuum.
13. TEA: If clear tea, use mild soap and clear water. If it contains cream, use Chlorothene.

CARPET CARE SUMMARY CHECKLIST

• Determine where and if carpeting should be used; it would be more successful above the first floor and, in hospitals, it would be best to avoid carpeting in patient and treatment areas.

• Choose the carpet carefully, avoiding solid colors, natural fibers, and cut pile. Preferably use a continuous-filament nylon, with a dense, short, loop pile, in a floral or tweed pattern. Use glued-down carpet to simplify repairs and provide a safer surface.

• Protect the carpet with soil entrapment devices, especially gratings and matting or runners.

• Obtain a stain removal kit, and remove stains and spots as quickly as possible.

• Regularly dry vacuum the carpet with a machine that contains a brushing action; change the collection bag when it is two-thirds full.

• Use surface brightening to improve the appearance of the carpet, such as by using the detergent-impregnated granule system, or the dry foam system.

• For each three surface-brightening cleanings, use water extraction for the fourth cleaning.

28

How to Clean During Construction

The natural tendency at a time of construction or renovation is to throw up your hands and "give up" on housekeeping until things settle down and it can be handled in a normal, routine fashion. This is the same attitude which is so commonly seen during periods of inclement weather, where snow, water, sand, or salt may be tracked onto the floors, making it very difficult to preserve the appearance of the floor.

But construction and renovation create conditions which require you to actually *intensify* sanitary maintenance activities, rather than curtail them. The fact is, for new construction, you must begin maintaining finished surfaces just as soon as they are in place; and for renovation or expansion, you must not only protect the new surfaces but also the existing surfaces nearby.

At such times, then, you are faced with the very real problem of housekeeping becoming considerably more difficult—both from a psychological and practical performance standpoint—while at the same time it becomes more important. Even the fact that it is only a temporary condition brings problems of its own.

HOW CONSTRUCTION AFFECTS MAINTENANCE WORK

During the rough construction or renovation period, where concrete is poured, steel beams put in place, bricks laid, and other rough construction work accomplished, you have no real interest in the housekeeping of that particular area for its own sake. It is up to the contractor to provide a level of housekeeping and orderliness within the construction area which will make the work efficient and safe. Nevertheless, you must protect adjacent or nearby areas from contamination by the soils and waste material produced in the construction.

There are several reasons for this:

1. Floors can become badly damaged by scratching or abrading by workers tracking in particles of concrete, metal chips, splinters, mud, etc. This is true not only of resilient floors, but also of floors normally thought to be quite durable, such as terrazzo or marble.
2. Air-borne waste particles and dust can damage mechanical systems, laboratory or office equipment, etc.
3. Comfort alone becomes an important factor. Tenants in a building will react strongly to unpleasant conditions caused by excessive dust and soil. Occupancy in a hotel will reflect directly the housekeeping level at such times.
4. In addition to comfort, cross-infection control is even more important and is extremely difficult when there is excessive dust and soil. Cross-infection control is especially critical in hospitals and schools.

6 WAYS TO ISOLATE CONSTRUCTION WORK TO FACILITATE BUILDING MAINTENANCE

Try to keep the renovation or construction work and its by-products completely isolated from that portion of the building which is not involved. Here are some effective steps to take in this direction:

1. Provide "air locks" in hallways, which will confine the dust to the portion of the building under construction or renovation. These should be constructed of two sets of fabric, canvas, or plastic curtains, which must be pushed aside in order to walk through, but yet which will fall back into place. They should be the full length and width of the hallway opening. Make sure these curtains are vacuumed regularly—at least once per week. From a safety standpoint, have at least one portion of them transparent. Be sure that the area between the curtains is well lighted.

2. Place plenty of rubber mats and runners in connecting areas to entrap as much dirt and soil as possible, so as to keep this dust and soil from being ground into the older flooring and tracked throughout the building. An ideal location for such mats and runners is within the "air lock" area. Make sure these mats are vacuumed daily and washed out at least once per week. The use of old carpeting is very helpful, and many organizations save discarded carpets to be able to use them for such a purpose.

3. Check the filters in the forced air system regularly so that they do not become "loaded" with construction dust. Keeping the filter system efficient will do much to cut down on air-borne dust.

4. Put up "off-limits" signs so as to prevent construction workers from tracking steel filings, stone chips, sand, dust and soil into areas not involved in their work.

5. Similarly, put up signs to prevent sight-seers from walking into the construction area and then tracking soil back into the original facility. To help enforce such a rule and to satisfy employees' natural curiosity, you might consider conducting a periodic tour of the new areas for interested persons.

6. Cover equipment or machinery that is located near the construction area that is not regularly used with polyethylene sheeting or other protective material so as to keep out soil and dust.

MAINTENANCE GUIDELINES DURING CONSTRUCTION

After you have taken all possible steps to isolate the soils created during construction from any existing facility, you must still maintain the non-construction areas. Despite the above precautions an additional soil load will undoubtedly be brought in. Here are some guidelines:

1. Wet-clean floors more regularly, as dry mopping will not remove many of the heavier particles which become embedded in floors and their protective coating. Use auto-scrubbing equipment daily for this purpose in open areas.

2. Vacuum walls, trim and furniture periodically to prevent permanent soils. This occurs when dust is permitted to form a hard material in association with water or other vapors, oils, etc.

3. Once the construction is complete, wash walls and trim thoroughly.

4. Apply a protective coating of floor wax, finish, or seal to all floors before construction actually begins. This way, as much of the wear as possible will be confined to the protecting material, eliminating permanent damage to the floor surface itself. To prevent hard materials from being ground through the coating down into the floor causing permanent damage, strip and reapply the floor coating much more frequently than normal. When the project is finally complete, strip and rewax all affected floors.

5. Pay special attention to the connecting areas between the existing facility and that under construction or renovation. These should be wet-cleaned daily.

6. Try to prevent the transfer of soils from one floor to another. This can be done by cleaning stairwells and elevators, as well as entrance areas more often. Elevators and entrance areas that are carpeted will be easier to clean.

HOW TO TEMPORARILY INCREASE MAINTENANCE STAFF DURING CONSTRUCTION

It is management's responsibility to provide the money necessary to be able to accomplish all the extra duties required at such times. You must recognize the economic import of the situation in terms of damage to surfaces, and its effect on everyone who uses and inhabits your facility.

If the construction or renovation program is an extensive one and lasts a long time, it will undoubtedly not be possible for your existing housekeeping staff to handle this extra work even if the original staffing were efficient. Here are some of the ways in which to provide the additional time required:

1. Use overtime for existing employees. This is undesirable except for a short period of time, since employees who become accustomed to overtime feel that something has been taken away when it is discontinued, and a severe morale problem ensues. Also in such cases, you must consider the fatigue factor.

2. Hire temporary employees either specifically for the extra jobs required, or else to permit a redistribution of work assignments.

3. Hire additional full-time permanent employees to handle the extra work-load. These employees can then become the cadre for the staffing of the new facility. This is very desirable from the standpoint of having a ready-made, properly trained work force which can correctly maintain the new facility from the day it is first occupied.

4. Hire a contract cleaning establishment to perform specific cleaning jobs at assigned frequencies, or to clean entire areas for a fixed period of time. In these cases, it is very important to specify materials, methods, frequencies, and supervision. See Chapter 7 for further discussion on contract cleaning.

HOW TO WORK WITH THE CONSTRUCTION CONTRACTOR TO KEEP DIRT AND DAMAGE TO A MINIMUM

If you approach the contractor (or the construction department where the work is done internally) before the work begins, you may be able to agree on construction methods that will considerably reduce the amount of soil created. Such methods can actually be written into the specifications and, in many cases, will not increase the cost appreciably. Where the cost is increased, you must consider the added amount in light of its value in terms of preventive rather than corrective action.

Some of these methods include:

1. Specify wet-grinding rather than dry-grinding for a number of operations, which will sharply limit the dust problem.

2. Instruct the laborers regularly to clean up accumulations of dirt and dust and to remove them from the building before they are spread around.

3. Remove the waste, particularly plaster and concrete waste, in such a way as to avoid clouds of dust. For example, laborers should avoid dumping the materials from upper levels. Keeping such waste wetted down also helps.

4. Use sweeping compound, treated dust mops and cloths, and vacuums in clean-up operations.

On the final building clean-up, the typical construction contract requires the contractor to finish floors with a protective coating. Generally, the material used is of a low grade and often may not even be compatible with materials which management may wish to use. In such cases, your contract should indicate the use of specific materials and methods of application. Again, some organizations prefer that this work be done by their own people—it is good training for new workers, for example—and receive an allowance from the contractor who is thus relieved of this responsibility.

Military doctrine gives us a clue as to effective housekeeping action during construction and renovation: when the situation becomes difficult, attack and attack again. Do not permit a retreat—it can turn into a rout. Face the problem and fight it!

Controlling Germs and Cross-Infections

The term "cross-infection" refers to infections which are actually acquired within an environment, as opposed to infections which may have existed in the particular individuals upon entering the environment. It should be noted that cross-infections are *not* the exclusive property of hospitals (though this is where the term originated) and that they may occur in any factory, office building or institution where there is close contact among people; in fact, the common cold is transmitted as a cross-infection. Thus, in any environment, reasonable sanitation precautions need to be observed to protect the people in that environment.

However, there are factors which invest the cross-infections in hospitals with far greater danger than those encountered elsewhere. First, the conditions which normally bring people to the hospital have developed strains which are resistant to antibiotics, so that it may not be readily possible to assist the already-weakened patient to fight the newly acquired infection.

While the germs causing cross-infections are introduced into an environment by people and are most readily acquired by person-to-person contacts, it must be assumed that they can be shed from the original carrier into the air, onto surfaces or objects in the environment, and then transmitted to the victim by these secondary agents.

Germs enter the body through inhalation, through the mouth and through lesions in the skin. Perhaps the most common means of infection, second to direct contact, is from air-borne organisms, and these are often present because floors and other surfaces are not cleaned properly. This is where sanitary cleaning procedures become relevant.

CLEANING AND DISINFECTING: THE FIRST AND BEST PROTECTION IN CONTROLLING GERMS

Obviously it would be desirable to destroy as many as possible of the germs which potentially may cause cross-infection while they are present on surfaces and before they are transmitted. This is especially true because of the resistance of these organisms to drugs, as already mentioned. Fortunately, there is no relationship between the drug resistance and resistance to disinfectants. It was once widely believed, even among physicians, that a thorough job of cleaning was sufficient to control germs on surfaces. Certainly, many of the micro-organisms are associated with and carried by soiled particles.

However, cleaning is not enough because it is never possible to remove all organisms. This is because the micro-organisms are so tiny that they can readily be concealed and protected in minute faults in surfaces from which they will not necessarily be removed by cleaning operations. Thus it is important to use disinfectants to destroy as many as possible of those remaining.

In actual practice, use cleaner-disinfectants in a single-step operation to remove soil and to disinfect surfaces. This permits the greatest operational efficiency. When cleaning and disinfection are done daily, soil cannot accumulate, which would interfere with the effectiveness of the chemical agents.

Admittedly, it has not been possible to formulate a cleaner-disinfectant providing adequate pathogenic destruction without some compromise in the cleaning ability which could otherwise be built into the product—a fact which also emphasizes the need for good cleaning techniques. Therefore, you should ensure that mopping or other cleaning solutions are changed frequently enough to prevent inactivation of the disinfectant by organic or inorganic matter picked up by the mop or other cleaning tool. It is impossible to state how often you should change your cleaning solutions; this depends on variations in the water supply, the types and amounts of soil, and other influences. As a general rule, it may be said that mop water should be changed after mopping about 1500 square feet of floor surface which is only lightly or moderately soiled; the solution should never be allowed to become obviously dirty. No matter how little area has been covered, a bucket of cleaning solution should never be taken from an isolation area to any other area.

For greatest assurance in cross-infection control in critical areas, the cleaner-disinfectant solution should be poured onto the floor without dripping the mop into the bucket, only using the mop to spread the solution and to apply pressure to scrub obvious accumulations of organic matter. A sufficient amount of solution should be put down to allow at least two full minutes of contact time, and the solution should then be removed from the floor with a wet vacuum. This procedure gives reasonable assurance of the emulsification and dispersion of organic matter and of sufficient time for the germicide to be effective in controlling pathogens. The relatively large quantity of solution being used minimizes the chance for the chemical to be inactivated by the excessive organic matter.

Avoid the use of sponges in cleaning, as these have been shown to promote the growth of bacteria. Use cloth instead.

THE FOGGING OR MISTING TECHNIQUE OF DISINFECTING CONTAMINATED AREAS

Numerous studies have been made of the fogging or misting technique and its effectiveness in disinfection of contaminated areas. Initially, it was hoped that this procedure could eliminate the need to wash walls and other surfaces after isolation or "dirty cases" and thus permit both earlier return of rooms to service and reductions in the staff required for cleaning and disinfection. However, the results of this procedure vary depending on the amount of disinfectant deposited on the surfaces; also there seem to be wide variances in the effect on air-borne organisms.

Consequently most authorities discourage dependence on this technique, although some of them agree that the procedure has limited merit in providing a significant initial reduction of organisms present. If the fogging technique is used, the decontamination must be completed by spraying or washing all surfaces with sufficient quantities of a cleaner-disinfectant solution to insure complete wetting and removal of significant soil accumulations.

The misting procedure offers a valid substitute for the time-honored but inefficient and inconvenient practice of airing out an isolation room for 24 hours prior to its cleanup and return to service. Application of the technique under varying conditions with presumably competent bacteriological monitoring has shown reductions in the total counts of organisms of a magnitude exceeding 95% in some cases. The technique achieves at least as much as the airing-out process in respect to gross reduction of organisms and provides a significant safety advantage as a preliminary to the terminal cleanup by properly trained, equipped and garbed housekeeping workers. Presumably all items already in the room during the isolation case would be left there during the misting procedure, but trash and linens would be transferred to plastic bags for later removal after the exterior surfaces of those bags had been substantially disinfected during the misting. The actual details of the procedure should be reviewed by the Infections Committee, and housekeeping personnel should then adhere to all specified steps.

THREE POPULAR CLEANER-DISINFECTANTS USED IN HOSPITALS

No chemical disinfectant formulation which is safe in use both for personnel and for surfaces involved can be expected to kill 100% of all the pathogens exposed to it. Many disinfectants are highly selective and do not destroy some of the most significant germs. Thus, it is important to choose one that does the job as completely as possible under practical use conditions. At the same time, the disinfectant should carry a minimum hazard of irritation or sensitization to personnel or damage to structural surfaces.

Synthetic Phenolic

A very popular type of cleaner-disinfectant for hospital use is the synthetic phenolic. These products are chemically related to phenol or carbolic acid, but the molecule has been modified by introduction of chlorine or oxygen or even benzene rings at various points in the structure of the basic molecule. The result of this modification is both greater disinfecting ability and reduced corrosiveness of the molecule. The individually modified phenolics have been found to differ, one from another, in their effectiveness on various micro-organisms. Consequently, the trend in formulation, in order to provide broad spectrum control, is to include more than one of the synthetics.

Other materials which are used in the formulation include certain alcohols, wetting agents and synthetic cleaners, and sometimes soaps. The presence of soap tends to reduce the likelihood of skin sensitivity on the part of using personnel, but it creates films with hard waters which may insulate the surfaces of conductive floors and thus create static and electrical problems. Hence, products based on synthetics rather than soaps are usually selected.

Well-formulated phenolics will readily destroy Staphylococcus aureus—the organism which first caused great alarm over cross-infections when they were found to develop resistance to the antibiotics. They will also destroy the principal causative organisms of tuberculosis and other difficult-to-kill pathogens such as the gram-negative Pseudomonas aeruginosa and Proteus vulgaris.

In the selection of a phenolic disinfectant (as in the case of any other type) primary consideration of claims should be based on the A.O.A.C. Use-Dilution Test. This should indicate the amount required for control of various organisms. Comparative evaluation of products with respect to cost can be made intelligently by relating the number of ounces per gallon necessary to provide control, along with the cost per gallon of the concentrate.

Quaternary Ammonium Chlorides ("Quats")

Quaternary ammonium chlorides can be formulated to kill a wide range of pathogens, but some of these products have not proved effective on the Bacillus tuberculosis, and many are not effective on Pseudomonas aeruginosa. Further, the formulations of synthetic phenolics seem to yield better cleaning results than those of the quaternary ammonium chlorides especially with oily soils. However, the "Quats" are excellent deodorizers, and are easy on the skin.

Iodophor

Another type of germicidal-detergent is the iodophor. Generally these should not be classified as cleaner-disinfectants because of the inadequacy of the detergent ingredients to cope with heavy organic soil; this difficulty arises in part from the fact that, for germicidal activity, the iodophors must be in acid solution (preferably below 6.0 pH) which fact alone reduces their ability to remove organic soils. Because of the acidity of iodophors, they are actually better in removal of mineral soils than are most neutral or alkaline detergents; but soils of that type tend to raise the pH of the solution (that is, reduce acidity) and thus detract from the germicidal activity. Another disadvantage of the iodophors is that, despite the use of wetting agents to "complex" the iodine and hold it in solution, there is a tendency for them to stain porous surfaces. As for germicidal potential, they are superior to the phenolics.

Other Cleaner-Disinfectants

No classes of products beyond the above three currently merit serious consideration for cross-infection control in hospitals. Some of those offered must be classed primarily as bacteriostats, rather than germicides, and are considered inadequate by hospital bacteriologists, since the primary aim is to kill as many of the pathogens as possible rather than merely to suspend their growth. The organo-metallic compounds fall in this class. Other materials are so toxic or so corrosive as to be considered unsafe for use; these include the mercurics and formalin.

Good formulations of synthetic phenolics appear to be the best choice for use in general hospitals or those specializing in treatment in tubercular cases. For schools and institutions which are the habitat of persons in normal health and which do not by design collect large numbers of persons suffering from communicable diseases or infections, less stringent control is required; hence, cleaner-disinfectants such as the quaternary ammonium chlorides should prove entirely acceptable, as long as they give adequate results in cleaning. One of the purposes of using cleaner-disinfectants is to control odors through the destruction of micro-organisms which might break down any organic residues to produce offensive aromatic materials—an objective which the quaternary ammonium chlorides will accomplish while giving the added advantage of effectiveness in tying up existing aromatics.

CLEANING PROCEDURES FOR ISOLATION AND CRITICAL AREAS IN A HOSPITAL

Hospitals will generally have specific rooms or wards which are designated for the isolation of patients suffering from highly communicable or infectious diseases. In some instances, the patient found after admission to be suffering from such a condition may be isolated in the area to which already admitted, rather than moved to a designated isolation area. The objective of course is to help prevent the spread of the organisms of the infected patient.

Another type of isolation patient is the one suffering from extensive burns or a comparable condition where it is desirable to limit the number of people contacting that patient and thus limit the possibility of entry of infectious organisms into damaged tissue.

Definite controls for isolation areas should be established by the Infections Committee. Restrictions should exist on the number of persons entering the areas and all such persons should be required to use protective devices such as gowns, masks, caps and shoe covers, which should be properly discarded upon departure from the room.

During the stay of the isolation patient, the amount of project cleaning must be limited, but certain daily procedures will need to go on. It is best for the fewest possible number of people to enter the isolation area; hence it is often desirable to assign one person to do all cleaning tasks, rather than to divide the work between a maid and custodian as is often done in normal patient areas. Standard procedures must exist for the complete cleaning of the isolation area once the patient or patients have departed.

The operating, delivery and maternity, and the emergency suites are generally considered to be critical areas. The infection resistance of occupants in these areas may be lowered by the shock of surgical procedures or childbirth and extra hazards exist when surgical incisions or wounds break the skin barrier.

In some hospitals, all cleaning in these critical areas is performed by personnel in the nursing services department. More commonly, building surfaces are cleaned by housekeeping personnel, while the equipment used by doctors and nurses is often left to the responsibility of nursing service personnel. In a rather large percentage of hospitals, agreements are reached between housekeeping and nursing services designating which equipment is cleaned by the one and which by the other.

Obviously, where there may be a division of responsibility, there should be close coordination between the supervisors of the departments concerned and—where housekeeping cleans specialized equipment—on the methods and frequency of cleaning. The housekeeping department should not be responsible for cleaning x-ray equipment, sterile storage cabinets (except when emptied by nursing service), suction units or stands carrying anesthesia equipment or bottles of anesthetic gas; however it is quite appropriate for housekeeping to clean the bases or legs of such equipment in conjunction with the cleaning of floor surfaces.

HOW TO CLEAN CONDUCTIVE FLOORS

Whenever explosive anesthetic gas is used in hospital operating rooms, obstetrics areas, or emergency suites, special floors providing controlled ranges of conductivity are a legal and practical necessity to guard against the possible explosion of these gases. For every type of explosive gas, there is a range of mixtures of gas and air which will be explosive. Since critical concentrations may quite likely be reached in operating rooms and similar areas, and since the mixture might be set off by an electrical discharge of any kind, precautions in the form of conductive floors are required. These floors are designed to provide a controlled amount of resistance to the passage of an electrical current or static discharge. At the lower limit of resistance, the passage of a current directly to the floor from electrically-wired equipment must not occur; at the upper limit of resistance, there must not be a sufficient difference in potential between a surface which tends to collect electrons by static buildup and one which does not, to cause electrons to be discharged through the air to produce an arc or spark. While we speak of "conductivity of the floor" the controls actually established are in terms of ohms of electrical resistance.

There are three common types of conductive floors: terrazzo, mosaic tile and vinyl tile. All of these achieve their conductivity through impregnation of proper amounts of certain conductive materials, and the floor must be laid in such a way as to be grounded to the framework and foundation of the building. With all types of conductive floors, it is important that no cleaner or finish used on the floor surface shall interfere with conductivity. Certain sealers are used by the contractor laying conductive terrazzo floors for the initial curing or sealing; also, there are some manufacturers who sell sealers and finishes alleged to be satisfactory for maintenance use on the surface. Most manufacturers of sealers and finishes do not recommend that their products be used on conductive floors, and most authorities on housekeeping are similarly wary of the application of *any* coatings once the floors have been put into use.

Synthetic cleaner-disinfectants seldom form hard-water films and thus, with proper techniques, should not produce problems on conductive floors. If there has been a loss of conductivity due to past practices, the first attempt to solve the problem should be a thorough cleaning with a chemical designed to remove any existing film. A good synthetic-base wax stripper will often serve this purpose. There are terrazzo cleaners based on chelating agents which are safe and often effective in scum removal. The use of acids should be avoided on any of these floors, especially on terrazzo and grouted mosaic tile, because of the probable deleterious effect of the acid on the Portland cement binder.

Conductivity can sometimes be improved in conductive terrazzo floors through the use of a standard calcium chloride solution in accordance with the procedure published by the National Terrazzo and Mosaic Tile Association. There is the possibility that past cleaning techniques may actually have removed the carbon black or other conductive component originally incorporated in the surfacing material. If this has happened, no cleaning procedure, however thorough, will permanently restore the conductivity.

If none of the suggested methods for restoring a defective conductive floor proves successful, advice should be sought from the original flooring contractor (if known) or from some other flooring authority. In the case of terrazzo, it is sometimes necessary that the floor be reground to remove that portion of the topping wherein the conductivity has been destroyed. This is an expensive procedure which should be performed only by experts.

30

How to Avoid Common Mistakes in Cleaning

There are a number of oft-repeated mistakes made in cleaning. Reviewing these common mistakes can be both interesting and instructive. Naturally, other mistakes are made, but these are the ones most often seen; and they are also the ones contributing to the greatest loss in time and quality.

DUST MOPPING

Operating Errors

A number of operational errors occur in dust mopping, the basic dry floor-care job. The purpose of the dust mop is to catch and hold soil on its porous and sticky surface. A good rule to follow in dust mopping is to keep the mop on the surface to be cleaned until it is necessary to vacuum or remove the mop head, and to lift it as little as necessary to avoid dropping some of the dust from the mop surface and causing it to become airborne.

Dust mops are best used in open areas, such as corridors, by moving the mop forward simply with a walking motion, with the handle of the mop stick held on or near the hip by one hand.

Don't swing the mop from side to side; this uses too many small muscles of the arm and torso, which fatigues the worker while throwing dust from the floor and from the mop into the air. You should utilize as few muscles as possible and favor the use of the larger rather than the smaller muscles.

The dust mop is one of six basic long-handled manual tools; each is to be used with its own proper motion for best results. Whenever a dust mop is used in the same way as one of these other tools, poor results will be achieved.

The swinging motion mentioned above, which uses so much effort and creates so much dust, is the characteristic motion for the *wet mop*.

A common error is to use the dust mop as if it were a *push broom,* with a sweeping, then lifting motion, ending with dropping the dust mop back to the floor near the beginning point. This lifting motion tends to throw a good deal of soil into the air, just like the swinging motion, but merely in a different direction.

Another long-handled manual tool is the *corn broom,* where working sideways, or at right angles, to the work simplifies the motion, but this merely makes dust mopping more difficult.

Another such tool is the *deck brush,* where the proper motion is a back-and-forth scrubbing motion using a good deal of pressure. When this technique is used with a dust mop, too much effort is used for the results achieved. Normally, the weight of the dust mop itself provides sufficient pressure, and one pass over the surface to be cleaned is sufficient.

Finally, the *floor squeegee* is best used by tapping it at the end of the forward stroke so that drops of water are not left in the dried area on the back swing. When the dust mop is handled in this way, the tapping motion does indeed release a good deal of the dust that has accumulated on the mop, but a great deal of this will go into the air, later depositing a film of dust on numerous horizontal surfaces, which will then require more cleaning than normal.

In congested areas, a *swivel-type dust mop* is best used, permitting the mop to slide around furniture legs and other obstructions with a simple twist of the mop stick. The size would range from about fifteen to twenty-four inches in length, depending on the congestion of the furniture, and the skill and strength of the worker. With good training, the worker can use the dust mop with one hand, leaving the other hand free for moving furniture and other obstructions.

Over-Treating or Under-Treating the Mop Head

Dust is held to the surface of the dust mop by the nature of the fibers; this natural retention is improved by the addition of an impregnating material. This impregnation is well controlled when disposable paper is used, or sometimes when a rental mop system is used. In the latter case, the impregnating material is charged into the washing machine at the final rinse cycle (this is the same material and method used for impregnating cotton matting).

In larger housekeeping departments, you can operate a washer-extractor so that this impregnation can be done on the premises; but in smaller operations, where the rental system is not used, the problems in mop treatment can be avoided by having all mops treated by one trained custodian at the end of his work shift, so that the fibers are properly impregnated and ready for use the next day.

Where each worker is permitted to treat his own mop, we often find two common mistakes being made: over-treating and under-treating of the mop head.

Over-treating will put so much liquid material into the mop head that it is not all absorbed, so that in use a film of treatment is actually deposited on the floor; this can cause slipping hazards, as well as a softening of the floor through chemical attack. The appearance of numerous indentations in the floor caused by furniture legs may indicate that the tiling has been softened by such chemical action.

Under-treatment means that the mop head will not hold the dry soil properly, and that a good deal of it will become airborne in one of a number of ways, so that the dust settles back down both on the floor which the worker was supposed to have cleaned, as well as on all other nearby surfaces.

Overusing the Dust Mop

There is a limit to the amount of soil which a dust mop can hold in its fibers and on its surface. On very heavily soiled floors, consider other types of cleaning—such as automatic scrubbing, or wet mopping, or dry vacuuming—so that the dry mopping does not cause an immediate "loading" of the dust mop, so that it is only efficient for a short while. As the dust mop begins to overload, it develops a slick, hard leading edge. The usefulness of the dust mop can be extended by exposing new surfaces, but it is a very common mistake to over-use dust mops far beyond their ability to retain soil. Naturally, the result is somewhat similar to wet mopping a floor with dirty water, since it causes a redistribution of soil.

In many facilities and areas where especially high quality is desired, such as hospitals and building lobbies, dust mops are collected, laundered, and retreated daily. This is a very desirable practice. With the disposable paper dry mops, these can be discarded at will as they become filled.

The day of the dust box, into which mops were shaken, seems to be gone; but an even more valuable method of removing soil to extend the usefulness of the dust mop, is by dry vacuuming, either with a tank-type vacuum or with a central vacuum system. Some of the central systems have built-in floor slots specifically for the cleaning of dust mops.

Mishandling the Mop

When dust mops are mishandled, they bump and rub furniture legs and baseboards, causing unsightly build-ups of soil that are later difficult to remove. In some organizations, everything from the floor level up is stained up to a height of about four inches. This problem can be resolved, of course, through a simple training program.

Wrong Mop Handle Size

Even if the dust mop is properly treated, the proper motion is used, and the worker well trained, productivity can be sharply limited, and the worker greatly fatigued, by providing a handle that is simply too short. Standard mop sticks—whether they be used for dust mops, wet mops, squeegees or the like—are only fifty-four inches long. Studies show that a mop stick when standing on the floor should reach to eye level, in order to be used without the worker bending over. When purchasing manual equipment that uses wooden or metal "sticks," it is desirable to purchase a stick that is at least five feet long, and in many cases six feet long. Where the handle is too long, it can be cut down to size.

Misuse of Matting

All dust mopping is a corrective operation to remove soil from floors once it has entered a building. A good deal of floor work, both dust mopping and other types, can be avoided through the use of matting used as a preventive device to trap soil at entrances and other locations before it can be tracked throughout a building.

As wage rates go higher and higher, the use of matting, both the rental type and the owned type, becomes more and more worthwhile.

A common mistake is made in the use of matting, however; unless it is policed regularly, it can become so filled with the very soil that one is attempting to trap, that this soil is picked up and transferred to other areas just as if the mat were not there.

Depending on the weather and the type of soil, the matting needs vacuuming, shaking, or otherwise cleaning, periodically—sometimes as much as every hour. Where the matting becomes so wet that it can no longer be cleaned, it should be replaced with dry matting. This attention must not be considered as time lost, because the more policing time given, the more soil will be eliminated from adjacent floor areas.

WET MOPPING

Probably more soil is removed from buildings every day through wet mopping than through any other cleaning technique. Yet, a number of common mistakes are made in this procedure that outwardly looks so deceptively simple.

Infrequent Changing of Mopping Water

The error most common to wet mopping is failure to change the mopping water frequently enough. (This is also probably the most common error in housekeeping in general—and it's one of the most significant mistakes.) Cleaning personnel will use mop water longer than desirable because they do not want to take the time or effort necessary to dump out the soiled water and prepare a new detergent solution; this is one reason why the location and equipping of custodial closets at strategic points is so important. Some workers will use mopping water too long through a misdirected attempt to save money by not wasting detergent.

A detergent solution can contain only a given amount of soil; the detergent molecules become "used up" in their envelopment of soil particles through the chemical phenomenon called micelles. The results of over-use of detergent are unhappy: the soil is redistributed, often with the display of the very swirl marks that put it down, to completely negate the efforts of the custodian. In terms of cross-infection, the mopping can create conditions worse than existed before the job was done!

Overusing Detergent

Just as failure to change mopping water represents an under-use of detergent, it is also quite common to over-use detergent. Untrained custodians feel that the more detergent that is used, the easier the job will be on them, expecting the additional chemical to do more of the work. It is too bad this is not the case, since it would vastly simplify the economics of cleaning. Overusing a detergent can cause so much suds that either an additional rinsing operation is necessary, or a sticky, unsightly film is left to attract and hold soil. Even where suds are not formed in excess, such a film may be deposited.

Consumption of detergents should be controlled through dispensing pumps, measuring cups, or proportioning devices, not so much to save the cost of the product—which, although of importance, is a relatively minor item—but rather to assure that efficient cleaning is performed.

Improper Baseboard Cleaning

One of the most difficult of all custodial jobs is the cleaning of baseboards. Special brushes, floor machine attachments, and even special machines have been designed for this purpose. Yet, most of the difficult part of baseboard cleaning can be avoided altogether through proper mopping techniques.

The untrained, unskilled worker falls into the correct pattern of wet mopping—moving backward while swinging the mop in a "figure S" movement—very easily; but if he is adjacent to a wall, at the end of each stroke his mop will come into contact with the baseboard or the lower part of the wall. Since the mop will be charged with a good deal of soiled water, this pounding action deposits the soiled water onto the baseboard. It has the same affect as if one would slap a wet, dirty rag against a wall. Frequent repetitions of this mopping technique create a soiled area that is very unsightly and difficult to clean. The problem is compounded further when wax is applied in the same manner; the finish deposited on the baseboard simply seals in the soil beneath.

This can be avoided through a simple technique known as "striping" or "lining" the baseboard. The mop is pulled along parallel to the baseboard, right alongside it, but preferably not touching, so that we have a mopped area that extends from the baseboard usually about eight or ten inches. Then, when the "figure S" pattern is started, the worker stays far enough away from the wall so that the ends of the strokes merely overlap the mopped area parallel to the wall, with the mop never actually touching the baseboard.

Inefficient Wringing Action

When wringing a mop, it is unavoidable that some of the soil will be left in the yarn, trapped by the fibers, especially near the head of the mop. But we can improve the effectiveness of the mopping procedure by reducing to a minimum the retained soil and water, so that on the next application to the floor we will not be putting both the soil and water back down.

With an inefficient wringer or inefficient wringing action, a good deal of the weight that is being lifted in the mop is never disposed of, and this becomes useless effort. Of course, when the mop remains relatively soiled as it goes back to the floor, some of the soil in the mop is transferred elsewhere.

The three principal types of mop "wringers" are the roller, side-pressure, and downward-pressure, in increasing order of effectiveness. With the roller wringer, it is difficult to squeeze near the head of the mop, and a good deal of physical effort is required with relatively poor results. The side-squeezer wringer is more effective, but has the disadvantage of spraying water on the floor at times. The downward-pressure wringer is by far the most effective type to use, also requiring the least effort.

Mops should be washed out in a washing machine or mop-washing device on occasion; this removes most soil and bacteria and, combined with the hanging of the mop so that it can receive sufficient airing, avoids problems with odors due to souring. The fan-type mop—which has a band of cut tape sewn near the bottom of the strands—washes well without knotting, and also performs well in use, as it lays flat on the floor without balling up. Some housekeepers will bleach their mops during laundering for better appearance.

Incorrect Placement of Mop

Mop handles can be dangerous weapons—when left projecting from a bucket. A number of accidents have been caused by such placement of mop handles, where people have walked into them and received serious facial or eye injuries. The same problem can occur when mops are carried carelessly. When mops are left in buckets, they should be kept next to walls or corners, and leaned against the wall if they are not self-supporting; when carrying mops, the stick should be held upright (one convenient method is to put the head in an empty bucket, with the bucket being held in one hand and the stick in the other).

Mops can also cause damage to surfaces by mishandling, the most common example of which is pulling the mop stick straight up from a roll type wringer in a low-ceiling area with resulting damage to the ceiling.

Using Too Much Water

Assuming that "if a little is good, a lot is better," custodians will use too much water in the mopping process. The flooding of floors, particularly if the area stays completely wet for some time, and this is repeated regularly, can cause severe damage to resilient floors. Eventually the tiles will curl at the corners and edges, due to the loosening of the mastic at the joints. This curling accelerates the loosening action, so that in time the tiles are completely lifted from the floor. This same action may be observed when water damage occurs through pipe leakage or from rain blown in through open windows.

SPRAY-BUFFING

The spray-buffing technique is one of the most productive floor care systems developed in recent times. Many of the organizations which have tried this technique have discarded it following apparently unfavorable results however. Usually this is not because the organization did not have suitable surfaces for this technique, but because one or more common errors associated with it had been made. As a result, many housekeeping departments which should be taking advantage of the spray-buffing technique are not doing so.

Spray-Buffing Heavily Soiled Floors

The most common error in the use of the spray-buffing technique is considering it as a *cleaning* method; thus, it is used on floors which are too heavily soiled; in this case, the floor often looks worse than before any action was taken at all!

In spray-buffing, a thick, loosely woven nylon or animal hair pad buffs a floor which has been wetted with a dilute polymer finish containing a small detergent component. This action tends to remove the abraded finish, impregnating it into the pad. Also removed are rubber heel burns, scuff marks, and a small amount of soil; the floor is left with a thin film of new finish. Where the floors are not overly soiled, the results become increasingly attractive, both from the standpoint of appearance and cost.

Using the Wrong Type of Floor Pad

It is a mistake to use any floor pad which may be handy for spray-buffing. Pads which are too coarse will scratch the floor; pads which are too fine will load too quickly. And pads which are too thin will curl or fold. Most pad manufacturers now provide products specifically made for the spray-buffing technique; color-coding helps to identify them.

If the pads are not given regular attention after use, the material removed from the floor can harden and becomes difficult to remove. A good deal of money has been wasted by throwing away pads which could have been restored to usable condition. Often a small section of pad (such as the material which comes from the center hole) can be used to brush off dryer soil. Pads can also be rinsed out after use. Where the build-up is heavy or hardened, a soaking in a stripping solution is helpful. For severe conditions, the pad can be run through a washing machine, or even steam-cleaned.

Once the pad has been torn, it can either be trimmed down for use on a smaller machine, or cut up to be used as scouring pads by hand, or tied to the heel of a wet mop for spot cleaning.

Clogging the Spray Device

A number of types of sprayers are available for applying finishes while spray-buffing; these can amount to anything from a small plastic finger-bottle costing less than a dollar, to a relatively expensive pressurized or electrical atomizing system. Some of these systems clog very easily, so that if the spray nozzle is not flushed out after each day of use, the finish hardens in the orifice, either making the sprayer very difficult to clean or causing it to be thrown away.

There are a few spray devices which do not clog after use; but where the slightest trouble is encountered, it should be made standard practice to clean the spray unit by spraying clear water through it following each day of use. Unfortunately, it's a common mistake for spray-buffing to be put in general use for a few days, then for custodians to revert to a former method of floor maintenance because all the sprayers have become clogged.

Spraying Too Far in Front

As the custodian moves his floor machine forward into the wetted area while spray-buffing, he continues to spray a few feet in advance of his machine. But he can make the mistake of spraying too far in front of the machine—which is easily done, particularly if the spray head has been adjusted to coarse droplets—in which case the solution may actually dry on the floor before the machine reaches it. Even though the machine then goes over the area later, it still leaves a pattern of dark circles on the floor, which require a good deal of effort for removal.

Care should be taken in the adjustment of the spray nozzle, and the length of the spray pattern.

Some workers make the error of spraying walls, furniture, the floor machine, and even the worker himself! Keeping the spray directed two or three feet away from walls or furniture is best, since the damp pad will move a quantity of the material to the sides, but properly in smaller quantities, since these areas are less heavily trafficked.

Overusing the Pad

As the pad rotates on a dampened floor surface, it will pick up a mixture of soils, including old wax material. A certain amount of this soil works its way up into the pad (hence the desirability of a thick pad), but a good deal, unavoidably, stays on the surface. After a time, the pad becomes "loaded"—actually the surface becomes slickened with the soil. Before this slick surface develops to any extent, the pad should be turned over or a fresh pad obtained. If the pad is used too long, very little cleaning action will take place, and the appearance and condition of the floor will be harmed rather than helped.

There is a final step to the spray-buffing technique that may be forgotten, with unhappy results: a dust mopping to remove the loosened finish and soil, to prevent its being tracked over the floor.

WAX STRIPPING

Overusing Stripping Solution

Just as in wet mopping, you can use too much detergent for the job. But while in wet mopping the over-use of detergent merely takes more time, in wax stripping it can cause damage to the floor. Most strippers are alkaline in nature, which gives them the ability to penetrate the finish or wax surface, and to break it up for emulsification and suspension in the solution. Too much of this caustic can leach the color from resilient tile—a process also known as "burning."

In order to remove a heavy build-up adjacent to baseboards, custodians have applied overly-concentrated stripping solutions, or even poured on or mopped on undiluted stripper; perhaps you have seen corridors that have been blanched or discolored down both sides, irrevocably damaging the floor, by just such a mistake in cleaning technique.

Leaving Water on Floor Too Long

Putting down too much water and leaving it down too long will permit the solution to work its way between the tile joints, and then under the tiles, causing them to loosen and curl, thus damaging the floors permanently as above, but in another way.

Allowing the Floor to Dry Before Stripping

After the stripping solution has been mopped on the floor and scrubbed, the resulting slurry of solution and old finish is removed preferably with a wet vacuum or with a mop. If the removal were not rapid enough, by the simple process of evaporation, the floor would dry out again—and this leads to probably the most common mistake made in wax stripping. If the floor is permitted to redry before the old material is removed, then the finish is redeposited on the floor, and forms a new film, which is just about as difficult to remove as if the floor had never been stripped at all. If removal is not attempted, this redeposited film leaves an unsightly blotch on the floor, which is then covered up with new wax, to create a very poor looking floor indeed.

Failure to Strip All the Wax Buildup

Blotches may be left by failure to strip completely, which most often occurs on a floor with a considerable build-up of numerous coats of finish, probably as a result of previous poor floor care techniques. Naturally, this is most often to be expected near baseboards and under and around furniture. With heavy buildups, it may not actually be possible to penetrate all the layers of finish with a single stripping, and successive applications of stripping solutions, scrubbings, and removals may be necessary in these built-up areas. It's always a good practice to examine the floor from a sharp angle to see if all the old wax has indeed been removed. Where only a few small spots remain, these can be removed individually without the need for an entire restripping of the floor (it's good to bear in mind the minimizing of the amount of water used on the floor).

Rewaxing Without Rinsing

When the floor has been completely stripped of old wax, it will still be covered with a fine film of detergent; this should be rinsed well with clear water, which would then be removed by vacuum (again preferably) or mop, so as to provide a truly clean floor surface to receive the new finish.

Where the mistake is made of rewaxing the floor which has not been rinsed after stripping, the detergent membrane can soften the finish from beneath, creating a tacky or slippery floor, as well as one which is dull in appearance. As usual, the error leads to more work than would have been done otherwise.

Cleaning Baseboards After Finishing the Floor

It is good practice to clean baseboards while stripping floors. For one thing, baseboard cleaning unavoidably causes solution to run on to the adjacent floor area, which would damage a finished floor. Then, the buildup of soil and finish on baseboards proceeds along with that of the adjacent floor area. It is not only expedient but simple to clean both areas simultaneously with a baseboard scrubbing machine or baseboard devices on regular floor machines.

The baseboard should be considered a part of the floor, since it is made of similar material that has been curved up the wall a distance of a few inches! And, custodians should not consider a floor properly stripped where the baseboard remains soiled.

WAXING

Applying Thick Coats of Wax

"If a little bit is good, then a lot is better"—this fallacious reasoning is applied to many aspects of custodial maintenance, and contributes to one of the most common problems in waxing. Custodians, by laying down thick coats of wax, feel that they are saving themselves work because of the additional protection to the floor, and believe they will not have to refinish the floor as frequently.

Unfortunately, thick coats of wax cause many problems, and generally lead to a great deal more labor than the application of two or three thin coats; in addition, it is almost impossible to get a high-quality appearance without the use of thin coats.

Thick films have the tendency to be much softer than thin films, and are thus more susceptible to scuffing and smearing. This softness also contributes to the retention of a good deal of soil, causing a dingy if not downright dirty appearance. Finally, the thicker film is apt to be a good deal slipperier than the thinner film.

The benefits of the spray-buffing system, which provides an almost microscopic film thickness, well demonstrates the values of the thinner coats of wax or finish.

The unsightly darkening of floors adjacent to walls and furniture, caused by a build-up of successive layers of wax and soil is more frequently seen than it should be. This condition is not caused by improper cleaning or stripping techniques; rather, it is caused by improper application of wax or finish. Because people don't like to rub up against walls when they're walking, footsteps are rarely made closer than six inches to a wall or other vertical obstruction. The principal reasons for waxing a floor are protection of the floor surface and appearance; both of these objectives can be achieved with a single coat of wax in untrafficked areas. Further, buffing (wet or dry) tends to spread the wax from the center of the floor to the sides, especially the softer waxes but also, to a lesser extent, the harder finishes.

After a floor has been stripped, only a single coat of wax or finish should be applied adjacent to a wall. Many custodians prefer to make this the second coat, therefore covering up the edge of the first coat that might begin several inches away from the wall. It's often helpful to follow the line of a resilient tile joint in applying wax away from the wall, which would put the coat about nine inches away from the wall.

Using New Waxes Without Testing the Floor

In an effort to improve quality or cost, purchases of custodial materials will change finishes and waxes from time to time. But this change should not be made without testing the new material with respect to its compatibility with materials which are already on the floor, if they are not to be stripped first. For example, where harder materials are put over softer materials, such as applying a polymer finish over a carnauba wax, cracking and powdering of the top finish would be likely.

Since floor finish materials are rather complex, containing natural waxes, synthetic resins, plastic polymers, emulsifiers, levelers, etc., the testing of samples before the purchase and application of large quantities can save a great deal of time and expense. Where there is any doubt, the new material should be applied following a complete stripping of the old.

Waxing an Improperly Cleaned Floor

A number of improperly maintained floors take on a very dingy cast, appearing grey, brownish, or sometimes taking on a mottled white cast. Although this can be due to failure to mop or sweep the floors regularly, very often the soil is actually sandwiched between two layers of wax, when wax is put down on a floor that has not been properly cleaned. After this has been done, dust mopping or wet mopping will not remove the trapped soil, and the floor will stay in very poor condition until a complete stripping is performed.

Applying a Second Coat of Wax Before First Coat Dries

Time is often a problem in the refinishing of floors, especially in those organizations which operate on more than one shift, or where floor finishing project teams have not been set up, and each area-assigned worker is responsible for his own project work near the end of his shift. This time pressure may result in the application of a second coat of wax over a first coat which has not properly dried. Drying time is affected by a number of factors: the humidity, temperature, air movement, the thickness of the wax film, the nature of the subsurface, formulation of the finish, etc. Normally, a twenty- to thirty-minute drying time is necessary if the material has been properly applied, and the first film must be at least dry to the touch.

Where the first coat has not dried properly, the application of the second coat will tend to re-emulsify the base film. At the least, the result would be a dingy looking surface. In addition, the re-emulsification would not be uniform, and a mottled or splotchy appearance might result, perhaps along with powdering, bubbling, etc.

Mishandling or Misusing the Mop

The mishandling of mops probably creates more work for custodians than any other improper technique. Just as in wet mopping mentioned previously, the striking of baseboards with wax will deposit material there, which causes an unsightly appearance as well as a difficult removal problem later on. The same thing can happen to furniture legs, file cabinets, doors and door jambs, and the like. The training of custodians in proper mop-handling technique is an indispensable part of a good housekeeping program.

In applying the wax film, a clean cotton mop is the best applicator with which to cover fairly large areas (some skilled custodians prefer the "string" typc of cotton mop, which has finer strands than the usual cotton mop; those who prefer lambs wool applicators, however, are not apt to get both good coverage and results).

To produce the proper thin film, the mop must be wrung fairly well, and moved in a "figure S" pattern, with the worker stepping backwards with each swing. Avoid rubbing the mop in a back-and-forth motion, since this causes a bubbling of the wax because of the emulsifiers which are part of its formulation. When the bubbles break, the wax dries with thousands of miniature craters in its surface, which collect soil and present a very poor appearance.

Reusing Wax That Was Poured from Its Original Container

Once a wax has been poured out of its original container, it should be considered as lost, since the pouring of a quantity back into the container may introduce bacteria-laden soil, which can spoil the entire drum or can. This spoilage may not be immediately noticeable, and a good deal of effort will be wasted in applying this faulty material to a floor, only to find out that it gives a very poor appearance.

Thus, when pouring out these emulsions for finishing or waxing, it is best to be conservative, obtaining a little less than might be thought needed.

It's always easy to pour out more, but any material remaining should go down the drain.

BUFFING

Buffing Without Cleaning

The first step in dry buffing is to prepare the floor by removing the loose surface soil with a dust mop. Of course, if the floor is quite soiled, a damp-mopping or wet-mopping would be necessary.

Where buffing is performed without this prior cleaning, the action of the floor machine will cause a scattering of the soil not only over the floor, but up into the air, from which it settles back down onto the floor as well as all other surfaces, such as furniture, ledges, equipment, and the like. Dust mopping is a very easy and quick job, and can save the expenditure of a good deal of effort later on.

Failure to Control the Movement of the Floor Machine

The single-disk floor machine is by far the most common type of floor care equipment in use. The rotation of the brush will tend to move the machine to the right or left depending on whether the handle is raised or lowered. Workers must be trained in the "feel" of the machines in open areas before they are permitted to use them in congested areas, to avoid bumping—which causes both damage and marking—of furniture, walls, doors, and other objects.

"Fighting" the Machine

Where an operator has not learned the proper handling of the equipment, he is apt to "fight" the machine rather than control it, and he can become fatigued very quickly before the performance of a reasonable amount of work. Proper training permits any custodial worker, regardless of his or her size or strength, to handle such machines with relative ease. Further, the development of vibrating and oscillating-type equipment also tends to limit this particular type of problem.

Using the Wrong Floor Pad

Synthetic floor pads are now being manufactured in four basic grades: stripping, scrubbing, spray-buffing, and dry buffing. It is a common mistake to choose the wrong pad for the dry-buffing technique. Naturally, the pad that is too aggressive will cut through the finish, and sometimes through the floor as well!

Fortunately, a number of manufacturers are beginning to color-code pads to indicate their use, and this greatly simplifies training custodial workers in proper pad selection.

In addition to these synthetic pads, other materials are available for dry buffing. These include steel wool (also available in various grades) and only the finer grades should be used for buffing; of course, steel wool should not be used on terrazzo because the small metal chips which might remain embedded in the floor can cause serious rust stains; and terry-cloth pads and lambs wool pads (which are used for ultra-sheen polishing).

RESTROOM CARE

Failure to Clean Under Lip of Toilet Bowl

Probably the most frequent omission in restroom cleaning is the cleaning of foreign matter hidden under the lips of toilet bowls and urinals. This is the source of a number of serious odor problems in restrooms, since bacteria multiply very quickly in such an environment (bacteria of course being the principal source of non-transicnt restroom odors). The custodial worker should be equipped with a small mirror, either of the pocket or dentist's type, for inspecting these hard-to-get-at surfaces. Normally an acid-type descaler, applied with a bowl mop, is required to remove these hidden soils and stains.

Too Frequent Descaling

Acid descalers do such a good job on soil removal that it becomes a common error to use it too frequently. Normally, descaling should be performed from once per week to once per month, depending on the hardness of the water (and thus the rate of scale deposit) as well as the amount of use the fixtures are given. More frequent flushing means more scale deposit.

It is not rare to find custodians using acid descalers on a daily basis, yet these materials can damage restroom fixtures and surrounding surfaces, particularly metals, and also represent a safety hazard in its application since the acids are corrosive to the skin and especially hazardous for the eyes.

Descaling should be set up on a calendar performance basis, and in some cases performed by special projects workers who are properly trained and equipped to do nothing but this in all areas.

Lack of Overhead Cleaning

"Tunnel vision" is a defect we all seem to have; it is not so much physical as it is psychological. That is, we tend to see things at eye level or below, since we often walk or stand with eyes downcast, but rarely do we notice things above our level. Thus, a restroom that might otherwise be perfectly clean might well have overhead areas that require a good deal more attention. Examples of such areas are ceiling diffusers, ventilation grilles, piping, light fixtures, and duct surfaces.

With a little attention, this problem is easily remedied, since overhead areas normally require only periodic (such as monthly) attention. As with many other aspects of housekeeping in all types of areas, in restroom cleaning we may be performing daily repetitive jobs adequately, but fail to give proper attention to those tasks which are performed infrequently. Good housekeeping requires attention to both!

Litter Due to Lack of Proper Policing

Perhaps you have walked into a restroom and found a crumpled-up paper towel balanced neatly on the top of a torpedo-type waste receptacle. Or worse, the receptacle surrounded by litter which has been thrown on the floor. And yet, on inspection, one finds that the receptacle is far from being full.

A still closer inspection reveals the cause; the push plate on the receptacle has not been properly cleaned or policed. A visitor washing his hands and then drying them simply does not wish to resoil his hands on a dirty push plate—it seems to be an invitation for him to avoid using the receptacle.

Much of the litter in wash rooms is caused by failure to provide proper cleaning and policing.

Failure to Clean Inside Stall Door

The custodian opens the door to a restroom stall, cleans the fixture, the walls and floor, and then closes the door; this results in the common error of failure to clean the inside surface of the stall door, and the wall area which it hides. This can only be done, of course, by the custodian entering the stall, closing the door, and then cleaning it.

Improper Deodorizing

The only positive approach to restroom deodorization is thorough cleaning and disinfection, thus removing and killing odor-producing bacteria. The substitute of odor-masking and perfuming devices such as drip fluids, deodorant blocks, wicks, and aerosols can be helpful in certain difficult cases where ventilation is not adequate, or where the public demands it merely through habit, yet the use of these devices should not represent an excuse to avoid good restroom cleaning practices.

ROOM CARE

Failure to Clean Beneath and Above Furniture

When a person sits down, he is able to see a good deal more of the floor under the furniture than when he is standing.

Cleaning personnel must be trained to clean underneath desks, tables, beds, and other furniture. This includes removing litter, dry mopping smooth floors (and wet mopping and buffing periodically), vacuuming and spotting carpeted floors.

Just as failure to clean *beneath* surfaces is a problem, so is a failure to clean *above* them. Thus we find collections of dust on the tops of lamps, over doors, on the tops of tall furniture such as cabinets, on high ledges, and the like.

Failure to Clean Door Hinges

If one area were to be selected as that which is most often missed in room cleaning, it would be the hinge area of the door frame at the entry to the room. Where this surface is omitted in the cleaning schedule, it is for one of two reasons. In one case, the door may normally stay closed, with the hinge area being covered by the door itself, and thus not seen either by the worker or supervisor. Interestingly, the hinge part of the door frame is invisible even when one opens and closes the door; it can only be seen when the door remains open.

The other possibility is a jurisdictional problem: where one worker is assigned to clean the room, and another to clean the corridor, the door frame becomes a "gray area," lying neither in the room nor in the corridor, and each worker leaves it to the other. It is usually best to assign the door frame to the person cleaning the room itself.

Insufficient Cleaning of Trash Receptacles

Waste baskets are often left uncleaned, with numerous unhappy results ranging from an unpleasant odor to the attraction of vermin.

This cleaning omission may occur for failure to schedule the periodic washing of the receptacles, since the person emptying the receptacle normally does not have the cleaning solution, sponge, and cloth necessary to do this job. An excellent way to wash waste baskets is with a mobile double-tank cart with drainage rack, which can move from one room area to the next, the baskets being cleaned on a project basis by two persons, one collecting and returning the baskets, the other washing them at the cart. Of course, waste baskets should be standardized so that it is not necessary to return any one basket to its original location.

Another approach to this problem, this time from a preventive rather than corrective basis, is the use of plastic liners. In quantity purchase, these cost little more than a cent each, and might last a week or more depending on the waste that has been put into the basket. They are particularly helpful where workers put wet waste—such as coffee—in waste baskets.

Ignoring Handprints on Walls

It is not uncommon for hand prints on walls, doors, partition glass, and other vertical surfaces to be ignored in the daily cleaning routine. Where "spotting" is done daily, it becomes a simple task to keep them removed, but where they are permitted to accumulate, the net result is often a complete wall washing or even painting job.

Further, human psychology being what it is, uncorrected problems of this type can amount to invitations for more of the same! Just as people do not hesitate to throw litter on top of other litter, but would hesitate to "throw the first stone," so they would not mind putting hands or feet on walls which are already marked and not receiving attention, but would hesitate to mark a clean area.

Improper Cleaning of Overhead Fixtures

Just as in rest room care, some custodians tend to ignore areas above the six-foot level. So, it is not uncommon to see rooms that are well cleaned in other respects but have dirty ceilings, grilles and louvers laden with soil, and pipes and ducts covered with films of dust.

Fortunately, the correction of these overhead problems is a simple matter, often most easily handled with a tank-type or pack-type dry vacuum equipped with an extension wand for reaching these overhead areas.

Difficulty Cleaning Cluttered Areas

Finally, a common mistake in room cleaning that is very difficult to cope with insofar as floor care is concerned, is one which is caused by management's failure to require the occupants of the room to keep areas clear. Floors which are cluttered with boxes full of business forms, old files, and the like, cannot be properly cleaned because no liquid can be used on these floors for fear of damaging the contents of the containers; it would even be difficult to dry-buff or spray-buff such an area. This leaves only dust mopping which, in time, will leave the floor in an unsatisfactory condition.

The same is true of cluttered furniture, where workers will leave tables and desks covered with papers and supplies. It is normally best to have custodial personnel clean only exposed areas of such furniture, since the restacking of papers and other materials may lead to a loss of time or information that could be rather costly. But even this is not wholly satisfactory, since the desk user, on clearing his desk, finds "islands" of uncleaned areas, which he may feel reflects on the custodial department rather than on himself!

Where management can do this, other than in executive areas, it is desirable to have personnel responsible for the cleaning of their own desk tops whenever there are any papers or other material on the surface; but where the user completely clears the surface (except for telephone, dictating machine, and such) this becomes a signal for custodial personnel to dust, and even wash or polish the desk if needed.

Many room areas provide this same problem on vertical surfaces as the horizontal surfaces just discussed. That is, walls and partitions may be cluttered with calendars, notices, cartoons, and lists which makes their spotting and cleaning very difficult. And the removal of these materials may actually take away part of the painted surface.

HOUSEKEEPING EQUIPMENT

Failure to Clean the Cord

The technological advance that is most often sought by cleaning management is the development of floor machines, vacuums, and other motorized equipment that can be operated with a light-weight battery. Meanwhile, we are required to use equipment with electric cords, and this leads us to a common mistake in the use of equipment: simply, failure to clean the cord.

The electric wire—whether it be for a vacuum cleaner or for a floor machine—is naturally pulled across soiled floors, and thus becomes well soiled itself. When the cord is moved about it transfers soil to furniture legs, door frames, clean floors, and other surfaces.

Workers should be trained, when putting the equipment away at the end of each use, to clean the cord while winding it onto the proper part of the machine. If this cleaning is performed after each use, it can be simply done with a cloth or sponge dampened in detergent solution. If it is done infrequently, however, it will be necessary to use stronger cleaning means, such as scouring powder or paste, or a warm stripping solution where some of the soil consists of wax or finish.

Inefficient Equipment Cleaning

The equipment itself suffers from this lack of cleaning attention, and although there is not the problem of soil transfer so much as is experienced with the electric cord, dirty equipment does provide a poor appearance and, in the case of wet vacuums, may led to foul odors.

If equipment cleaning is done after each use, as with cleaning the electric wire, the time requirement is rather slight. Workers should be instructed that no equipment should be put away until it has been properly cleaned, and of course the short time required for this should be allowed.

Equipment cleaning should involve manual equipment as well as power equipment; buckets, wringers, mop handles, and other such items should be cleaned after use.

Where such a program is first started, and a considerable build-up of soil must be removed first before beginning a daily cleaning program, the use of soak-tanks charged with a warm stripping solution, or the use of a steam-cleaning machine, can be very helpful.

Failure to Clean Custodial Work Stations

It is natural to judge workers by the tools that they keep and the work station where they prepare for their jobs. It is often possible to judge the morale of the custodial workers and the effectiveness of supervision by simply examining custodial closets to see whether they have been properly organized and kept clean.

The walls should be washed, the floors scrubbed, the door louvers vacuumed. The room should be kept lighted. Tools with handles should be hung in a vertical position, while chemicals and small supplies should be neatly stored on shelves. Empty boxes and litter should be removed.

A simple example of care of tools in the custodial closet is the washing of mops; where wet mops are merely thrown into a corner while still containing soil, they may "sour," the result of the multiplication of millions of bacteria. If the mops are washed out after each use—and of course it is desirable to have removable mop heads rather than the old "stick" type—and hung up to dry, this problem can be avoided. In some places where the appearance of the mop is important, it is even desirable to bleach them so that they will have a whiter appearance.

Equipment that is kept clean, stored in a closet that is also kept clean and is well organized, will help develop pride of performance and improve the public image of the custodial department.

Index

D

E

F

G

H

I

J

K

L

M

Y